CONTAMINATED LAND
Problems and Solutions

CONTAMINATED LAND
Problems and Solutions

Edited by

T. CAIRNEY
Consultant
W.A. Fairhurst Environment Division
and
D.M. Hobson
Environment Division Director
W.A. Fairhurst and Partners
NEWCASTLE upon TYNE

E & FN SPON
An Imprint of Routledge

London and New York

Published by E & FN Spon, an imprint of Routledge
11 New Fetter Lane, London EC4P 4EE

Simultaneously published in the USA and Canada by Routledge
29 West 35th Street, New York, NY 10001

First edition © 1993 Chapman & Hall
Second edition ©1998 E & FN Spon

Typeset in 12pt Times by Pure Tech India Ltd., Pondicherry
Printed in Great Britain by St Edmundsbury Press, Bury St Edmunds, Suffolk

British Library Cataloguing in Publication Data
A catalogue record for this book is available from the British Library

ISBN 0 419 23090 4

Contents

Preface to the second edition   xii

Acknowledgement   xiv

Contributors   xv

1   International responses   **1**

1.1   Introduction   1
1.2   Differences in national responses to land contamination   2
    1.2.1   Various national viewpoints   2
    1.2.2   The complexity of soil systems   3
    1.2.3   Identification of land contamination and necessary clean-up standards   5
    1.2.4   Variation in legislation and controls   7
1.3   Signs of an international consensus   8

2   Risks from contaminated land   **10**

2.1   Introduction   10
2.2   What is 'contaminated land'?   11
2.3   Estimating the amount of contaminated land   12
2.4   Standards and criteria to establish if land is contaminated   13
2.5   Risk assessment of ground contamination   17
2.6   The development of legislation, policy and practices in the UK   20
2.7   Netherlands policies and practices   25
2.8   Summary   26

3   Rational site investigations   **28**

3.1   Introduction   28
3.2   Purposes of site investigation   29
    3.2.1   Risks to land users   30
    3.2.2   Effects on the environment   31
    3.2.3   Level of detail required   33
    3.2.4   Topics of interest   34
3.3   Site investigation strategy   35
3.4   Preparation   36
    3.4.1   Data collection   36
    3.4.2   Recorded physical conditions   39
    3.4.3   Previous uses and contamination potential   40
    3.4.4   Preliminary conceptual model   44
    3.4.5   Site reconnaissance   45
    3.4.6   Defining objectives   47
3.5   Designing an investigation   48
    3.5.1   Information required   48
    3.5.2   Sampling strategy   49
    3.5.3   Non-targeted sampling   50

3.5.4	Targeted sampling	53
3.5.5	Sampling depths	54
3.5.6	Background conditions	54
3.5.7	Groundwater and surface water sampling	55
3.5.8	Gas monitoring	58
3.5.9	Analytical testing	59
3.5.10	Sample type and size	60
3.6	Exploratory techniques	62
3.6.1	Hand methods	62
3.6.2	Trial pits	62
3.6.3	Light cable percussion boreholes	64
3.6.4	Power auger drilling	64
3.6.5	Dynamic probes and small diameter boreholes	65
3.6.6	Soil gas surveys	65
3.7	Implementation	66
3.7.1	Specification	66
3.7.2	Site work stage	66
3.7.3	Recording of data	68
3.7.4	Sample preservation	69
3.7.5	Safety precautions	70
3.7.6	Testing programmes	70
3.8	Reporting and interpretation	72
3.8.1	Presentation of data	72
3.8.2	Characterizing zones	73
3.8.3	Assessment of targeted testing	75
3.8.4	Assessment of water testing	75
3.8.5	Report recommendations	76
3.9	Conclusions	76

4	**Reclamation options**	**78**
4.1	Introduction	78
4.2	Reclamation choice in the United Kingdom	78
4.2.1	Evolution of the UK's emphases on land contamination	78
4.2.2	Challenges to current UK emphases	81
4.3	Reclamation choices when health and environmental concerns dominate	84
4.4	Engineering-based (broad spectrum) techniques	87
4.4.1	Excavation and disposal	87
4.4.2	On-site encapsulation	88
4.4.3	Dilution of contamination	89
4.4.4	Clean covers	90
4.4.5	Soil dry screening	90
4.4.6	Summary	91
4.5	Innovative (narrow spectrum) techniques	93
4.5.1	Introduction	93
4.5.2	Remediation of oily contamination	93
4.5.3	Removal of cyanides and heavier hydrocarbon contamination	96
4.5.4	Treatment of metal contaminated sites	97
4.6	Summary	98

5	**Clean cover technology**	**99**
5.1	Introduction	99
5.2	Basic design decisions	100
5.2.1	What does a cover have to do?	100
5.2.2	How long does a cover have to remain effective?	101
5.2.3	What materials can be included in a cover?	102
5.2.4	Detailing the necessary material properties	103

5.2.5	Quantifying the cover design	107
5.2.6	Identifying possible failure mechanisms	108
5.2.7	The time before any failure becomes apparent	111
5.2.8	Ensuring that the client appreciates the design methodology and any liabilities this may imply	112
5.3	Simple covers	112
5.4	Covers intended to combat upward migration of contamination	113
5.4.1	Designing a capillary break layer cover	113
5.4.2	Example	116
5.5	Covers designed to minimize groundwater pollution	119
5.5.1	Introduction	119
5.5.2	Rainfall infiltration and cover design	119
5.5.3	Encapsulation covers	124
5.6	Soil cappings	126
5.7	Appropriate and non-appropriate applications of clean covers	126

6 In-ground barriers   128

6.1	Introduction	128
6.2	The requirements for a cut-off	130
6.3	Gas permeability	131
6.4	Potentials tending to cause flow	132
6.5	Types of vertical barrier	133
6.5.1	Driven barriers	134
6.5.2	Injected barriers	134
6.5.3	Cut-and-fill type barriers	134
6.6	The slurry trench process	135
6.7	Types of slurry trench cut-offs	136
6.7.1	Soil–active clay cut-offs	136
6.7.2	Clay–cement cut-offs	136
6.7.3	Clay–cement–aggregate cut-offs	137
6.7.4	Cut-offs with membranes	138
6.7.5	High-density walls	139
6.7.6	Drainage walls	139
6.8	Slurry preparation	139
6.9	Requirements for the excavation slurry	140
6.9.1	Excavation stabilization	141
6.9.2	Slurry loss	141
6.9.3	Bleeding and settlement	142
6.9.4	Setting time	142
6.9.5	Displacement	142
6.10	Compatibility of materials	142
6.10.1	Clays and water	142
6.10.2	Cements and cement replacement materials	144
6.10.3	Mix water	145
6.11	Sampling slurries	145
6.11.1	Sample containers, storage and handling	146
6.12	Testing slurries	146
6.12.1	Density	146
6.12.2	Rheological measurements	147
6.12.3	The Marsh funnel	148
6.12.4	Other parameters	148
6.13	Testing hardened properties	149
6.13.1	Unconfined compression tests	149
6.13.2	Confined drainage triaxial testing	149
6.13.3	Permeability tests	150
6.14	Specifications for slurry trench cut-offs	150
6.14.1	Mechanical performance: permeability	151

6.14.2	Stress–strain behaviour	152
6.14.3	Durability	153
6.14.4	The design compromise	153
6.15	Overtopping and capping	154
6.15.1	Capping	155
6.16	Future developments	156
6.16.1	The bio-barrier	156
6.16.2	Active barriers	156
6.17	Conclusions	156

7   Reclaiming potentially combustible sites

158

7.1	Introduction	158
7.2	Combustion processes	158
7.2.1	General combustion	158
7.2.2	Spontaneous combustion	161
7.3	Combustion tests	162
7.3.1	Introduction	162
7.3.2	British Standard tests	162
7.3.3	Direct combustion testing	162
7.3.4	Loss-on-ignition tests	163
7.3.5	Calorific value testing	163
7.3.6	Fire Research Station test method	165
7.3.7	Combustion potential test	165
7.3.8	Air permeability test	171
7.4	Use of the combustion potential test	173
7.4.1	Case study 1	173
7.4.2	Case study 2	174
7.4.3	Case study 3	174
7.4.4	Case study 4	174
7.4.5	Case study 5	175
7.4.6	Case study 6	175
7.5	Conclusions	175

8   Gas and vapour investigations

177

8.1	Introduction	177
8.2	Principal subsurface gases and their properties	177
8.3	Gas and vapour monitoring	178
8.3.1	Monitoring equipment	178
8.3.2	Gas monitoring observation works	179
8.3.3	Gaseous properties of importance	186
8.4	Establishing most hazardous gassing conditions	188
8.4.1	Introduction	188
8.4.2	Factors affecting landfill gas (and other degradation product gas) results	188
8.4.3	Factors affecting the measurements of gases not generated by degradation	191
8.4.4	Conclusions	193
8.5	Gas and vapour survey procedures	194
8.5.1	Introduction	194
8.5.2	Selecting observation hole locations	194
8.5.3	Including more hazardous gassing conditions in gas and vapour surveys	195
8.5.4	Data consistency	197
8.5.5	Gassing categories	197
8.6	Solutions to subsurface gaseous contaminants	199
8.6.1	Introduction	199
8.6.2	Gas source identification	201
8.7	Legislative controls and official advice on redevelopment	202
8.7.1	Introduction	202

8.7.2	Main restrictions	203
8.8	Gas emission predictions	204
8.9	Case studies	204
8.9.1	Glasgow greenbelt	204
8.9.2	Paper mill site, central Scotland	205
8.9.3	Former steel mill site, West Midlands	207
8.9.4	Midlands foundry site	209
8.9.5	Leisure development, London Docklands	211
8.10	Summary	213

9 Establishing new landscapes   215

9.1	Introduction	215
9.2	Plant requirements	216
9.2.1	Sunlight	216
9.2.2	Anchorage	217
9.2.3	Water availability	219
9.2.4	Plant nutrients	221
9.3	Soil cover	222
9.3.1	Evaluation	222
9.3.2	Depth of cover	222
9.3.3	Soil ameliorants	224
9.4	Soil fertility	226
9.5	Site preparation	227
9.6	Establishing grass cover	228
9.7	Establishing trees and shrubs	230
9.8	Maintenance	232
9.9	Species selection	233
9.10	Natural regeneration	234
9.11	Conclusion	235

10 UK legal framework   236

10.1	Introduction	236
10.2	Sources of law relating to contaminated land	236
10.2.1	Legislation: primary and subordinate	236
10.2.2	Case law	237
10.2.3	Common law	237
10.2.4	Guidance Notes and Codes of Practice	238
10.2.5	European Union regulations, decisions and directives	238
10.3	Law relating to contaminated land	239
10.3.1	General outline	239
10.3.2	Principal legislation	239
10.3.3	Common law	246
10.4	The Environment Act 1995	247
10.5	Other relevant legislation	256
10.5.1	Introduction	256
10.5.2	Government overall policy	256
10.5.3	Building regulations	257
10.5.4	Waste management powers	257
10.5.5	Summary	259
10.6	Conclusions	260

11 Introduction to waste management practices in the United States of America   261

11.1	Introduction	261

11.2	Regulatory impetus for remediation	261
	11.2.1 Background on regulations	262
	11.2.2 The landbans	264
	11.2.3 BDATs (Best Demonstrated and Available Technologies)	265
	11.2.4 Characteristic wastes versus listed wastes	266
	11.2.5 Phase II disposal role – Universal Treatment Standards	268
11.3	Regulation escalation and backlash	268
11.4	The remediation process under CERCLA	275
	11.4.1 Preliminary assessment/site investigation	277
	11.4.2 Assignment to National Priority List	277
	11.4.3 Remedial investigation/feasibility study	277
	11.4.4 The remedy selection process	279
	11.4.5 Remedial action selection	280
	11.4.6 Remedial process selection	281
	11.4.7 Remedial design/remedial action	281
	11.4.8 Post closure	281
	11.4.9 Public participation	281
	11.4.10 Cost	282
	11.4.11 Reforms needed	282
11.5	Trends in remediation technology	282
11.6	Remediation under RCRA	283
11.7	Radioactive wastes	284
11.8	Remediation markets	285
11.9	Remediation business outlook	286
11.10	Conclusions	286

12 Netherlands' reclamation practices — 288

12.1	Introduction	288
12.2	Legal and administrative controls	288
12.3	Private-sector initiatives	290
12.4	Decision-making on contaminated sites	291
	12.4.1 Introduction	291
	12.4.2 Current practice	291
12.5	Designing a risk assessment	294
	12.5.1 Introduction	294
	12.5.2 Use of CSOIL model	294
	12.5.3 Ecotoxicological assessments	295
	12.5.4 Contaminant migration	295
12.6	Remediation procedures	296
	12.6.1 Introduction	296
	12.6.2 Remediation plan and specifications	297
	12.6.3 Regulation of excavated soils	297
12.7	Soil remediation practices	298
	12.7.1 Introduction	298
	12.7.2 Current trends	298
	12.7.3 Recent initiatives	299
12.8	Summary	300

13 Effective management of contaminated land reclamations — 301

13.1	Introduction	301
13.2	The pre-site work period	301
	13.2.1 Introduction	301
	13.2.2 Information to contractors	302
	13.2.3 Insurance cover	304
	13.2.4 Contact with external bodies	304
	13.2.5 Conclusion	305

13.3 Health and Safety at Work Regulations and Legislation 305
    13.3.1 Introduction 305
    13.3.2 Breaches of Health and Safety Regulations 305
    13.3.3 Consequences of information limitations 306
    13.3.4 Regulatory requirements 308
13.4 The site investigation period 308
    13.4.1 Introduction 308
    13.4.2 Information to be collected 310
    13.4.3 Conclusion 311
13.5 Quality assurance of reclamations 315
    13.5.1 Introduction 315
    13.5.2 Quality systems 315
    13.5.3 Appropriate quality systems 322
13.6 Summary 328

Appendix I Soil guidelines (UK) and Dutch Intervention Values 329

Appendix II Semi-quantified risk assessment methodology 335

Appendix III Other useful guidelines and standards 337

References 358

Preface to the second edition

There has been a marked and abrupt change in professional reactions to land contamination problems from the early 1990s, when the first edition was written.

Throughout the previous decade and a half, contaminated land had been treated with very considerable caution. Few practitioners were involved and these found frequent national/international conferences to be essential opportunities for the exchange of their incomplete experiences and information. In this rather esoteric atmosphere, unspecialized organizations (i.e. most site investigation firms and consulting civil engineering practices) either avoided contamination commissions or specifically enrolled one of the few available specialists when contaminated sites had to be investigated or remedied.

Today the situation is very different. Large numbers of organizations and individuals confidently tackle land contamination problems, specialist conferences are less frequent and attract smaller audiences, and the remediation of contaminated land often seems to be no more than a minor extension of normal geotechnical and civil engineering practice.

In many ways, this new confidence is welcome, since there never was any need for most contaminated sites to be seen as the preserve of a specialized elite.

However, there is real concern, and that is whether or not best environmental solutions are being implemented to reduce future risks to human health, other life forms, construction materials and the wider (particularly the aqueous) environment.

A pedestrian application of the limited guidelines and standards, set up by various national governments, to ensure that land contamination risks to future generations are adequately minimized, is not in itself enough to give necessary safeguards. For example, demonstrating that the soils beneath a site do not exceed the Threshold Limits, established by the UK's ICRCL advice, can mean very little: these Threshold Limits do not cover all the contaminants of importance; they are currently phrased in terms of total concentrations of solid materials; they give no note to solubility/leachability/mobility potentials; and they ignore time variabilities which can give rise to greater risks in some future conditions.

Necessary safeguards are achieved only by thorough evaluations of contamination factors, in the light of the specific conditions on each different site. This, of course, is a call for rigorous and defensible contamination risk assessments and a caution to those less aware practitioners who view contamination evaluations as no more than comparing sampled soil contamination concentrations to the numerical values cited by regulatory bodies.

This necessary emphasis on evaluating future risks is the primary reason for the production of this second edition.

T. Cairney
D. Hobson

Acknowledgement

The constructive criticism and advice of our colleague Stuart Walker in the preparation of Chapters 10 and 13 is gratefully acknowledged. Without access to his specialist understanding of recent legislation and regulations, our task would have been far more onerous.

T. Cairney
D. Hobson

Contributors

Original contributors to the first edition

1 International responses: T. Cairney, W.A. Fairhurst and Partners, 1 Arngrove Road, Newcastle-upon-Tyne, NE4 6BD.

2 Land contamination: M.J. Beckett, Land reclamation consultant and formerly Secretary to the ICRCL, 1 Falcon Court, Alton, Hampshire, GU34 2LP.

3 Rational site investigations: D.M. Hobson, W.A. Fairhurst and Partners.

4 Reclamation options: M.J. Beckett and T. Cairney.

5 Clean cover technology: T. Cairney and T. Sharrock, W.A. Fairhurst and Partners.

6 In-ground barriers: S.A. Jefferis, Golder Associates, 54–70 Moorbridge Road, Maidenhead, Berks, SL6 8BN and visiting professor in the Department of Civil Engineering, Imperial College of Science, Technology and Medicine.

7 Reclaiming potentially combustible sites: R.H. Clucas (Liverpool John Moores University, Clarence Street, Liverpool, L3 5UG) and T. Cairney.

8 Landfill gases: M.V. Smith, Stranger Science and Environment, Louisa House, 92–93 Edward Street, Birmingham, B1 2RE.

9 Establishing new landscapes: G.S. Beauchamp. W.A. Fairhurst and Partners.

10 Quality assurance: T. Cairney.

11 UK legal framework: G.J. Longbottom, W.A. Fairhurst and Partners.

12 Introduction to US waste management approach: M.K. Meenan, Halliburton NUS Environment Ltd, Thorncroft Manor, Dorking Road, Leatherhead, Surrey, KT22 8JB.

13 Health and safety: S.A. Simmons and W.K. Lewis, HASTAM, Aston Science Part, Love Lane, Birmingham, B7 4BJ.

Second edition

1–5, 7, 8, 10 and 13 revised or rewritten by T. Cairney and D. M. Hobson.

6 Stephan Jefferis, Golder Associates.

11 Richard J. Ayen, Vice President (Technology), WMX Technologies, Anderson, S. Carolina.

12 A Van der Horst, J.P. Okx, R.M.C. Theelan, W.J.F. Visser and J.M.C. de Witt, Tauw Milieu Consultancy, Handelskade 11, Deventer, Netherlands.

1
International responses

1.1 Introduction

When land contamination was first identified in the late 1970s, in a few of the industrialized nations, almost nothing was known of the hazards this could pose. The governments concerned, in the United States, The Netherlands, West Germany and the United Kingdom, thus had to react in a state of partial information. Other industrialized countries, which had not had the problem thrust upon them, found the partial evidence of environmental hazards unconvincing and chose not to treat land contamination as a problem of real concern. Some still persist in this view.

This situation is in sharp contrast to the universally recognized concerns over air and water pollution, whose effects have been obvious even to casual observers for centuries. Indeed, water pollution worries were recorded as far back as the time of imperial Rome, when the effects of the city's untreated sewage on the River Tiber were recognized as dangerous to human health. Likewise, the identification of air pollution as a concern has a long history. The smoke clouds that covered Elizabethan London were clearly related to various human ailments, and quite serious (though unsuccessful) efforts were then made to ban the use of coal for domestic fuel [1].

The growth in industrialization in Western Europe, in the 18th and 19th centuries, gave even more convincing proof of the obvious links between air and water pollution events and such declines in the quality of life as the loss of fish in some rivers adjacent to industrial centres, the destruction of trees and crops lying downwind of smelters and industrial chimneys, and the reduced life expectancies likely to occur in many industrial employments [2]. These persuaded the more enlightened administrations of the day into enacting increasingly effective controls over air and water pollution. As this process was gradual, neighbouring nations had the opportunity to absorb the lessons that had been learned and to gain access to the relevant scientific and engineering information.

When the influential environmental movements appeared in the 1960s (largely as a result of Rachael Carson's [3] apocalyptic predictions of the world our children could inherit if controls were not enforced), a considerable technical background existed on which to base more effective environmental conservation measures. Thus, while national differences still persist on some aspects of air and water quality protection, the overall need for such conservation measures is not subject to dispute.

In contrast, little or no concern over land contamination and the fate of spillages and wastes on industrialized land appeared in earlier years. To a large extent this remains the case, though recently announced policies for sustainable development and controls over wastes obviously imply that land contamination is gaining a higher official recognition. Despite this recent trend, other environmental concerns, e.g.

- over ozone layer depletions;
- on the decline of fish and mammal populations in the North Sea; and
- of radioactive leakages in the Arctic Circle from sunken Soviet Navy nuclear submarines

still have much more obvious impact on public perceptions.

In this situation of limited technical guidance and of generally low key public pressures, it is not surprising that different national administrations adopted very different responses to the identification of land contamination.

1.2 Differences in national responses to land contamination

1.2.1 *Various national viewpoints*

That potential hazards can arise from the addition of chemical substances to soils is probably not a matter of dispute. There is, after all, a widespread knowledge that altering the chemical condition of a soil will affect plant life and that substances deposited on land can later migrate to streams and rivers, and then enter the human food chain.

The problem comes when it is necessary to prove that actual risks have occurred, or are likely to take place and so justify particular control measures.

Some nations have found no difficulty in being absolute in their stance on this, and senior Dutch spokesmen have expressed such definite views as: 'soil pollution is a gigantic problem with serious consequences for man, plants (including crops), animals and the abiotic environment' [4].

The Netherlands' authorities have persisted in this view and have recently [5] enacted new soil quality legislation. This introduces Soil Intervention Values (at which clean-up is mandatory) to replace the earlier established and well known Dutch 'C' values. The fact that these Intervention Values are generally stricter (for soils) than the earlier standards reflects continued Netherlands' concerns over ecotoxicological and human health risks, and a national willingness to remove the heritages of past industrial activities almost irrespective of financial burdens.

In contrast, other nations like the United Kingdom have been less certain of the scale of the likely hazards and initially chose to discuss land contamination only in the context of the safe redevelopment of land, e.g.

Contaminated land is that land which, because of its former uses, now contains substances that give rise to the principal hazards *likely* to affect the proposed form of development and which requires an assessment to decide whether the chosen development may proceed safely, or whether it requires some form of remedial action, which may include changing the layout and form of development [6].

This British caution has continued.

The latest UK element of environmental protection law (the Environment Act 1995 [7]) seeks clearly to establish an optimum balance between environmental protection and the wider economy, and so introduces the concept of 'significance'.

Contaminated land is thus defined as 'any land which appears to the local authority in whose area it is situated to be in such a condition, by reason of substances in, on, or under the land that

- 'significant harm is being caused or there is a significant possibility of such harm being caused; or
- 'pollution of controlled waters is being, or is likely to be, caused.'

This stress on the word 'significant' will remove a large proportion of former industrial sites from the contaminated category and will focus attention on those few sites (probably rather less than 10% of the total) of real concern. This arguably is a more realistic approach than the attitudes developed in The Netherlands.

Other industrialized nations have generally adopted stances intermediate between the extremes of British and Dutch practices (Table 1.1), though some (e.g. Italy) have yet to recognize that a problem could exist.

The reason for this variation in possible viewpoints is easy to identify.

1.2.2 *The complexity of soil systems*

Soil–water–air environments are extremely complex; different soil fractions and constituents give rise to diverse reactions when anthropogenic chemicals are introduced. In many cases, contaminants become bonded, particularly to clay and organic content particles, and so are unavailable to present future risks [8]. In other situations introduced chemicals may remain unbonded, or can be remobilized by changes in soil acidity or redox potential, and then become far more able to create risks. While retention bonding is much commoner than contaminant leaching, this need not always be the case. Thus the simplicity of 'cause and effect' relationships – such as those noted in air and water pollution incidents – is usually absent. One result of this lack of unequivocal evidence of harm being created by soil contaminants is

Table 1.1   Different responses to the problem of contaminated land

Supernational/national grouping	Response
European Union	No specific soil quality directives. General aim to ensure sustainability and preclude uncontrolled waste disposals. Strict waste control directives. Strict groundwater quality guidance.
Belgium	No specific soil quality laws. Waste control, environmental permit and water quality laws used to enforce some land remediations.
Denmark	No specific soil quality laws. Chemical Waste Sites Act 1983 used to enforce clean-up of some more hazardous sites.
France	No specific soil quality laws. Waste regulations enforced to prevent more contaminated land being formed. Tax on waste disposal being used to fund clean-up of worst known sites.
Germany	Soil Protection Act planned. Processes of ranking sites in hazard order developed. Registration of sites undertaken. Residual Pollution Law enforces soil remediation. System varies from state to state, as do the soil criteria which are employed.
Netherlands	Soil quality legislation in force. Multifunctionality remains national policy of clean-up standards. Highest soil clean-up requirements of any nation.
United Kingdom	Soil quality legislation recently introduced (1995). No soil quality standards produced, instead a risk assessment approach is required. Multifunctionality rejected as a national policy. Controls exerted largely by local authorities.
Australia/New Zealand	No soil quality legislations. Recognition of the problem very recent.
Canada	No federal soil quality legislation. States have very variable emphases. Some (e.g. Quebec) mirror Netherlands attitudes to a degree. 'Polluter Pays' policy favoured. Interim soil quality standards exist.
USA	Soil quality standards devised, but vary from state to state. Federal legislation severe and enforced. Federal clean-up of problem sites in progress.
Other industrial nations	No official recognition of land contamination as a problem.

that national governments lack certainty over any need for controls and protective actions.

Good examples of the complex behaviours of contaminants in soils are provided by such workers as Thornton [9], who showed that vegetables grown in soils, which were 10 times more contaminated than usual urban soils, still appeared to generate no adverse effects to the health of people consuming these garden crops.

Thus it is not surprising that demonstrable cases of harm from contaminated soil are almost totally absent in the technical literature. For example, Dutch studies [10] on a population exposed to chlorinated benzenes and phenols from a household garbage dump failed to prove that any individuals had in fact suffered, despite long-term exposure to what are obviously dangerous substances.

This lack of actual proof that harm has occurred does not, of course, mean that risks are non-existent or negligible, but merely indicates that they are difficult to identify and probably very slow to appear.

However, national administrations, forced on by local public pressures, cannot delay decisions for decades until harm has actually appeared. Thus intuitive and emotional responses have tended to typify the various national reactions.

1.2.3 *Identification of land contamination and necessary clean-up standards*

In Europe, The Netherlands has been most emphatic in recognizing land contamination as a major concern. Other nations and states (e.g. Germany, Austria, Norway) have followed the Dutch lead though with less commitment, while a surprising number of governments have yet to decide whether or not a cause for concern exists (e.g. Wallonia (Belgium) and Greece) (Table 1.1).

In the light of this variation, the UK's position, that risk assessments are essential to identify those more hazardous sites, is less idiosyncratic than is sometimes claimed.

The most recent tabulation (Table 1.2) of the scale of clean-up necessary [12] in different countries, makes clear the obvious differences in emphasis between those nations which have succeeded in devising registers of contaminated land. For a small country of the size of Holland to have identified more than twice as many sites requiring clean-up as its much larger and more heavily industrialized German neighbour is obviously not credible, unless – as is the case – the Dutch have chosen to apply a more rigorous definition of land contamination and have had the national will to accept the financial and other consequences. Also, the remarkably small number of

Table 1.2 Estimated scale of the necessary clean-up of contaminated land (after [12])

Country	Likely number of suspect sites	Likely number of contaminated sites	Estimated clean-up costs (US$ $\times 10^9$)
Netherlands	600 000	110 000	25[a]
Germany	200 000	50 000	10–60
United States (Superfund)	35 000	1 200	30
Denmark	6000–10 000	Not known	Not known

[a] The Netherlands clean-up costs have also been estimated at $27 billion, or $270 million per annum for the next 100 years [12].

US sites requiring remediation is obviously indicative of a very different national emphasis from that in The Netherlands.

The Dutch lead position resulted from the 1983 adoption of 'A-B-C' soil quality criteria. These had the logic of establishing the

- chemical contents of clean soils (the 'A' background levels)
- higher concentrations (the 'B' levels) at which further investigation was required, and
- still higher concentrations (the 'C' values) at which soil clean-up was mandatory.

These standards can, however, be criticized for advocating single contaminant concentrations as precise indicators, and for ignoring the variations which soil matrix differences inevitably generate. It simply is not defensible to give one concentration of a particular contaminant a unique significance. If the contaminant were in permeable sands it is evident that mobility opportunities would be much greater than in peaty clays where fixation by cation exchange and humic acid bonds would be probable. That The Netherlands authorities embarked on this unjustifiable adoption of single contaminant concentrations to represent 'clean' and 'contaminated' soils is very surprising, since Dutch knowledge of background chemical variations, in different soil types and in areas well away from any prior industrial activities (Tables 1.3 and [13]), was much greater than was available elsewhere.

Recently[5] the new Dutch soil quality criteria (Appendix I) have corrected this unsatisfactory situation and now permit higher chemical levels in soils with more clay and organic matter content. Thus the allowable content of mercury in a 'standard' Dutch soil (with 10% organic matter and a 25% clay content) is less than 10 mg/kg. However in peaty clays, clean-up intervention need not take place until a mercury content of 27 mg/kg is exceeded.

Nevertheless the latest Dutch criteria still fail to allow for any consideration of contaminant mobility. For example, metalliferous slags typically display high concentrations of zinc and other metals, but these usually prove to be so insoluble that harm to health or to the environment cannot reasonably be suggested.

Table 1.3 Comparison of some background values in nature reserve topsoils and the ABC guidelines

Element	Concentration range (mg/kg) in			A value in guidelines (mg/kg)
	Sandy soils	Loamy soils	Clay soils	
Chromium	11–43	33–75	99–117	100
Cobalt	0.3–2.0	2.4–10	14–16	20
Zinc	6.4–43	28–150	135–153	200
Arsenic	1.4–18	5.1–13	19–21	20
Barium	80–266	168–416	466–525	200
Bromium	2–18	5.7–42	14–16	20

It is also apparent from Table 1.3 that the Dutch have chosen to accept the costs and difficulties of cleaning up old industrial land to a condition similar to that of a nature reserve. No other country has so far been willing to accept the financial burdens and technical difficulties that this choice presents.

The standards adopted by other countries are equally easy to criticize. In the United Kingdom, the pitfall of ascribing a single contaminant concentration that makes clean-up mandatory has been avoided, but at the cost of leaving the interpretation of the guidelines open to a large element of professional judgement. This can give rise to very different clean-up standards being used on adjacent sites and makes it difficult to justify any claim of a national standard.

In those countries where federal and state authorities coexist, the situation can be more complex. In the United States, the Environmental Protection Agency has developed a comprehensive and evolving set of clean-up criteria, yet a state like New Jersey (with a particularly rigorous view on environmental clean-up) makes use of its own informal guidelines for the soil and water concentrations that must trigger site clean-up activity [14]. Likewise in Germany, the federal authorities have established a national strategy to remove contaminated land hazards, but the implementation is necessarily left to the various Laender authorities, which still lack any uniform and agreed evaluation system on which contamination hazards are judged [15].

This degree of variation is unsurprising: political judgements in an area where scientific certainty is absent will depend on public attitudes and perceptions. The consequence has been a wide diversity in the chemical concentration levels at which soil clean-up is required; cadmium (for example) does present real risks to human health and so should not be allowed to remain at high concentrations in soils, but different national agencies have set the clean-up point at concentrations (for land to be re-used for domestic gardens) which vary from 1.5 mg/kg to 20 mg/kg. Similar variations are apparent for almost all soil contaminants.

1.2.4 *Variation in legislation and controls*

The differences in national emphases obviously have governed the legislation and controls that are seen as necessary.

The Dutch view of soil multifunctionality logically requires contamination clean-up to a very high standard. This, in turn, has led to The Netherlands Interim Soil Clean-up Act (1983), the Soil Protection Act (1986), and the 1994 Soil Protection Law, all of which permit local authorities to set constraints on the environmental impacts of industry. Without doubt, this is a particularly comprehensive set of land contamination controls and is supported by a national programme of research and development to prove the effectiveness of innovative clean-up technologies. A recent effect of these policies is that the state has found it impossible to organize, fund and execute all the clean-ups that are seen as necessary [12] and efforts are in progress to

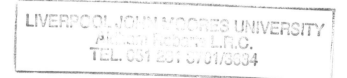

encourage industry voluntarily to clean up some sites. Also, technical difficulties of attaining the required clean-up level, in some cases, have led to criticisms of the national standards, and the recent modifications of the Dutch
Building Act (which permit local authorities to include soil contamination as
a factor governing whether or not a building permit is granted) look likely to
harm the necessary inner city rebuilding programme, since a great many sites
have been levelled with dredging spoils that are more contaminated than the
Dutch mandatory clean-up levels. Thus the Dutch position looks to be one
where a laudable emphasis on environmental conservation is increasingly
coming into conflict with other equally desirable national aims.

The United States situation is, in some ways, analogous to that in The
Netherlands and environmental controls have been enforced to a level that
is surprising, given the usual US emphasis on constitutional freedoms. The
Comprehensive Environmental Response, Compensation and Liability Act
of 1980 and the Superfund Amendments and Reauthorization Act of 1986
give the Federal Environmental Protection Agency (EPA) very considerable
powers. Funding to support technology trials and demonstrations is also
provided on a scale much greater than in most other countries. The EPA
can force abatements of environmental threats and damage and can compel
responsible persons to pay the costs of rectifying environmental degradation.
Since these powers can be applied retro-actively (unlike those of UK or EU
legislation), many commentators [16] have seen the system as unfair, more
likely to involve expensive litigation than land clean-up, and a direct cause
of the near impossibility of insuring against environmental liabilities. Other
writers have noted that the system has led to an increasing number of derelict
(and possibly contaminated) sites, which have been abandoned by bankrupt
companies. Thus, the required environmental protection might actually be
being reduced by the severity of the legislative controls.

Other European countries have avoided the severity of the Dutch and US
legislation, though most have gone some way to imposing environmental
restraints on land contamination, and on the use of such land.

What has been learnt in the past decade seems to be that land contamination is simply one factor that should be of concern in a modern industrialized
society and that an overly rigorous legislative emphasis on it can, in itself,
adversely affect other desirable environmental objectives.

1.3 Signs of an international consensus

Some differences in national responses to the identification of land contamination will inevitably and very properly persist. Geological, population
growth and land scarcity factors alone make this inevitable.

In small, heavily populated countries where shallow groundwater is used
for public water supplies (such as The Netherlands) it is reasonable to seek

to have as high a soil clean-up standard as can be afforded. Less thorough remediations, suitable (say) for light industrial re-uses could give rise to future problems if land for housing were unavailable.

Elsewhere, the mere presence of a thick cover of poorly permeable glacial clays removes much of the concern which could be voiced over pollution of groundwaters. This is the position in the UK where deeper aquifers, used for water supplies, have generally been protected from the ill-effects of former industrial activities.

Where land is still abundant (as in much of the USA and Australia) the need to recycle disused industrial land can be less pressing, and adequate safeguards can be gained by partial remediations designed only to protect water resources from pollution effects.

Remediating contaminated land is an expensive business, particularly when thousands of old industrial sties are identified. The apparent simplicity of 'making the polluter pay' has often failed to place the cost burdens on the guilty parties, and instead, has given rise to a lucrative legal defence industry. Thus nations have to weigh the significant costs of land remediation against other pressing social needs.

Despite these reasonable differences, the first signs of an international consensus have begun to emerge. This includes:

- the recognition that only when contaminated sites are located and examined can the hazards which contaminants in the ground pose be identified, some form of registration is thus essential;
- the agreement that scarce national resources ought to be targeted only on problem sites;
- the necessity of preventing the formation of still more areas of contaminated land by enforcing increasingly strict waste disposal and integrated pollution control policies; and
- the pressing need for scientifically precise rankings of the hazards which contaminants in the ground can pose. On-going exchanges of technical information, such as [17] offer the best hope of achieving the factual base which today is still elusive.

Since the publication of the first edition of this book, only slow progress has been apparent in the convergence of differing national attitudes. It is to be hoped that the next decade will be productive.

Thus, the next decade offers the hope that the distinctions between national attitudes may reduce, that a sounder basis for establishing the actual hazards from land contamination may appear, and that misapplication of scarce national resources can be minimized.

While the bulk of this book is based on UK information and methods, there is, in fact, no reason to see this as a parochial attitude. Contamination of soils from past industrial uses does not differ technically from one country to another; all that changes are the national policies.

2
Risks from contaminated land

2.1 Introduction

The literature on land contamination is dominated by accounts of different investigation and remediation methods and their effectiveness in various site conditions. This is unsurprising. Most authors and their readers trained as scientists or as engineers and naturally find the solution of complex technical problems especially interesting.

In contrast, policy issues and the methods of devising suitable regulatory systems find very few publicists, probably because they seems less exciting and important. Yet it is precisely these more 'boring' subjects which primarily determine whether clean-ups of contaminated land are necessary at all, the soil quality levels which have to be achieved in reclamations, and the ultimate remediation costs which the state, private organizations or individuals will have to bear.

To understand more fully the differences in the various national reactions to land contamination's identification, and so appreciate why particular advice is given on site investigation good practices (Chapter 3) and on contamination remediation choices (Chapter 4), it is helpful to consider why basic policy and regulatory systems differ so widely. In particular, answers to three salient questions, i.e.

- what actually constitutes 'contaminated land'?
- what chemical concentrations levels distinguish 'clean' soils from those which are 'contaminated'? and
- what risks – if any – can arise from the presence of anthropogenic contaminants in soils?

are worth considering.

One recent effect of greater professional understanding of land contamination's effects is that the expenditure of scarce national resources has to be targeted preferentially on sites which pose the most acute risks to public health and the environment. Other pressing social needs – e.g. job creation, education and training opportunities, and the extension of health care – obviously compete for whatever funding is available and can have more compelling cases.

2.2 What is 'contaminated land'?

This seems almost a silly question, since it is well understood that it is the introduction of higher than natural concentrations of chemical substances (resulting from prior uses of land) which distinguish 'contaminated' from 'clean' soils.

However, the difficulty in answering this seemingly simple question is central to the differences which occur today between policies in different countries.

In the UK, as long ago as 1984 [1] a clear distinction was drawn between 'contamination' and 'pollution'. This persists to the present day as the core belief in the UK's environmental protection policies. 'Contamination' was recognized as being a result of 'the introduction or presence of alien substances or energy', but – in the absence of adequate scientific information – this could not be judged to be 'liable to cause damage or harm'. An example of what might be termed soil contamination would be the presence of a thin peaty clay, at substantial depths (say 2.8 m) in an otherwise clean soil profile. The presence of highish concentrations of mercury and cadmium (at – say – concentrations of 3.4 mg/kg and 12.3 mg/kg respectively) would certainly show that metal levels are much higher than would be normal in clean soils, but do these metal concentrations, in this particular setting, present risks of future harm? Without a good deal of additional data it is not possible to answer this question.

In contrast for 'pollution' no such uncertainty exists. The definition again specifies that it is the 'introduction of substances or energy' which are of concern, but goes on to identify that these will be 'liable to cause hazards to human health, living resources and ecological systems, damage to structures and amenity, or interferences with legitimate uses of the environment'. In terms of soils, an example of a polluted site might be one adjacent to, and topographically higher than, a surface water course, and where the soil layers are heavily saturated with diesel fuel spillages. This land would inevitably be seen as polluted, since it is obvious that a high probability exists of the oils migrating to the nearby river to cause water quality deterioration.

This distinction between contaminated and polluted ground obviously implies that some form of risk assessment will have been undertaken to establish that the hazards, created by the presence of subsurface contamination, could result in probable harm. The metallic concentrations in the thin peaty clay layer, noted above, certainly pose chemical hazards, since both mercury and cadmium adversely affect human health and the well-being of other life forms [2], but for these hazards to create problems, and so be raised to the higher status of risks, calls for a target to become affected by contaminant migration along discernible and predictable pathways. Without an unbroken chain from pollution risks – on the UK's policy – cannot be claimed. An everyday example should make the distinction clearer – metals

a hazard $\rightarrow$ pathways permitting contact $\rightarrow$ a target coming into contact
with the hazard,

such as lead and uranium certainly are hazardous to health, particularly if
they exist at very high concentrations, yet good-quality crystal drinking
glasses (with up to 10 000 mg/kg of metals introduced to give different colorations) are used without any concern. The metals are, of course, so locked
into glassy silicate compounds that they remain totally insoluble and immobile, and so cannot harm the users of the crystal glasses. Similarly the presence
of higher than normal concentrations of chemicals in soils certainly poses
hazards, but this alone is not enough to suggest that risks will actually result.

In stark contrast, in nations such as The Netherlands, no such subtle distinction is drawn. There the discovery of contaminant hazards in soils is deemed to
be enough to necessitate land clean-ups, and 'hazards', to a large extent, are
equated with 'risks', and contaminated land is equated with polluted land.

This basic difference in perception itself is the essential cause of the discrepancies in the numbers of contaminated sites which have been identified
in different national programmes (see Table 1.2).

Contaminated land is seldom visually obvious to a layman except in most
polluted cases. The self-evident distinctions between other categorizations of
land types (i.e. between arable and industrial land) are absent. Equally contaminated land is not always marked by the sort of industrial dereliction
which typifies disused railway yards and abandoned iron foundries; some
green pleasant areas are underlain by significantly contaminated soils, and
derelict industrial premises can prove to be quite uncontaminated. So it is
necessary to define land contamination in terms of those specific concentrations of anthropogenic chemical substances which might pose risks. The difficulty is choosing these contamination levels.

2.3 Estimating the amount of contaminated land

Because of the inherent difficulty in defining what constitutes land contamination, it follows that accurate estimates of its presence are also hard to
make, and that comparisons between the areas identified in various countries
(e.g. Table 1.2) have to be interpreted with caution.

Two methods have been used for estimates of the extent of land contamination, i.e.

- those based on surrogates for contamination (e.g. dereliction); and
- those derived from actual surveys and land registrations.

In the UK where no national contaminated land registration has get taken
place, the first approach has necessarily been used. This – while consistently
indicating the existence of significant amounts of land contamination – has

given rise to widely varying estimates, which cannot be regarded as reliable. Most recently, it has been suggested that up to 100 000 contaminated sites might exist in England and Wales alone [3], though it is highly likely that only a small fraction of these will pose a threat to public health or the environment. This lack of factual certainty was recognized as a limitation on policy by the UK government [4,5] and led to the abortive attempt to set up registers of land 'which might be contaminated', under Section 143 of the 1990 Environmental Protection Act [6]. This intention, however, had to be abandoned in 1993 [7] after land owners and developers objected to the policy, which would have permitted sites to be registered on the suspicion that contaminated soils could exist, yet would not have allowed the removal of site registrations even if later investigations showed no contamination or that the land had been remediated to entirely acceptable levels. Currently, revised forms of contaminated land registers (but only of land *proved* to be contaminated and to such a level that remediation notices – specifying necessary remedial action to protect people and the environment – have been served on land owners or occupiers) are being introduced under the powers of the 1995 Environment Act. The means of determining whether land is contaminated to the extent that a Remediation Notice could be issued is dependent on guidance from the Secretary of State. As this has not been issued at the time of writing it will be a considerable time before it is possible to specify with certainty how much contaminated land (i.e. polluted land on the Royal Commission's definition) actually exists in Britain.

Most other European and North American nations are further advanced with the formal registration of contaminated sites, and so are able to supply better estimates of its abundance. However, since the criteria used to define 'contamination' obviously differ, these estimates are not directly comparable. As Table 1.2 indicates, The Netherlands makes use of criteria, which in the UK's view would indicate contamination but not actual pollution, while the USA's identification processes are more in tune with UK approaches and so have identified only a very small number of problem sites.

2.4 Standards and criteria to establish if soils are contaminated

Once any nation decides to ensure that land is safe for re-use, it becomes necessary to establish lists of chemical concentrations which distinguish between 'clean' and 'contaminated' soils and the water and gases which occur within these.

On first sight this might not seem to be too difficult a task, particularly since so many such listings have been devised (Table 2.1). The method to be adopted would (say for risks to human health) probably be to identify what ingested weight of a particular contaminant would be likely to result in risks to human health, judge how much ingestion would

Table 2.1   Examples of soil quality criteria currently employed in Europe

Denmark	proposed limit values
Finland	polluted soil values
Germany	Federal draft law values
Germany	Berlin land re-use levels
Germany	Eikmann–Kloke values
Netherlands	Intervention values
United Kingdom	ICRCL

be probable in normal living conditions, and then relate this to particular measured contaminant concentrations in the soils at land surface or at shallow depths. In this way it would seem feasible to establish which critical contamination limits are of probable concern (Table 2.2).

Table 2.2   Typical human health exposure assessment

The risk likely to occur from eating garden crops grown in contaminated soils may be gauged as follows:

Maximum concentration (as dry weight) of contaminant known to exist in garden produce (from site investigation)	$= X$ (mg/kg)
Likely maximum daily intake of the garden produce (proved by survey)	$= G$ (g/day)
Likely period of crop intake (proved by survey)	$= D$ (days)
Total likely annual intake	$= X \times G \times D$ (g)

If the intake dose equals or exceeds national or international health risk guidance, then remedial action will be necessary. Focus, in such health risk assessments, is usually on more sensitive human subjects (children and older people) with limited body weights, and a greater susceptibility to health risks.

However, it takes only a brief consideration to realize that matters are far from so simple.

- Many contaminant elements occur as a range of compounds, some of these are more soluble, available for biological uptake and toxic, while others are far less hazardous. For example, arsenic is a very common chemical addition to soils under older industrial sites and exists in various organic and inorganic forms. Toxicity is known to depend on the actual compound's oxidation state, yet chemical analyses of suspect soils seldom reveal any such detail, and usually only indicate the total metal concentrations present. Thus assumptions have to be made, by those bodies devising standards and criteria, as to which actual compounds might exist and the risks which might arise at various contaminant concentrations.
- Metals, in particular, occur as a wide range of possible species (Table 2.3) and are bonded to a greater or lesser extent. Solubilities and bioavailabilities thus differ and a concentration of a metal as one particular species might have no risk significance, yet precisely the same concentration (if it occurred in another of the feasible species forms) might pose substantial risks to people using the land. As noted above, when chemical analyses of samples of soils reveal only total metal concentrations, it is not possible to establish whether

Table 2.3 Speciation possible for commoner contaminants (zinc as an example)

Simple hydrated ion	(e.g. $Zn(H_2O)^{6+}$)
Simple inorganic ion	(e.g. $Zn(H_2O)_5Cl$)
Stable inorganic complex	(e.g. Zn_5)
Simple organic complex	(e.g. Zn in composts)
Adsorbed on inorganic colloids	(e.g. Zn in clays)
Adsorbed on organic colloids	(e.g. Zn in humus)
Free element form	(e.g. metallic zinc dusts in factories)

metals are available or not, and so – again – devisers of numerical concentration lists have a real problem.

- Soils are not simple environments. Factors such as pH, redox potentials, and the amount of clay minerals and organic matter present affect whether or not contaminants are locked away or are easily available for uptake to create risks. Processes in soils which affect the availability of inorganic contaminants are reasonably (though still incompletely) understood [8] while those for organic compounds are still subject to considerable doubt. Thus it is extremely difficult to specify any unique concentration value for a specific contaminant which (if exceeded) implies that land is contaminated, unless this is done in terms of *all* the soil environment variables and interactions which are feasible. Obviously this would be a major task, which no national contaminant concentration listing has yet been able to include.

- To make matters even more complicated interactions occur in soil environments between different contaminants. Sometimes these increase the potential toxicities of the various contaminants involved (synergistic effects) while in other cases competition between individual contaminants actually reduces the overall toxic potentials (antagonistic effects). In yet other situations, no interactions occur at all, and so simple additive effects of several discrete contaminants can be identified. As if this were not complex enough, studies on simpler life forms have shown that the interactions differ depending on the species exposed to the effects of combinations of contaminants (Table 2.4). As toxicological trials cannot ethically be conducted on human subjects, it is obvious that the same knowledge cannot easily be established for risks to human health, and so very real uncertainty persists.

Table 2.4 Interaction effects of different metallic contaminants

Metals which occur	Interactions possible	Targets affected
Nickel and zinc	Synergism	Bacteria
	Antagonism	Fungi
	Additive effects	Cyanobacteria
Nickel and lead	Antagonism	Fungi
	Additive effects	Algae
Cadmium and nickel	Synergism	Bacteria
	Antagonism	Algae

Contaminant concentrations in soils are invariably low compared to the known cases where accidental exposures have proved that chemical hazards have caused very real harm, so industrial health and safety experiences offer little in the way of added insight into which soil concentrations indicate 'clean' conditions, and which others imply unacceptable contamination, i.e. pollution [9].

The situation is even less clear when risks of harm to the aqueous environment have to be considered. Obviously this calls for assessments of migration pathways and contaminant mobilities, which can be affected by various physical and/or chemical conditions. Any suggestion that a single acceptability criterion could ever adequately describe the possible risks of water pollution occurrence could not be defended.

Thus despite the best efforts of different national organizations, no demonstrably accurate and comprehensive listings have yet been devised which adequately distinguish between 'clean' and 'contaminated' conditions in all types of soils. Because of the uncertainties involved, the fact that different suites of contaminants affect different groups of targets, it is in no way surprising to find that the allowable concentrations of soil contaminants vary to a real extent in the different national listings. For example, the concentration of cadmium, above which the soil which contains this metal is deemed to be contaminated, varies (for land to be used for crop production) from 1 mg/kg to 50 mg/kg, in different national listings [10]. Despite this, such listings are essential; otherwise how could remediation practitioners judge whether or not enough soil quality improvement had been effected? In some nations, the listings have legislative force and are deemed to be soil quality standards. In others, such as the UK and Denmark, only advisory guidelines are offered, and have to be applied with professional judgement (i.e. estimation of the predictable risks likely).

One final point, which is worthy of consideration, is whether or not critical chemical concentration values from another nation's standards or guidelines should be used when one's own country's listings provide no advice. From what has been stated above it should be obvious that in a situation of scientific uncertainty and very partial knowledge, different states have produced guidelines or standards which primarily satisfy their unique political, economic and public perception framework. No national guidelines or standards have yet been derived from incontroversial technical evidence. Thus to employ contamination criteria, devised in a country where official attitudes differ markedly from those in which the contaminated site occurs, should only be done with great caution. The trend in the UK (where contaminant guidelines encompass relatively few chemical substances) for use of the much more complete Netherlands' standards to supplement the ICRCL advice [11] often seems to be especially ill-judged. If practitioners do have to use other nations' contaminant concentrations it would be more reasonable if they employed criteria from states, e.g. from Quebec [12] where, at least, the geological and hydrogeological conditions resemble those of Britain.

2.5 Risk assessment of soil contamination

Note has been made, above, of the variations in the definitions of contaminated land, the effect of these on national policies, and of the current situation in which no available contaminant guidelines or standards can be seen as other than best available estimates.

Given this situation, it would not be logical to accept that the presence of measured concentrations of contaminants in soils, in itself, is enough for an investigator to be convinced that risks to various targets will arise, and thus show that site remediation is imperative.

Instead it seems much more reasonable to advise that thorough risk assessments always have to be conducted, once site exploration has revealed the distributions and concentrations of contaminants under the site and which pathways for contaminant migration and contact exist.

This, of course, is a re-affirmation of the Royal Commission on Environment Pollution's views[1].

Several different types of risk assessment can be used, i.e.

- qualitative;
- semi-quantitative; and
- quantitative assessments.

In the qualitative process, an assessor judges, from the site's subsurface information, whether or not risks will be probable and then ranks these simply as 'high', 'medium' or 'low'. This is by far the quickest and cheapest method of assessing soil contamination risks, but it is one in which the biases and experience (or lack of it) of the assessor can overly dominate what should be a rigorous and defensible evaluation process. Because of the freedom allowed to assessors in qualitative judgements significant errors can often be identified (Table 2.5) [13] and it is sometimes very difficult to accept that such assessments are other than personal statements of opinion.

More satisfactory are those semi-quantified risk assessments [13] in which formal judgemental protocols are included and the biases of assessors are restricted. For such assessments, critical concentrations of particular suites of contaminants are used, and these necessarily differ as different possible risks are considered.

This should not be surprising; the contaminants likely to attack concrete and construction materials simply are not those which might harm human health or affect the qualities of surface and underground water resources.

An apparently much more comprehensive and complete risk assessment is possible if a fully quantified assessment process, such as that designed by the

Table 2.5   Abstracts from some qualitative risk assessments

Site	Claim	Reality
A	'No risks from landfill gases are probable. Methane was not detected and carbon dioxide levels seldom exceeded 2.1% by volume.'	The judgements were made on data collected in mid-winter over two weeks. Later gas measurements in the following hot summer revealed a far riskier situation with explosive methane levels rising under positive gas pressures.
B	'Insurance cover will be difficult to obtain, since high concentrations of metals occur at shallow depths. This will pose likely future and serious health risks.'	Metal working slags did exist at depths of 3 m and below the site's groundwater table. However, these metals were so insoluble that groundwater quality remained high. As all soils above the slags were demonstrably clean, the risk (cited) is over-stated.

US Environment Protection Agency and used by workers such as La Goy [14], is utilized. This was devised to establish the risks to human health from contaminated soils, and is founded on an impressively long listing of different contaminant potency estimates. These (expressed in units of lifetime risk) when divided by the critical human target's body weight, and then multiplied by the daily exposure dose of the contaminant the human subject is likely to encounter, provide numerical risk assessments. Using this approach various workers (for example [15]) have been able to suggest what order of benefit (in terms of reduced risks to human health) can be gained by undertaking site remediation to a greater or lesser extent. To European readers, such numerical assessments can appear extremely impressive, but a closer examination of the contaminated potency estimates, on which the USEPA method rests, soon reveals that these are subject to essentially the same criticisms as those made earlier on the various national guidelines and standards. Many of these estimates necessarily were obtained from animal studies and may not be relevant to human health judgements, dose–response relationships have been included which owe as much to political as to scientific views, and there is doubt that the assumed daily ingestion doses are actually realistic estimates of the contaminant weights retained in human bodies to pose health risks. Despite these comments, what is undeniable is that the USEPA approach is rigorous and not as subject to individual assessor biases as are qualitative risk assessments.

Given the still-current uncertainties over the fates, potencies and interactions of the contaminants which can exist in soils, it is difficult to accept that the use of the USEPA risk assessment methodology is justified in European conditions, particularly since it fails to address the greater number of different risks which can occur when contaminated land is to be re-used (Table 2.6). Likewise, the cheap and rapid qualitative risk assessment method is too often unconvincing for its continued use to be advocated (see Table 2.5).

Table 2.6   Risks possible when contaminated land is re-used for sensitive purposes

A. *Risks which are breaches of UK statute law*
- Risks of polluting surface waters
- Risks of polluting underground waters
- Risks of creating area-wide air pollution
- Risks of contaminants migrating into adjoining properties or land

B. *Risks which could affect future users of the land*
- Risks of gases/vapours entering enclosed spaces
- Risks to the integrity of construction materials
- Risks to plants established in contaminated soils
- Risks to human health, particularly from ingestion and inhalation

For these reasons, semi-quantitative risk assessments seem to be as much as is justifiable in the current information situation; they also have the real advantages of being comprehensive in the range of risks which can be evaluated and in providing consistent outcomes, even when different assessors evaluate the same site's information. A summary of the assessment method advocated is given in Appendix II.

When a contaminated site is to be re-used for some productive use (as is the commonplace result in the UK because of population growth pressures and land scarcity) several different risks could be feasible (Table 2.6). Some of these, should they arise, would be breaches of the UK's statute laws and so would attract legal challenges, and probably fines or penal sentences in worst cases. Others would primarily affect future users of the site and could lead to these individuals seeking legal redress for any damages for nuisances suffered.

Table 2.7   Contaminants of greatest importance to different risk targets

Risk of	Contamination suites of significance
Surface water pollution	Liquids (oils, leachates), water soluble metals and sulphates. Very acidic or very alkaline soils.
Groundwater pollution	As surface water risk listing.
Area-wide air pollution	Dusts rich in asbestos, metals, or organic compounds. Large-scale gaseous emissions.
Off-site pollution migration	As above.
Gases/vapours entering dwellings	High concentrations of flammable, toxic or asphyxiating gases in soil profiles, where these gases are under enough positive head to permit upwards flow to the site surface.
Construction materials being degraded	Highly acidic soils containing water soluble sulphates (or sulphur wastes) and chlorides. Damage to steel requires both air and water to be present. Organics affect PVC pipes and fittings.
Plants being harmed	Extremes of soil pH, together with the presence of water soluble phytotoxic metals. Gaseous concentrations which cause soil oxygen levels to fall will also be a risk.
Human health	Dusts rich in metals, asbestos or organic particles. Extremes of soil pH harm human skin and permit contaminant entry into the body. Toxic metals at high concentrations can harm 'pica' children.

In the case of each risk, the effects of a unique and distinct suite of contaminants (at whatever critical concentrations are deemed to pose potential hard) (Table 2.7) on the future well-being of the selected targets has to be assessed. Ideally such assessments should be conducted before any reclamation is undertaken (when they usefully indicate which risks are of the greatest significance in the context of a particular site and its planned re-use), *and* after reclamation has been completed, to ensure that the reclamation work has satisfactorily achieved the necessary freedom from future risks.

While such risk assessments are implied in the 1984 Royal Commission Report [1] and have been part of the UK's ICRCL approach to contaminated land evaluations since 1983 (section 2.6), it is only in recent years that governmental [16,17] and professional [18] publications have included descriptions of the assessment methodologies which might be employed. This reflects the increasing concerns of funders [19] and developers, in a period of increasingly rigorous environmental protection legislation, that they do not become open to future liability claims or legal actions.

2.6 The development of legislation, policy and practices in the UK

Restoration of land dereliction commenced in the UK after the 1966 Aberfan colliery spoil heap disaster. The aim was to remove the worst effects of older industrial activities by providing financial grants. Contamination remediation, while obviously included in this, was not a primary aim, even after 1976 when soil contamination was identified as a concern. Instead the purpose of such funding schemes as the Derelict Land Grant was to convert unusable land to new productive uses and so create employment, provide better housing and urban environments, and increase public open space amenity opportunities. Other funding initiatives – the Urban Programme, City Grant and the Urban Development Corporations – have had the same primary objectives, though all have contributed to the reclamation of contaminated sites, where contamination has been a salient restriction on the possible re-use of areas of land.

In parallel with state-funded land clean-ups (which have been supplemented by a large and effective private-sector reclamation industry in those areas where demand exists for housing and industrial developments) legislation to reduce environmental damage has become more and more effective. However, until the passing of the 1995 Environment Act no specific legislation addressed the problem of soil contamination. Instead other wider legislation was used to force necessary land clean-ups and to prevent the continued formation of still more areas of contaminated land. Of greatest importance in this process have been (in England and Wales):

- the Town and Country Planning Act 1990

- the Town and Country Planning (Assessment of Environmental Effects) Regulations 1988
- the Building Regulations 1991
- the Environment Protection Act 1990 (Part III on statutory nuisances)
- the Health and Safety at Work Act 1974
- the Clean Air Act 1993
- the Environmental Protection Act 1990 (emissions from prescribed processes)
- the Water Resources Act 1991
- the Environmental Protection Act 1990 (Part II on Waste)
- the Waste Management licensing Regulations 1994.

A similar body of legal constraints exists in the different legal systems which apply in Scotland and in Northern Ireland.

In general terms, the use of these various items of legislation has been successful, e.g.

- local authority planners when considering development applications should treat contamination as a salient issue, and so should require appropriate remedial actions in cases where future risks seem likely;
- the separate building control system should enforce the production of risk-free conditions below building foundations, and insist (in particular) that gaseous entries into structures should be precluded;
- the Environment Agency for England and Wales and the Scottish Environment Protection Agency, as statutory consultees under the planning process, and as the organizations enforcing the Water Resources Act, should demand that site clean-up is to a standard high enough to avoid any pollution of surface or underground waters, and should not merely be those necessary to make site surfaces safe for planned re-use; and
- the Waste Regulatory Authorities should insist on waste disposal practices which are suitable to ensure that further areas of land do not become contaminated.

However, the intrusion of so many separate regulatory bodies, each with its own particular interests and emphases, has on occasion created difficulties. Planning authorities which have a presumption in favour of development, and in attracting employment to their local areas, can ignore the advice of statutory consultees and so permit reclamation standards inadequate to prevent on-going pollution of water resources. Building control inspectors can, after planning approval has been granted, require extra remediation work which the Planning Authority did not see as necessary, and a Waste Regulatory Authority's decisions – on the disposal of contaminated soils from sites which are in the process of reclamation – can be contrary to those agreed with other regulators.

Because of this type of difficulty, the most recent element of the UK's environmental protection legislation – the 1995 Environment Act – has found it necessary to combine the water, inspectorate of pollution and waste regulatory functions into a single Environment Agency for England and Wales. A similar reorganization has also been introduced in Scotland. The hope is that, by this, conflicting regulatory requirements will be reduced and the particular difficulties associated with land contamination will be governed by a single element of law. Local authorities, which still retain the critical planning development powers, will be required to identify areas of land contamination, and must do so on a risk assessment basis, using guidance provided by the Secretary of State for the Environment.

Use will continue to be made of the advisory ICRCL guidance on safe contaminant concentrations in soils, though a revision of this is imminent. The ICRCL approach (Figure 2.1) has been to specify two contaminant trigger concentrations:

- a lower Threshold Value, below which soils can be considered to be uncontaminated;
- an upper Action Value (not established for other than a very few contaminants) above which the soils are taken to be contaminated (i.e. polluted and likely to create risk); and
- an intermediate range of contaminant concentrations in which there is a need for professional judgement of the risks which could exist, and how (if at all) these need be reduced.

It should be remembered that contaminants not included in the ICRCL listing may exist and could create significant harm.

This approach faithfully mirrors the Royal Commission on Environmental Pollution's distinctions between contamination and pollution, and includes the essential need for a convincing assessment of predictable future

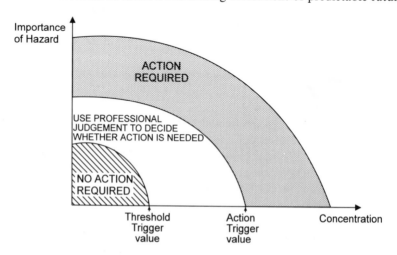

Figure 2.1   Trigger concentration zones.

risk before clean-up is necessary. However, while logical and clear, this guidance has been widely misunderstood by regulators, who – all too often – have viewed chemical concentrations (in soils) only slightly above the specified Threshold Values as clearly requiring land quality remediation. In very many cases, this lack of appreciation of the distinction between contaminated and polluted soils has resulted in unnecessary expenditure on reclamation, analytical and proof-testing work. To end this malpractice finally is obviously a primary aim of the requirements set out in the 1995 Environment Act. These will force local authorities to undertake risk assessments *before* they can issue Remediation Notices and so identify land as actually being contaminated.

Good practice on identifying whether land is contaminated has been developed since 1980 by the ICRCL, and can be cited as a clear systematic approach, i.e.

- *What is the site's history?* As Chapter 3 makes clear, a knowledge of the prior uses to which land has been put is extremely useful in indicating where soil contamination of different types might occur. This permits initial site investigations to be focused on the localities which probably are of most importance.
- *What is the intended re-use of the land?* Different forms of development vary in their sensitivity to the presence of contaminated soils. Once initial information is available on the subsurface occurrence, types and concentrations of contaminants, it can be possible to select a land re-use which, at lowest costs, will be safe. Insisting on placing a very sensitive re-use (housing and domestic gardens) on a very polluted site will inevitably be extremely expensive, and may not be ultimately as safe a land usage as is required.
- *What contaminant types, concentrations and distributions exist?* This is the purpose of site investigation (Chapter 3), which is only carried out so that the suitability of land for a planned re-use can be assessed with sufficient confidence. One factor has to be stressed, since it often is overlooked, and that is that a well-designed site investigation should clearly establish which specific pathways exist to allow site contamination to affect targets or receptors.
- *Do contaminants pose predictable risks?* To answer this question calls for a convincing assessment of each of the possible risks (Table 2.6), which are feasible in the light of the planned re-use of the land and of the contaminants which exist in the site's soils (Table 2.7) and of the likelihood that pathways will permit targets to be put at risk. The methodology outlined in Appendix II can be helpful in judging the likely scale of predictable risks.
- *What form and type of site reclamation is necessary?* This question is best left unanswered until each possible risk has been assessed, since assessment scores will highlight the most important risks, and so indicate

which type of reclamation is necessary. Not all possible reclamation methods will be appropriate (Chapter 4), as some are quite ineffective in reducing particular risks.

- *What proof-testing is needed to confirm that reclamation has achieved necessary standards?* Proof-testing is always helpful, particularly in today's climate of concern over future liabilities and legal challenges. Additionally proof-testing gives reclamation practitioners the information to design future reclamations more effectively, and to predict possible failures at any early stage.

In the UK's policy, trigger concentrations are *not* substitutes for soil quality standards and *must not* be used as such.

Vastly greater amounts of scientific evidence will have to established before any nation poses such defensible quality standards, and even should these be established the applicability of standards against site-specific factors will inevitably have to be considered.

The fact that risk assessment is a necessary element in deciding if contamination is severe enough to necessitate remedial actions implies that site investigations have to be comprehensive enough to provide necessary basic information for assessment purposes. There simply is no sense in attempting to assess partial and incomplete information. This indicates that while guidance such as that given by British Standards Institute (BSI DD 175 [20]) should be followed where appropriate (Chapter 3), analytical testing of samples ought to be appropriate to establish which risks actually are of concern (Table 2.8). Pedestrian adoption of the 'standard analytical suites' (Table 2.9) offered by many testing houses, will not be enough.

Knowing the total concentrations of (say) cadmium and polyaromatic hydrocarbons in a soil is a quite inadequate set of measurements, if the concern is that the site should not pollute nearby water resources. When such

Table 2.8 Types of chemical investigations likely to identify riskier soil conditions

Risk	Analytical bias advocated
Pollution of waters	Emphasis on measuring water solubility and contaminant mobility.
Pollution of air	Emphasis on chemical composition of dust-size soil particles and on the flow rates and concentrations of a range of gases and vapours.
Gas/vapour entry	Emphasis on measuring the concentrations and flow rates of a range of flammable, toxic and asphyxiating gases.
Material degradation	Emphasis on water solubility and mobility measurements and on soil and groundwater pH.
Plant death/plant concentration of contaminants	Emphasis on plant-available measurements of phytotoxic contaminants and on soil and groundwater pHs.
Human health	Emphasis on the chemical concentrations of dust size soil particles and on acid-soluble determinations. Extremes of soil acidity/alkalinity important to judge risks of dermal contact.

Table 2.9 Typical 'standard' analytical package offered by laboratories

'Total' elements	Anions	Other determinations
Arsenic	Chloride	pH
Boron	Sulphate	phenols
Cadmium	Sulphide	Toluene extractables
Chromium		
Copper		
Lead		
Nickel		
Zinc		

Possible additions

Coal tar determination (indirect estimation)
Cyanides
Mineral oils (indirect estimation)
Polyaromatic hydrocarbons (PAHs)

risks appear to be feasible then the 'standard analytical packages' necessarily must be supplemented by leachability tests [21].

One final cautionary note is essential, and that is that even the most thorough of site investigations only permits a very small fraction of a site's soils to be sampled and analysed. On the basis advised by the British Standards Institute [20], for example, only some 0.6% of a site's surface will be trial-pitted or drilled. This indicates that sampling has to be carried out to a consistently high standard, if chemical analytical measurements are to be taken as representative of a large volume of soils.

2.7 Netherlands' policies and practices

While a large number of countries have developed their individual policies and approaches to land contamination, it is useful to consider Dutch practices, not because they are better or more comprehensive than those which exist elsewhere, but because they depart so markedly from those which apply in the UK.

As indicated earlier, the Dutch authorities have never shown any doubt that contaminated soils pose significant threats (Chapter 1, section 1.2.1), or accepted that it is reasonable to distinguish between soil 'contamination' and soil 'pollution'. As a consequence, Dutch practices have stemmed from a uniquely rigorous soil quality protection legislation and from clean-up standards which have legislative force.

Initially these standards were the well known (A)-(B)-(C) values (Chapter 1, section 1.2.3) and a very considerable part of the Dutch nation's resources was diverted to cleaning-up old industrial land to the conditions which typify the entirely unspoiled soils under nature reserves. This quite unique emphasis on a multifunctional standard for all reclaimed land is understandable given the size of Holland, the population pressures it has to resolve, and the need to ensure that shallow groundwater resources remain as unpolluted as possible.

However, the costs associated with the policy have been high and the Dutch state has encountered difficulty in funding all the soil reclamations it sees as necessary. Additionally it has proved sometimes to be technically impossible to clean up some contaminated soils to the standards which are required [22].

Perhaps as a consequence of these factors, Netherlands' policy and practice has recently evolved. The (A)-(B)-(C) soil values have been abandoned and been replaced with simpler Intervention Values (Appendix I). Additionally far more severe Target Values have been derived, though these are indicators more of ultimate national remediation hopes than of any practical achievement which might be possible in the short or medium term. The former (B) soil values, which were used to identify the point at which the discovery of slightly enhanced chemical concentrations called for additional testing and investigation (to determine if meaningful risk might occur because of these contaminant concentrations), have been replaced by the arithmetic average of the sum of the Intervention and Target Values.

While Netherlands' policy appears to remain far more rigorous and expensive than that existing in the UK, it is apparent that a shift towards the sort of distinctions promoted by the Royal Commission on Environmental Pollution[1] has taken place, i.e.

A distinction has been made in the policy for soil remediation between the seriousness of soil contamination and the priority for remediation. If Intervention Values are exceeded there is serious soil contamination. Subsequently it needs to be assessed whether dealing with the soil contamination has priority. This depends on the actual risk for human health and ecosystems at the contaminated site as well as the risk of spreading of the contamination. These risks will depend *on the use of the contaminated site* [emphasis added][23].

2.8 Summary

The significances which should be awarded to the occurrences of anthropogenic chemical substances in soils are still debatable. This is unsurprising. Interactions in the complexity of soil environments make it near impossible to establish the simple cause and effect relationships which can be so apparent in less complex situations.

Thus various nations and states have devised contamination criteria which differ to a greater or lesser extent. As these criteria do not address all the risks which might arise from the presence of soil contaminants (and instead are usually restricted to possible ill-effects to human health), their use in all risk assessment conditions will not always be appropriate.

Despite these scientific uncertainties, contaminated land is increasingly being put back to productive use, and often for quite sensitive purposes where harm could arise at some future time. Thus it is essential that probable risks be assessed on each site individually. Site-specific differences – in the types, distributions and concentrations of anthropogenic chemical substances which have been added to a site's soils *and* in the geological or man-made pathways available for contaminants to come into contact with targets – are such that no universal risk assessment approach is defensible.

However, to assess contaminant risks calls for a thorough knowledge of subsurface conditions in a site, and this is the reason why site investigation is necessary for contaminated land (Chapter 3). It would be unjustifiable ever to attempt to assess soil contamination risks without as complete a knowledge of site conditions as is possible.

3
Rational site investigations

3.1 Introduction

Site investigation is a widely used term and is often taken to mean physical exploration on site, such as the excavation of trial pits or the sinking of boreholes. It is, perhaps, for this reason that all too often investigators charge onto site with no real understanding of what they are looking for or why. Such investigations can often end up as nothing more than the examination of a number of unrelated points of detail with no coherent strategy or means of interpretation.

For an investigation to achieve its purpose, a rational approach must be taken to define its objectives, determine the level of information required, and properly plan the various stages necessary. In other words, an investigation must be designed to meet the specific needs of the project to which it relates. This requirement is of greater importance when applied to contaminated land, which rarely follows the comparatively logical sequences of geological soils. The British Standards Draft Code of Practice DD175 [1] defines site investigations as follows:

> The planning and managed sequence of activities carried out to determine the nature and distribution of contaminants on and below the surface of a site that has been identified as being potentially contaminated. These activities comprise identification of the principal hazards; design of sampling and analysis programmes; collection and analysis of samples; and reporting of results for further assessment.

Another increasingly important reason for advocating that site investigations are planned well enough to yield truly indicative and comprehensive information is the current necessity to carry out risk assessments. These are essential before it can be claimed that any area of land is actually contaminated, and so likely to harm people or environmental targets [2]. It would be futile to commence any risk assessment with incomplete information, particularly if the scale of hazards, which might occur, were inadequately defined. In the context of land contamination, this translates into a need for a clear understanding of a site's subsurface physical conditions, complemented by truly diagnostic chemical analytical results, which adequately typify contamination occurrence, distribution and type.

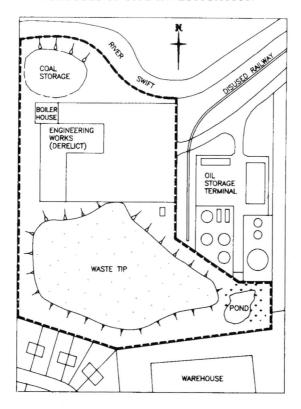

Figure 3.1 Example site. (- - -) Site boundary.

Throughout this chapter the process is described by reference to an example project, which is, in reality, a composite of a number of actual schemes chosen to illustrate the various procedures. A plan of the site is shown in Figure 3.1. The site chosen comprises a riverside industrial area on which is situated an engineering works, now derelict, with associated coal storage and waste tip. An operational oil storage terminal lies to the east and residential housing and industrial warehousing to the south.

3.2 Purposes of site investigation

Investigations of contaminated land are relatively expensive, and so are only undertaken when there is a need to obtain practically important information.

The critical questions which have to be answered by site investigation, i.e.

• does subsurface contamination occur?

- if so, what targets could be harmed by the contaminants which are present? and
- do subsurface geological and hydrogeological conditions allow deeper contamination to adversely affect the land re-use which is intended, or the wider environment?

are, in fact, the basic steps in any assessment of probable risks. The presence, distribution and concentrations of various contaminants must be seen as the specific hazards this parcel of land possesses. The routes via which these hazards could come into contact with probable targets are equally important, since no risk can be claimed unless it is probable that target–hazard contact will occur.

This theme of stressing both the occurrence and type of soil contamination and the geological/hydrogeological conditions within which it occurs is constantly re-emphasized throughout this chapter. Otherwise it will not be possible to produce a reasoned assessment of future risk – and this *essentially* is the prime reason for undertaking site investigations.

3.2.1 *Risks to land users*

While it is obvious that a good number of discrete risks (see Table 2.6) might be generated by the occurrence of contaminants in the soils below land surfaces, traditionally the main focus of concern was to ensure that future inhabitants on, or users of, suspect land should not be exposed to undue risk. This could arise because:

- direct dermal contact with contaminants occurred (say in garden soils);
- inhalation or ingestion of contaminated dusts took place in areas where soils were open to erosion;
- people ate garden or allotment crops which had abstracted and concentrated more soluble contaminants;
- the integrity of homes, or buildings, became affected by soil contaminants degrading construction materials, and, in worst cases this caused structural collapses; and
- gases or vapours entered enclosed spaces and posed fire, explosion, toxicity, or asphyxiation risks.

Because of this human-centred emphasis, the UK's government advice [3] is to relate contamination occurrences and concentrations to various different land re-uses (Table 3.1). There is an obvious logic in this approach – some insensitive re-uses (e.g. hard cover) effectively preclude most contaminant contact routes and so can be developed over relatively high contaminant concentrations. In contrast, very sensitive land uses (e.g. domestic housing with private gardens) offer much greater prospects of a human target being in

Table 3.1 Categories of land re-use recognized by UK government guidance

(Most sensitive)	Domestic gardens
	Allotments
	Parks, playing fields, public open space
	Landscaped areas
	Buildings without gardens or substantial landscaped areas
(Least sensitive)	Hard covered areas

contact with soil contaminants, and so only distinctly lower (and safer) contamination levels should exist in these conditions.

The short-term risks of reclamation workers coming into contact with contaminants always have to be addressed irrespective of the land re-use.

A number of other countries and states (e.g. Ontario, Canada) have accepted similar land use categorization, though The Netherlands (uniquely) permits no such variation. Holland is both small and heavily populated, and this makes it defensible there to require the same high quality of land remediation in all cases.

When making use of published hazard guidelines or standards, it is important to be sure that the cited critical concentrations are those relevant to the risk which is being considered. A large range of risks is possible but authors of contaminant listings never cover all these situations and, instead, preferentially emphasize one risk category (usually human health) to the exclusion of all others. As a single example, inhaling or ingesting copper salts certainly can be toxic if high enough concentrations exist, but the 130 mg/kg value given in the UK's current advisory guidelines has no relevance for human health at all. This concentration, instead, is that at which plant populations can just begin to experience growth difficulties, if the plants are rooted in thin, acidic soils which lack appreciable clay or organic contents.

The investigation will therefore need to identify the presence and nature of contaminants within the surface layers and to a depth consistent with its accessibility to future targets. If major earthworks or deep foundations are envisaged, information at greater depth will be required than if only superficial disturbance of the ground is anticipated. It will be necessary to determine the nature of the ground and define its ability to transmit contaminants laterally to other parts of the site or upward by rising groundwater or the capillary movement of soil moisture. Possible hazards during construction work or during the investigation itself must also be considered.

3.2.2 *Effects on the environment*

Of equal importance is the potential effect of more easily mobilized contaminants on the wider environment. Soluble substances at 2 m depths in a sandy soil profile will be of little consequence for users of a site, but could represent very serious threats to deeper aquifers or to adjacent surface water courses.

Owners of land which causes (or is deemed likely to cause) environmental degradation can be penalized under various statutory legislation [2,4,5] and may be compelled to initiate expensive emergency clean-ups. It is far easier and cheaper, if the redevelopment of a site is being planned, to carry out all the remediation needed for environmental protection as part of the land reclamation. This will reduce concerns over environmental liability in future property transfer transactions and accordingly protect land values.

It will therefore be necessary to identify the spatial distribution of contaminants, determine containing strata and barriers, and define permeabilities of the ground to water and, if present, to gases. The hydrogeology of the locality may need to be investigated to an appropriate degree. It will also be necessary to identify the primary sources of contamination that may lead to the release of contaminants into the environment, and to evaluate the ability of the site to produce noxious liquids or gases.

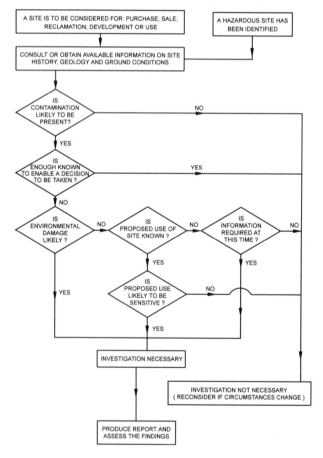

Figure 3.2 Determination of the need for site investigation: decision tree.

BSI DD175 provides a decision tree illustrating a stepwise approach to determine whether an investigation is necessary. This has been redrawn in Figure 3.2 with the consideration of environmental effects added.

3.2.3 *Level of detail required*

Site investigations can be needed for a range of very different purposes (Table 3.2) and so inevitably will differ in scope and necessary detail.

Table 3.2   Various reasons for undertaking investigation of contaminated land

Pre-purchase (to establish likely costs and liabilities)
For development (planning) application
For remedial design (to determine which risks are significant and how these may be tackled and at what costs)
To prove that a chosen remediation method will be adequately effective
To establish remediation success post-reclamation (proof testing)
To establish if breaches of environmental protection legislation occur

For instance, if the purchase of the example site (Figure 3.1) were being considered, a prudent buyer would need that information which directly would influence the purchase price. Questions such as:

- is the land contaminated?
- does contamination give rise to significant liabilities and remediation costs?
- what scale of remediation cost seems likely?
- is the seller's price reasonable in the circumstances?

might call only for a very restricted investigation.

Consulting readily available information sources (sections 3.4.1 and 3.4.2) and establishing past usages of the land, to predict which types of contamination would be likely to occur (section 3.4.3), should be enough to devise a preliminary conceptual model of the site's subsurface conditions (section 3.4.4). This could then be rapidly checked by excavating only a few trial pits, or drilling one or two deeper boreholes.

In contrast, the information needed to ensure that a chosen remediation method will be adequately effective can be more expensive and take much longer to acquire. If preliminary site investigations had revealed that the waste tip, on the example site, was contaminated not only by heavy metals in the ashes which had been tipped there but was also a source of migrating landfill gases, one remediation choice might be to screen out the finer ashes (which usually contain the bulk of heavy metal contamination) and – hopefully – take out, at the same time, the organic fine matter which preferentially generates landfill gases. To confirm that this approach would be successful would call for a trial screening of large and typical samples of the fill materials, chemical analyses of the various particle size fractions produced by the screening process to identify which screening is best, and gas generation measurements on the coarser screened fractions to ensure that these could be safely re-used to fill the excavated tip. Obviously this would be a far more complicated matter than the excavation of a few trial pits or the drilling of some boreholes and would call for a good deal of expensive chemical analyses.

In reality, investigation of contaminated land is rarely completed in a single stage, but usually occurs in several phases during which the results of previous work are revised to produce an increasingly refined 'model' of the site. This is of particular importance in development projects where expenditure must reflect the degree of certainty that the project can proceed and the amount of time that the costs will be outstanding [6].

3.2.4 *Topics of interest*

Having established that an investigation is necessary, its purpose and the level of detail required, it is possible to identify the specific items which are of interest:

1. Define the physical site condition.
2. Identify likely contaminants.
3. Quantify the extent and severity of contamination.
4. Assess the effects and constraints of future users.
5. Assess the potential for environmental harm.
6. Assess the hazards for restoration or construction workers.
7. Establish which type of remediation is likely to be suitable.

A summary of topics is listed in Table 3.3.

Table 3.3   Topics of interest for site investigation of contaminated land

---

Physical site conditions
Natural geology and topography
Soil types and physical properties
Disturbance and alterations caused by man's activities
Structures, buildings and underground features
Surface water drainage
Groundwater regimes, locations, depths, directions of flow

Likely contaminants
Previous uses and contamination
Activities on and adjacent to the site
Background contaminants

Extent and severity of contamination
Chemical concentrations in soils and their distribution by area, depth and soil type
Chemical concentration and distribution of surface and groundwaters
Background quality of soils and of water entering the site
Presence, nature and concentration of gases on the site
Temperatures within the ground

Effects on users
Nature and level of soil contamination with reference to user threshold values within the depth
  likely to be of relevance
Contamination levels in soils, which might be removed, to allow assessment of waste disposal
  arrangements
Surface water quality
Potential for gas emissions
Potential for combustion of soils

Potential for environmental harm
Nature and level of soil contamination to the maximum depth of penetration for identification of
  primary contamination sources
Presence of contained contaminants subject to future leakages (e.g. drums, tanks)
Quality of groundwaters with reference to relevant guidelines and background levels
Identification of water contamination phases (free, suspended, dissolved)
Identification of non-aqueous polluting liquors
Definition of aquifers
Permeability of soils and identification of containing strata and barriers
Discontinuities and migration pathways for water and gases
Quality of surface waters
Soil contamination in surface layers likely to become airborne as dust

Hazards during construction
Contamination of soils and waters with reference to exposure limits
Presence of drums, tanks or buried containers
Unstable conditions

3.3 Site investigation strategy

In order to get the most out of an investigation, it must be approached in a systematic way, always bearing in mind its purpose, the level of detail required and the topics of interest. At the outset the following questions should be addressed [7]:

- What is known about the site?
- What is not known about the site?
- What needs to be known?

The process essentially involves the construction of a conceptual 'model', which can be used to assess the condition and behaviour of the ground, and the mechanisms and processes that lead to predictable risk. An initial model must first be constructed using whatever is already known from existing records, previous investigation work (if any), and from what is visibly evident. An investigation programme must then be designed to refine the model to the extent that is necessary and provide specific information required by the project. The design must take into account physical constraints and other limitations, such as cost, access, etc. It must, above all, lead to the collection of data that are capable of interpretation in a logical manner.

Physical exploration work on the site and subsequent laboratory testing must be implemented to a designed plan, with sufficient flexibility to allow adjustments to be made in reaction to conditions encountered.

Table 3.4   Site investigation strategy

Stage	Activities
Preparation	Define the actual objectives. Consult available information. Carry out a visual inspection. Establish an initial conceptual model.
Investigation planning	Decide which information is needed to prove/disprove the model. Identify constraints such as access, services and financial limitations. Select the investigation techniques. Decide the necessary sampling and testing strategies.

Investigation
(often conducted in
several distinct phases)
	Carry out the planned exploratory, sampling and testing work.
	Ensure that complete records (geological logs, strata descriptions,
	    photographs, test results) are collected and related to known
	    locations/depths.
	Continue to revise the conceptual model as information comes
	    available.

Reporting	Analyse results and refine the site model. Produce separate factual and interpretative reports.
Risk assessments	Identify, in a defensible way, the contaminant implications for the planned re-use of the land.

Finally, the interpretation of the results must follow a rational basis, which can be clearly understood. The results should produce an adjusted and more accurate model of the site, an assessment of its implications, and information on the various topics of interest.

The basic strategy that should be adopted in summarised in Table 3.4.

3.4 Preparation

The first stage in any site investigation requires the establishment of what is already known, what can be observed, what can be deduced and what conditions can reasonably be expected. This process is described by BSI DD 175 as the *Preliminary Investigation.*

It will, in the first instance, provide the means of constructing a preliminary conceptual 'model' which is based on the first assessment of site conditions. The Preliminary Investigation is in two parts: (i) study of documentary material; and (ii) site reconnaissance.

3.4.1 *Data collection*

This is the collection and assessment of data that already exist, are available and are relevant to the investigations. The information should be grouped into categories and it is most important that data are collated in a logical manner. To this end it is useful to refer to the topics of interest listed in section 3.2.4 and Table 3.3.

Sources of information are considerable and a comprehensive listing is provided in the BSI DD 175. However it is not necessary, or desirable, that all such sources are used. Once the accessible and most promising sources have been consulted, the likely value of further data that might be obtained from more obscure sources must be weighed against the effort involved in obtaining them. Since the site investigation under consideration may be a later stage of the phased process described in section 3.2.3, much of the original data collection may already be available in a previous report. This should be reviewed to establish whether more detailed data are now justified as a result of the more advanced stage of the project. The main sources that should normally be consulted are listed in Table 3.5

Table 3.5 Main sources of information

Local library	Maps
	Books, journals
	Newspaper records
Ordnance Survey	Current and superseded maps
National map libraries	Various maps
British Geological Survey	Geological maps and memoirs
	Well and exploration records
	Hydrogeological records
Planning Authority	Records of past use
Environmental Health Officer	Private water abstractions
	Animal burial pits
The Coal Authority	Mining records
Minerals Planning Authority	Mineral extraction records
Environment Agency (SEPA in Scotland)	Licensed waste disposal activities
	Surface water run-off records
	Outfall details
	Abstraction points
	River details
Public utilities	Location of services
Present and previous owners, occupiers and users	Details of activities and processes carried out
	Plans and photographs
Drainage Authorities	Surface water drainage
Aerial photographs	Historical and modern photography

At the start of the data collection, a preliminary inspection of the site should be carried out. This need not be in detail, but should be sufficient to allow the investigator to become familiar with the general topographical conditions, main features of the site and the surrounding land. General panoramic photographs should be taken at this stage. A more detailed inspection will be undertaken after the collected data have been analysed.

The natural topography should be established from superseded maps, which should preferably trace the site back to 'greenfield' conditions. Examination of subsequent editions of maps will provide a series of snapshot views of the site at specific points in its development. From these it will be possible to deduce previous modifications to the natural ground structure, topography and drainage. Unfortunately, few historical maps contain accurate ground levels. Examination of smaller scale Ordnance Survey plans, such as 1:10 000 scale maps, for which superseded editions are available, will give coarse details of ground elevations. Old ground surveys provided by site owners are invaluable in providing more detailed information. When using this information it must be remembered that the Ordnance Survey datum was changed in 1940, and the Ordnance Survey must be consulted to obtain the relevant conversion.

Examination of a series of old maps will also provide an indication of the history of the activities that have taken place on the site, and names and details, which can be further researched in library records.

Older maps are not always readily referenced to the modern site and this makes assessment difficult and often leads to misinterpretation. When comparing such maps, therefore, it is useful to produce a composite plan based on the most modern map showing the main points of historical detail. To do this it may be necessary to produce a series of intermediate plans, starting with the second earliest, onto which details of the oldest map are plotted. These details should then be transferred to the following plan in the series and the sequence continued up to the latest. Common points of detail that occur on several maps should also be cross-checked to obtain the most accurate fit.

A composite plan for an example site is shown in Figure 3.3. From this it can be seen that the river channel was realigned after 1850 and that the Engineering Works was formerly a brickworks. The modern waste tip has clearly been placed into an old quarry, and the area used for coal storage was previously an ash tip, presumably from the adjacent boiler house. Railway links into the waste tip indicate the possibility of imported wastes. The sources of these may be determined by examination of adjacent maps. The oil storage terminal appears to have undergone at least one stage of redevelopment and the names of the quarry and the brickworks are provided.

The names of features and places obtained from maps or other sources can easily be researched in local libraries and in other archives. This will often provide references in historical books, journals and newspaper cuttings, which give useful data about the site. The depth of worthwhile research will depend on individual circumstances and will be a matter for judgement.

Aerial photographs are another useful source of historical information. When available, as stereoscopic pairs, they can be used to obtain an indication of ground levels. This will be particularly useful in determining the depth of the filled quarry in the example site. They might also give an indication of the type of fill being placed and the method of filling. Surface photography will also assist in such details.

While Figure 3.3 depends to a large extent on information from old maps, other data sources were consulted to ensure that the historical map information had not omitted briefer periods in which particularly contaminating land uses had taken place. It has to be remembered that the Ordnance Survey produced map editions only at 20- to 30-year intervals. The tendency of many investigators to consult only past editions of Ordnance Survey maps (because this entails only the effort of contacting the Ordnance Survey or one's local reference library and paying a small fee) thus can be risky (Box 3.1).

Box 3.1 Over-reliance on a singe data source

An executive housing development was planned in Central Scotland in an especially attractive area.

Past OS maps (1857, 1897, 1912, 1957 and 1981) all showed the site as farmland. No industrial or contaminating land uses were suggested.

Site visits revealed very normal grassed pasture land, so the land was purchased as 'greenfield' site.

What had not been shown on the maps was that deep quarrying (to 13 m) had taken place in 1961 and that the void was later (1969) filled with 'inert' wastes. The loosely compacted wastes were actually contaminated with fuel oils and tars and posed significant problems to the housing development.

Local electricity board and waste disposal records – which had not been consulted – did reveal the quarrying, its depth and the amount of tipping which took place.

3.4.2 Recorded physical conditions

The natural geology of the site may be established from geological maps and memoirs, and from previous investigations. This process is well documented elsewhere and forms a normal part of any geotechnical investigation. Its importance to contaminated land is that it establishes the framework onto which the details of the disturbed site can be constructed. In particular, details of the depth to rockhead, drift materials and aquifer locations are of interest. Useful information may also be available about the hydrogeology of the area. More detailed information will be obtained if previous site investigation records can be consulted.

Details obtained from historical maps and other records will indicate modifications that have occurred due to man's activities. For instance, on the example site (Figure 3.3), the waste tip area had been both excavated and then filled. It must, of course, be borne in mind that, since historical records are not a continuous record, they may not show all modifications.

Information obtained from British Coal and the Minerals Planning Authority will give details about mining and mineral activities, and Waste Disposal or Regulation Authorities can provide information on licensed

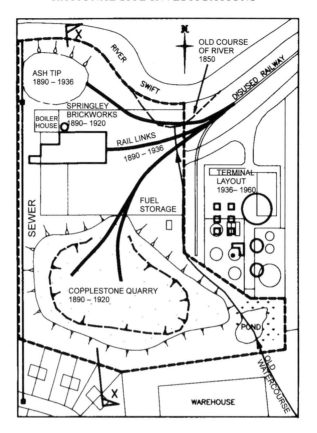

Figure 3.3   Example site: conposite plan. (- - -) Site boundary.

landfill operations. These all represent modifications to the natural conditions.

The evaluation of the drainage system on the site is of particular importance since original drainage channels and water courses which have been piped, often remain as pathways even after filling. A separate drainage plan should be prepared which indicates historical as well as modern drainage. This will be made up of information from old maps and from information provided by drainage and river authorities.

3.4.3 *Previous uses and contamination potential*

Knowledge of previous activities is essential to the understanding of contamination potential. This information is normally obtained firstly from records and secondly from the detailed questioning of previous users. The importance of the latter depends very much on whether the previous user carried out the most contaminating activities.

However, it can often be the case that the most recent land uses were the most intensive and contaminating, partly because only a relatively short time has elapsed and rainfall has not yet been able to leach out more mobile contaminants.

General profiles of the range of contaminants produced by various types of industry are listed in BSI DD 175[1], and ICRCL Guidance Notes [8 to 11] provide greater detail for a few industrial activities. Especially useful are chemical industry encyclopaedia [12] which contain particularly complete advice on the raw materials and waste products of various industrial processes, and so allow probable soil contamination to be estimated (Table 3.6)

Table 3.6   Prediction of soil contamination on a former paper mill site

---

General production process
The production of dispersed wood fibres in a water medium. Pressing and dewatering of the fibre pulp. Improvement of the paper by pulp bleaching, introduction of additives to increase mechanical strength, improvement of appearance by adding fillers, brightening agents and dyes.

Additives employed
Kaolin, china clay, titanium dioxide, calcium carbonate, zinc and barium sulphides/sulphates, various waxes, starches, glues and asphaltic emulsions. Water soluble organic dyes (additives invariably very fine grained; i.e. <1µm in diameter).

Pollution concerns during production
Air emissions (particles of sodium sulphate and carbonate, oxides of sulphur and nitrogen, various hydrocarbon emissions). Water pollution (BOD from pulps, phenols, chlorides and micro-organisms); sludges and ashes (of dewatered wastes) generally are the most acute pollution concerns.

Predictable soil contamination
Landfill gas emissions and subterranean smouldering risks (from sludges). Metallic contaminants (from ash wastes). Water soluble metals (from vats of additives, colorants, process chemicals which have been demolished). Extremely alkaline soil conditions. Local areas of reduced soil oxygen.

Source: after [12].

The UK Government has prepared profiles of processes and activities with the potential to contaminate land, which also provide some useful information [13].

To make a site-specific assessment, it is necessary to consider all the possible ways in which contamination may be brought onto the site and the mechanisms that could result in its deposition within the ground. Operational practices for the relevant industries and, where possible, for the specific plant concerned, must therefore be researched. Tables 3.7 and 3.8 summarize the main sources of contaminating materials and processes. By careful analysis of a particular industrial process, it is often possible to predict with some accuracy both the contaminants that will be present and their

Table 3.7   Sources of contaminating materials

Raw materials
Products and by-products
Rejects
Transitional products (materials part way through the manufacturing process)
Wastes
Ashes
Lubricants and coolants
Fuels
Building materials (e.g. asbestos, paints)
Vehicle and plant consumables
Cleaning materials
Laboratory chemicals

Table 3.8   Contaminating processes

Storage methods
Leakage and spillages
Transport
Processes
Disposal of wastes
Water treatment
Maintenance
Research and testing
Demolitions and redevelopment

most likely locations. This, of course, is rarely comprehensive, since during the life of any industrial site there will be many thousands of unrecorded activities which might lead to contamination.

Information from previous users, if available and willing, will provide details of the activities that occurred during their occupation of the site and permit a much better assessment of contamination potential. Initially a questionnaire should be prepared detailing all the information required. This should be followed by an interview in which the operation of the site should be discussed in detail. The most productive approach is to trace the progress of all raw materials entering the site, from delivery and storage to dispatch as finished products or wastes. Figure 3.4 illustrates this technique on an oil terminal site.

One particular benefit which this type of enquiry can yield is that the chemical analytical strategy, for the main investigation, can be 'tailored' to include contaminant forms which otherwise would be overlooked. For instance, if the last manager of the engineering works had advised that enamelling of metal fabrications (using hydroscopic cobalt, chrome and zinc salts) had been a major activity, it would be prudent to include water leachable tests for these metals on soil samples. Merely analysing samples for the 'total' metals, which are offered by most analytical houses, could provide far less useful information and could lead to a predictable risk to local surface water qualities being overlooked.

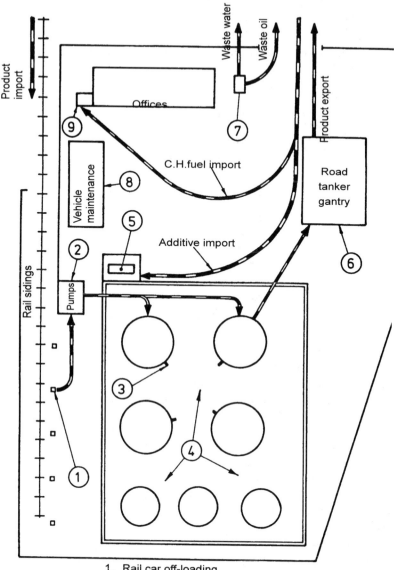

1. Rail car off-loading
2. Pumps
3. Tank drainage
4. Tank compound
5. Additive storage
6. Road tanker loading
7. Oil interceptor
8. Vehicle maintenance
9. Central heating oil storage

Figure 3.4 Establishing contamination potential by auditing production activities.

3.4.4 *Preliminary conceptual model*

Having completed the research of the known site history, the first conceptual model should be produced. This should be represented as plans, indicative sections and descriptive text summarizing what is known. The model should define:

- natural geology and topography;
- modifications, mining and other alterations;
- filled and distributed areas;
- locations of potentially contaminating activities;
- historical and modern drainage paths;
- services and other constraints.

The main elements for the example site would comprise the composite plan in Figure 3.3, deduced sections, such as shown in Figure 3.5, and drainage and service layouts. The natural topography of the example site is a gently sloping valley side falling northwards to the River Swift, but also eastwards towards an old watercourse, which originally ran northwards through, what is now, the oil terminal. Natural geology comprises alluvial sands and gravels adjacent to the river, overlying a succession of laminated and stiff boulder clays and weathered shales. Modifications identified are:

- filling of old river channel;
- tipping of ash;
- excavation of quarry;
- filling of waste tip;
- regrading of brickworks/engineering works and oil terminal.

Contaminating activities identified are:

- disposal of ash – coal storage area;
- coal storage – coal storage area;

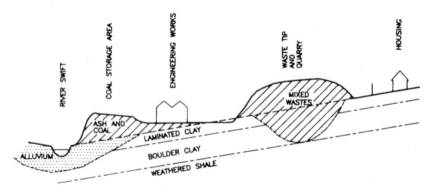

Figure 3.5 Deduced section X–X for the example site.

- chimney – brickworks;
- metal machining, hardening and finishing – engineering works;
- vehicle maintenance – south end of engineering works;
- fuel storage – south-east corner of hardstanding;
- wastes from engineering works – waste tip;
- imported wastes (foundry ash, possible chromium wastes from nearby chemical industry, domestic refuse) – waste tip;
- oil contamination – adjacent to oil storage terminal.

The site should be divided into zones of similar type, background or use. This will allow further analysis, observation and investigations to be taken in a logical manner.

Subdivision of the example (Figure 3.6) might comprise:

1. coal storage area;
2. main engineering works buildings;
3. boiler house;
4. vehicle maintenance areas;
5. hardstanding area;
6. area to west of engineering works;
7. waste tip;
8. unfilled land around the waste tip;
9. ponded area.

Other investigators could increase the number of contamination zones – in particular one encompassing the infilled watercourse on the site's eastern flank could be argued, since this might be acting as a linear channelway for more mobile contaminants. Whether or not this correction is necessary would become apparent once subsurface investigation commenced; if oily liquors were found in a trial pit within the old river course, and extending northwards to the River Swift, it would be obvious that the zonal configuration (Figure 3.6) requires modification. This would be a minor matter of correction, and the benefit of delineating zones (with which soil contamination is relatively homogeneous in type) is the possession of a clearly understood site model into which items of investigation detail can be inserted. No site model is likely to be precisely correct in its initial configuration, and personal emphasis will result in different investigators adopting somewhat different zonal boundaries. This, however, is not of importance, so long as a model is devised and then corrected whenever necessary. Only in this way can it be ensured that the significance of one fact, out of the thousands revealed by even a small-scale site investigation, is not overlooked.

3.4.5 Site reconnaissance

The purpose of a site reconnaissance is to check the information obtained from documentary evidence and to add further detail. The inspection should therefore be structured to ensure that all aspects of the conceptual model and topics of interest are addressed, and it should be carried out by reference to the areas into which the site has been subdivided.

The reconnaissance should, wherever possible, be conducted on foot and it is usually best to walk around the perimeter of the site first, before inspecting the central area and points of detail. This gives an understanding of the overall scale of the site and allows landmarks to be easily located.

A reasonably large-scale plan of the site, with points of interest marked, should be used to guide the inspector and to allow additional points to be recorded. The recording of detail as numbered points on the plan with descriptive text entered into a notebook is the best approach. Photographs should be taken of points of detail to supplement the more general photography taken during the preliminary inspection.

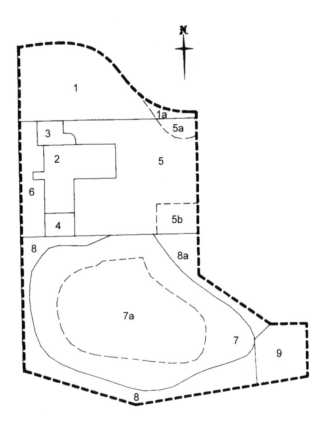

Figure 3.6 Zoning of the example site. (1) coal storage area; (1a) filled river course; (2) main engineering works buildings; (3) boiler house; (4) vehicle maintenance area; (5) hardstanding area; (5a) filled area; (5b) fuel storage; (6) area to west of engineering works; (7) waste tip; (7a) former quarry; (8) unfilled land around waste tip; (8a) adjacent to oil terminal (9) ponded area.

During the reconnaissance, the following main observations should be made with reference to each area of the site:

- surface materials, condition and appearance;
- materials exposed in excavations and side slopes;
- vegetation type and condition;
- unusual colours, fumes, odours and obvious contamination;
- presence of tanks, drums and other containers (details of any identifying marks and signs of leakage or spillages to be recorded);
- lagoons, ponds, pits depressions or swampy areas;
- tip materials, extent and description;
- signs of heating or combustion;
- drainage, outfalls, streams and watercourses;
- drainage interceptors and water treatment plants;
- appearance of surface waters;
- accessibility for exploratory investigations.

At the same time, or more practically during a separate visit, simple on-site testing and sampling can be undertaken. This may include collection of surface soil and water samples, superficial excavation to expose near surface materials, and testing for gases and vapours using portable meters.

Having completed the site reconnaissance, the observations should be properly recorded by reference to each of the zones into which the site has been subdivided. The description of the site 'model' should be adjusted and refined according to the observations made.

Leaving the site reconnaissance visit until after the initial site model has been devised is both more productive and safer. Knowing that a boundary between site zones is likely to occur in a particular area focuses the mind, and most investigators then find it easier to identify minor soil colorations or vegetation changes which could more accurately define the boundary's position.

Greater safety is gained by not disturbing or digging into particularly hazardous materials; if library research had suggested that the final disposals into the waste tip were of asbestos insulation, from an adjacent naval dockyard, it would be imprudent for an unprotected investigator to disturb the tip's soil cap and possibly release asbestos fibres into the air.

3.4.6 Defining objectives

Armed with a 'model' of the site, based on documentary records and the site reconnaissance, the next step is to define the objectives of the required exploratory work. This will be dependent on the amount of data already available and the purpose of the overall investigation. In some cases sufficient information may already have been acquired and no further investigation may be necessary.

Primarily, the objective will be to provide specific details of the conditions in each area of the site, the level of contamination and quantitative data to permit an analysis of its implications.

3.5 Designing an investigation

3.5.1 *Information required*

The site investigation may initially be to determine whether an area of land is contaminated and to what level, but it will usually be required to determine what remedial works and precautions are needed. The site investigation should therefore be designed with this in mind. A result which indicates that the site is generally contaminated with a range of contaminants is of little value in specifying remedial action. The problematic parts of the site and the areas where particular contamination occurs will need to be known. In addition, those parts of the site that are not contaminated will need to be identified and this will require establishment of background conditions.

The investigation should therefore be designed to determine the extent of each of the different materials making up the site within each sub-area and to describe their properties. The materials will be identified by:

- physical properties;
- chemical properties;
- location;
- depth.

The main physical properties to be determined are visual appearance, smell, temperature, soil classification and compaction. The investigation may also require the evaluation of various geotechnical properties at the same time. Permeability testing will provide information about the ability of the ground to allow migration of liquids and gases.

The evaluation of these properties is best achieved by a combination of observations made on site and subsequent testing, either in-situ or on samples collected for laboratory analysis.

Good practice is to analyse samples of soil and groundwater *not only* for the contaminants suggested by the site model, *but also* for

- commonly occurring contaminants, such as the arsenic and cadmium additions (probably from past air pollution) which so typify soils around many industrial centres;
- contaminants to which the planned land re-use would be especially sensitive (e.g. landfill gas emissions below areas intended for domestic housing); and also

- unusual contaminants, whose existence was not suspected. Checking that radioactive wastes had not been put into the waste tip could be an example of this.

This listing is not intended to suggest that all, or even a large proportion of, samples should be analysed to this degree. Obviously this would be far too expensive ever to justify, but extending the analytical strategy on a few well-chosen samples is certainly worthwhile, and indeed is an important part of proving and refining a site model.

While attention has to be given to contamination occurrence and distribution, the importance of pathways for contaminant migration should always be borne in mind. The material and physical properties which could permit pathways to occur will only be visible when trial pits and excavations are open, and investigation designs should allow for this. The problems which can arise if physical pathways are not evaluated are all too commonplace; possessing only the knowledge that enhanced concentrations of (say) cadmium (17 mg/kg) and mercury (5.4 mg/kg) occur at depth in an otherwise clean soil profile really does not help an investigator to decide whether or not a risk to human health is predictable. If the metals exist only in a thin peaty band within stiff clays and at depths beyond probable excavation by future residents, then risks will be trivial. There is well-established literature [14] to demonstrate that humic acid bonds so fix and hold metals that they are unlikely to be demobilized. However, if the metals exist at or near ground level in a porous sandy profile, the same judgement could not be made.

The investigation of contaminated land and the refinement of the conceptual 'model' is an evolutionary process. This often leads to the necessity of carrying out site and testing work in more than one phase. Initial tests, for instance, may show unusually high contamination at a single location, which warrants further work to delineate its extent.

3.5.2 Sampling strategy

The design of the investigation should ensure that observations are made and samples are obtained which are representative of the various materials making up the site. In the Department of the Environment's publication on *Problems Arising from the Redevelopment of Gas Works and Similar Sites* [15] it is stated that the sampling process should aim to:

'Identify the highest levels of contamination' and
'Describe the distribution of contamination both spatially and vertically'.

Preparatory work will already have provided a degree of understanding of where the different materials are located and what their properties are likely to be. Using the 'model' prepared during the preliminary investigation, all materials making up the ground within the site can therefore be subdivided into zones. Each will consist of materials considered to have characteristic and relatively uniform properties related to soil type, method of deposition, and exposure to contaminating processes.

For instance, the coal storage area in the example site will almost certainly comprise a coal-rich surface layer overlying boiler ashes with natural alluvial deposits below forming the original sloping river valley side. These will in turn be underlain by natural clays and rock. The investigation of this area would therefore seek to quantify the physical and chemical nature of each of these materials and delineate their extent. The degree of subdivision is dependent on the level of detail necessary and the minimum size of area to which the representative parameters need to be assigned.

In addition to the investigation of the solid materials, it will also be necessary to obtain information on the location, nature and quality of ground-waters and soil atmospheres. These should initially be described with reference to the materials in which they are located.

An appropriate exploratory pattern, density and total number of samples in each of the individual segments must be specified.

In considering this section, the reader is reminded that as well as subdivisions of the site by area, the ground in any area may contain a number of zones at different depths. The ashes in the coal storage area in our example site will need to be assessed as a separate segment to the underlying alluvial deposits within the same area (Figure 3.5).

The strategy to be applied to each zone may need to be different. There are two main categories:

1. zones within which no particular pattern or variation in conditions are anticipated, e.g. waste tips and ground into which contamination may have been deposited in a haphazard manner;
2. zones where contamination is related to known points of origin, e.g. ground adjacent to the doorway of a contaminating process building.

3.5.3 Non-targeted sampling

The majority of zones will fall into the first category. For these an appropriate spatial distribution of sampling locations must be selected. At each sampling location sampling depths must also be decided to ensure that representative samples are obtained for each of the anticipated layers. Samples of ash fill, alluvium and laminated clay would be needed from sampling locations within the coal storage area shown in Figure 3.5.

Where industries have occupied a particular area of land for scores of years, it has to be remembered that:

- plant layouts and activities varied from time to time;

- new plants were often built over the buried remnants of older works, and that tipped wastes (used to level the ground) often fill older cellars, vats and tanks;
- production wastes were usually dumped on whatever unused areas existed; and
- local hardcore (e.g. slags, ashes, foundry sands) were employed to fill low-lying ground without any consideration of the soil contamination thus being brought in.

The net results are often subsurface variabilities far greater than those ever likely to exist under 'greenfield' land.

This poses the practical problems that very dense investigations are extremely expensive, and that it can never be guaranteed that every small subsurface abnormality can be located no matter how closely spaced the investigation holes are. To locate all subsurface anomalies really calls for all soils in each zone to be excavated, examined, sampled and chemically analysed and this, obviously, is beyond any conceivable investigation budget. Site investigation of contaminated former industrial land itself has to be seen as an exercise in risk assessment.

If the intention is to interpret the data in a strictly statistical manner, a random pattern of sampling should be adopted. Unfortunately this has several disadvantages: sampling points will be unevenly distributed within the segment and constraints on the site surface will inevitably prevent completely random selection. A large number of samples are required to provide a high level of confidence and, to be truly random, samples at varying depth must be taken from separate excavations.

The most widely used method of locating sampling points is on a grid basis. This has the advantage of even distribution throughout the segment and can easily be set out on site. It is a more systematic approach to which subsequent investigations or restoration activities can easily be related. The size of the grid to be used is very much a matter of judgement. BSI DD 175 recommends minimum numbers of sampling points for different site areas.

A method of calculating the probability of locating a given size of contaminated area related to the numbers of randomly located samples is reported by Bell et al. [16]. Application of the binomial theorem gives the following simple relationship:

$$P = 1 - (1 - a)^n$$

where P is the probability of obtaining at least one contaminated sample, a is the proportion of the area of the site that is contaminated and n is the number of samples taken. This can also be expressed as follows:

$$n = \frac{\log(1 - P)}{\log(1 - a)}$$

Bell *et al.* [16] note that, although this formula is derived specially for randomly located sampling points, probabilities will not be significantly different if sampling locations are set out in a uniform grid pattern. (This only applies to relatively high levels of confidence.)

Using this method Table 3.9 gives the minimum size of contaminated area for which the sample numbers recommended in BSI DD175 would achieve 95% probability of locating. From this it can be seen that relatively large areas of contamination need to exist before this level of probability of finding them is achieved. Smaller occurrences are less likely to be located by non-targeted testing.

Table 3.9   British Standard draft guidance on the numbers of sampling points

Area of site (ha)	Recommended number of sampling points	Implied spacing of investigation points	Area (m²) of contamination needed to give rise to one contaminated sample
			(95% confidence level)
0.5	15	18 m	905
1.0	25	20 m	1129
5.0	85	24 m	1732

Source: after [1].

ICRCL 18/79 [10] recommends that the grid should be no larger than the largest area of contamination that could be handled without difficulty if it were not found during the investigation, but only discovered during the development. For small sites, a grid spacing of 10–25 m is suggested and for larger sites, 25–50 m. Leaving aside the problem that there is no certainty that any subsequent development would involve activities that would enable such contamination to be discovered, the statistical method described above can be applied to this philosophy. On a 1 ha (1 ha $= 10^4 \text{m}^2$) site a 10 m grid would give a probability of only 63.4% of finding a contaminated area of 100m^2 (i.e. the size of the grid).

Ferguson [17] used computer modelling to suggest that a better probability of detection could be achieved using a herringbone pattern of investigation points rather than a uniform grid. His analysis examined theoretical detection rates for different shaped areas and obtained a marked improvement on elongated areas of contamination when the herringbone pattern was used. The approach involved the assumption that contamination occurrences are spatially correlatable rather than being statistically independent hotspots. It is very doubtful whether this assumption can be justified as practically correct. For example the proven existence of asbestos fibres in ashy fill (encapsulated in a trench dug into boulder clay) gives no prediction as to the asbestos contamination of soils 20 m from the encapsulation. In such cases, contaminant 'hotspots' are anomalies with no statistical relationships with the surrounding soil qualities. A statistical relationship might occur if mobile or soluble contaminants had migrated from a central hotspot or if uniformly contaminated materials had been spread over a specific area of the site. Even if the more optimistic 'herringbone pattern' detection rates are accepted, the areas in the right-hand column of Table 3.9 are only reduced by less than 50%.

It can be seen that detection of contamination by non-targeted sampling alone will yield only poor levels of confidence. It is therefore essential that the initial research to identify likely variation in conditions is as thorough as possible and that the subdivision of the site into zones and layers is based on that knowledge. Non-targeted sampling patterns and densities must then be selected for each zone taking into consideration the likely variability and distribution of contamination within that part of the site. The results of testing *all* samples from a particular zone within the site must be treated as representative of the conditions within the zone (e.g. arsenic content in a particular zone varies from 5 to 150 mg per kg with 4 out of 10 samples exceeding 40 mg per kg). No further conclusions can be drawn about a particular zone from a non-targeted sampling regime. However, by adequate zoning of the site as a whole, following examination of test results, different conditions can be attributed to different zones which have had an identifiably different site history or are known to contain different materials.

Smith and Ellis [18] reported, from a gas works investigation, that investigation points separated by 25 m gave results which were not significantly different to those provided by much closer (6.25 m) holes, and this has sometimes been interpreted as suggesting that more closely spaced investigations are unnecessary. However, the results observed probably simply reflect the proportion of the site that was contaminated and the distribution of the contaminated materials in that particular case.

The density of investigation points which is finally selected will often be determined by external factors such as financial constraints. Whatever density of sampling locations is finally accepted it is most important that the limitations in detection rates are reported along with the results.

3.5.4 *Targeted sampling*

In segments where the contamination source is known, the sampling strategy should be targeted around that source. Normally sampling points should be located at regular distances along lines radiating from the contaminant source. The spacing selected should be based on the anticipated distance over which the contamination may have spread. The method of translocation must also be considered in locating sample points. Drainage ditches may have carried contaminants some considerable distance from their source. Surface spreading, however, would normally be much more local to the point of deposition. A staged process may be appropriate, with a wide spacing used initially and closer spacing subsequently applied between those points between which contamination falls from a high level to an acceptable level. This philosophy can also be applied to a sequence of layers below a large contaminating source when the overall quality of each layer is assessed using the technique described in section 3.5.3.

Provision should also be made in the investigation for the collection of additional samples of small isolated pockets of material which are visually suspect. In such cases, samples of the surrounding matrix of 'normal' material, representative of the zone as a whole, should also be taken.

3.5.5 *Sampling depths*

The depths at which samples should be taken will depend on the variability of materials encountered and the level of detail required. Exploratory holes may pass through a number of individual layers of different material, which will represent different zones. These may have been predicted by the preliminary investigation or may be unexpected. In any event at least one sample should be taken in each zone encountered in the hole. Zones that extend to some depth will require multiple samples taken at different levels.

An essential feature is to ensure that samples are collected at sufficient depth to prove the presence of clean natural ground [19]. In many situations contamination has percolated from surface fills and wastes and collected in, or on, underlying natural strata. Remediation of the quality of the uppermost layers of these natural soils may thus prove necessary and this can only be decided if investigation samples have been taken well into natural ground.

BSI DD 175 recommends a minimum of three samples from each excavation, one at the surface, one at the greatest depth and one at a random point in between. The cost of taking samples is relatively cheap compared with the cost of excavation or subsequent testing; therefore it makes sense to take too many rather than insufficient samples. The testing strategy can then be determined following detailed consideration of the whole investigation.

Where it is important to establish the variability of materials with depth, samples at 0.5 m intervals should be taken, although it is very unlikely that these will all need to be tested. Testing strategies are described in section 3.7.7.

3.5.6 *Background conditions*

In addition to establishing the concentration of contaminating substances within the site, the investigation should be designed to gain an indication of the general background level of these substances in the local environment. These can be expected to vary considerably from place to place and a number of sources give typical ranges of concentrations [20, 21].

A number of sampling points should be located in typical materials away from possible contaminating influences. This may require excavation outside the site and the co-operation of adjacent landowners. Alternatively it may be possible to establish areas of the site which are clearly not affected by contaminating processes. Representative samples from such areas should be taken.

The reason for suggesting this is that background chemical concentrations vary very substantially in different geographical and geological locations and chemical concentrations distinctly higher than those cited in various national guidelines and standards can occur. For example, in the Bristol area and adjacent Avon Estuary, cadmium contents in surface soils (in areas well away from any prior industrial activities) are generally higher than the 3 mg/kg threshold value cited in UK guidance. Remediating soils, on a site in such an area, to chemical limits which are lower than the natural background is scientifically unjustifiable and an unnecessary expenditure.

3.5.7 Groundwater and surface water sampling

The determination of groundwater qualities is an essential part of contaminated land investigation, since contamination transfer is largely controlled by water movement, and groundwater qualities reveal useful evidence of contaminant leaching rates from solid materials. As surface water bodies and aquifers are generally very sensitive to pollution effects, recent legislation (Environment Act 1995) rightly makes the water polluting features of contaminated soils an important element in their identification.

It can thus initially be surprising how relatively few groundwater analyses are contained in most contaminated land investigation reports. Cases where more than 2000 chemical analyses of soils are accompanied by only four or five groundwater analyses are common enough not to attract comment.

The main reason for this obvious information imbalance is that groundwater entry into trial pits or boreholes often takes more time than is available in many routine site investigations. The site investigation industry is very competitive, and the emphasis is often to excavate trial holes as rapidly as possible and then backfill these to remove risks to local inhabitants. Additionally the monitoring works needed, to provide meaningful groundwater levels and representative samples for chemical analysis, call for careful and specific design, and that information will seldom be available until – at least – a preliminary site investigation has been completed. Because of these factors, it has become normal for groundwater investigations to be predominantly carried out as a second stage of investigation.

The benefit of groundwater quality information in the assessment of contamination risks is beyond doubt. For instance, the nineteenth-century reclamation of a tidal marsh by tipping slag wastes, to raise the ground levels, is likely to give enhanced chemical concentrations in solid samples (Table 3.10). But do these analytical measurements signify predictable future risks? Leachability tests will often show that almost all the tipped fills are essentially

Table 3.10 Comparison of soil and groundwater chemical qualities. River Tees. Middlehaven Docks

| Sample location | Zinc content | | PAH content | |
	Solids (g/kg)	Water (μg/l)	Solids (μg/kg)	Waters (μg/l)
1	0.7	5.0	10.0	15.0
2	3.3	205.0	21.0	LT 1.0
3	13.0	<5.0	66.0	1.0
4	0.5	10.0	13.0	213.0
5	5.2	7.0	12.2	7.0
6	0.1	<5.0	35.0	5330.0
7	0.6	10.0	9.0	3330.0

Table 3.11 Leaching test results on samples which exceeded 10 times ICRCL threshold levels

| | Zinc | | |
TP ref. no.	Soil (mg/kg)	Aqueous extract (mg/l)	Leached contaminant (mg/kg)
C5	8830	<0.1	<1
C5	8340	<0.1	<1
E4	4200	<0.1	<1
E4	6630	<0.1	<1
E4A	3210	<0.1	<1
E6	12500	0.14	1.4
F4	4050	<0.1	<1
F5	4720	0.38	3.8
G2	9800	<0.1	<1

Sample weight = 100 g
Leachant volume = 100 ml
Leached contaminant (mg/kg) = 10 × leachate concentration (mg/l)

non-leachable and that actual risks are insignificant (Table 3.11). However (Table 3.10, PAH results for samples 6 and 7) localized areas may usefully be revealed where this certainly is not the case and where remedial action is needed to prevent pollution of controlled water resources.

It is essential therefore that adequate samples of water are taken, usually wherever encountered and Table 3.10 demonstrates the variability of water quality. The results of water testing will also give invaluable information about contaminant mobility and rate of dispersion.

To obtain meaningful groundwater levels calls for a borehole into which a water monitoring well is installed. This usually comprises a well casing with a specifically designed slotted intake section surrounded by a gravel filter pack. It will be necessary to 'develop' the borehole (i.e. pump or surge the water column in the hole to pull in finer clogging particles from the surrounding strata) to improve hydraulic continuity between the water-bearing horizon and the observation hole. A more rudimentary system can be more cheaply constructed for shallower groundwaters by excavating a trial pit, installing a slotted casing and backfilling the hole with granular fill into which waters can enter without significant head losses (Figure 3.7).

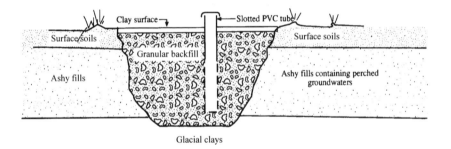

(a) Simple trial pit converted to a groundwater monitoring installation

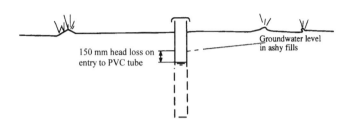

(b) Unacceptable high head losses due to poor development
of the trial pit installation

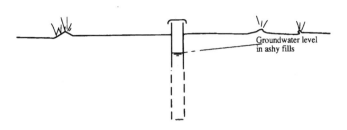

(c) Accurate groundwater level after installation has been developed

Figure 3.7 Simple trial pit installation

Ensuring that the slotted portion of the borehole's casing is only in the horizon where water levels and samples are required, seems too self-evident for specific mention, yet it is quite common to find that this is not the case.

Groundwater flow rates generally are very slow and typically are no more than a metre or so per day in most contaminated sites. Thus it is possible for stagnant water – which need not be typical of groundwaters in the surrounding strata – to remain in a borehole tube. Good sampling strategy is thus to pump out three borehole volumes before any sample is taken for chemical analysis. In this way (Box 3.2) sampled groundwaters are drawn a significant distance from within the surrounding water-bearing strata and are more likely to typify actual aquifer water qualities.

Box 3.2 Effect of complying with the groundwater sampling protocol [22]

Borehole diameter	50 mm
Saturated thickness of aquifer	7 m
Aquifer porosity (say)	33%
Volume of water contained in the borehole	13.75 litres

Removal of 3 borehole volumes calls for the abstraction of 41.25 litres of water before any sample is collected.

Because the aquifer's porosity is much less than 100% the sampled water can be shown to have been drawn from an average horizontal distance of 0.12 m from the centre of the borehole.

Surface waters should also be sampled and analysed to show whether the area of contaminated land is currently affecting surface water qualities. Samples can be taken from pools, ponds, drains and water courses, indeed wherever it seems probable that a distinction might be drawn between waters entering the contaminated site and those leaving it. In the example site (Figure 3.1) sampling the quality of the R. Swift both upstream and downstream of the development area would be good practice, and it would be prudent also to obtain groundwater samples at the southern extremity of the old water course (on the site's eastern boundary) and compare this to a sample of groundwater from the old water course where it drains into the R. Swift.

3.5.8 Gas monitoring

Examination of the soil atmosphere is an essential part of the investigation of contaminated sites. Fills in the ground may contain degradable materials, which can produce flammable gases, carbon dioxide and depleted oxygen. The presence of carbon monoxide may indicate combustion processes within the ground. Spilled liquids may produce volatile vapours that may be toxic or flammable, and the presence of such vapours may indicate contaminants that have not been located by exploratory excavations.

The investigations must therefore provide for monitoring of gases during the exploratory works and the installation of monitoring wells at locations at which gases might be expected. The ash tip and filled quarry in the example site would need to be monitored for landfill gases, and the area adjacent to the oil terminal checked for volatile hydrocarbons.

A preliminary gas survey using subsurface probes and portable equipment will give an early indication of likely problem areas and enable the planning of gas monitoring well locations prior to the main work.

A number of other techniques can be used to investigate gases and this subject is dealt with more fully in Chapter 8.

3.5.9 Analytical testing

The investigation will include testing of materials both in-situ and in the laboratory. In-situ testing will largely be restricted to the investigation of gases and measurement of soil temperature. However, several techniques are available, including the use of portable gas chromatography. These can be used to provide quantitative information as the investigation proceeds and enable reactive adjustments to be made to the investigation programme.

Great care should, however, be taken and major amendments avoided since there will rarely be time properly to consider all the implications. Despite this, in-situ tests are extremely useful in directing the main investigative effort to the most important areas and may avoid the need for further investigation phases when, for instance, carrying out the targeted sampling described in section 3.5.4.

A laboratory testing programme, which will provide the appropriate level of accuracy and cover the range of contaminants required as discussed in sections 3.3 and 3.5.1 must be prepared. There is limited published guidance on methods of chemical analysis, and often different laboratories will use different methods. BS 1377 [23] gives methods for tests normally undertaken as part of geotechnical investigations such as acidity and sulphates. BS 6068 part 2 [24] gives methods for testing of wastes and associated materials. BS 1016 [25] gives various methods of determination of calorific value and carbon content. The Health and Safety Executive provides guidance on methods of determination of hazardous substances [26] and further information is available from ADAS [27] and the HMSO *Blue Book* series [28]. In the United States, guidance is available on a range of analytical techniques [29, 30]. There are two published methods available for the evaluation of combustibility of soils. The BRE gives details of a method that provides a qualitative assessment. A test described by Cairney et al. [31] gives more detailed quantitative information and enables combustibility to be related to compaction and air flow as well as material type.

Where possible, it is preferable to obtain analytical measurements which can be directly compared to official guidelines and standards. Clients and regulatory bodies obviously find comfort in this, and there seldom will be any dispute that a soil is 'clean' if all its chemical contents are lower than published threshold values. Because of this very understandable reason, many investigators in the UK ask laboratories to analyse solid samples only and for the relatively few determinations cited in the ICRCL's guidance [3].

However, this approach does give rise to very real difficulties when risk assessments (the essential reason why site investigations are undertaken on contaminated sites) have to be undertaken, since ICRCL guidance (and that of almost all the equivalent national soil standards) is expressed only in terms of 'total' contamination concentrations. A consideration of the chemical conditions, which could give rise to the risks listed in Table 2.6, soon shows that 'total' contaminant concentrations are seldom reliable indicators. Thus it will often be necessary to 'tailor' analytical determinations more closely, and (as examples).

- obtain water leachability results [32] in those cases where risks to surface waters or groundwaters are possible, and
- have the plant-available phytotoxic metals (copper, nickel, zinc) measured in preference to total metal content for situations where an area is to support gardens or vegetation cover.

Table 2.8 (Chapter 2) indicates the type of analytical bias which can often prove useful.

To an extent, this approach has become normal practice; Building Research Guidance [33] is much more commonly followed today when sulphate contents of soils and groundwaters are measured, and water leachability testing is now a routine inclusion in situations where water resources are vulnerable to the influence of a contaminated site.

Analytical costs – which are often the most expensive element in a contaminated land investigation – have to be borne in mind. These can be reduced by employing cheaper and simpler indicator tests (such as the Toluene Extraction test for organic contents) and only incurring the costs of more detailed testing (such as the polyaromatic hydrocarbon, oil types and total petroleum products tests) when some chosen maximum concentration is exceeded.

3.5.10 Sample type and size

Although national guidelines and standards cite critical contaminant concentrations, there is very little guidance on the sample size and type. Their selection is probably based on the assumption that only 'spot' samples (taken from one specific band and only over an area of a few square millimetres) have been submitted for chemical analysis.

'Bulk' samples, taken as a cut through a vertically thick horizon or over a meaningfully long horizontal distance, will not reveal maximum peak contaminant concentrations, but instead will provide something closer to an average contamination level. Thus their use is generally seen as poor practice.

However, this restriction is not always sensible, as there are particular circumstances where bulk samples are usefully indicative, e.g.

- when excavated materials are stockpiled and have to be judged suitable/ unsuitable for re-use, and
- when soils are imported to raise site levels or fill areas where contaminants have been removed.

In these cases, the need is to obtain truly representative samples. This – while easily specified – is actually quite difficult to attain and will call for

- the collection of a good many subsamples (at least 10 in number and each at least 1 kg in weight) covering every available side and level in a stockpile;
- careful mixing of these subsamples on a clean board to give as uniform a material as possible, (mixing of samples is not suitable where unstable contaminants such as volatile substances are of interest); and
- quartering the mixed materials to abstract the sample for chemical analysis.

Even when such care is taken, unacceptable analytical variations can be found (Table 3.12) and may represent inadequate sampling/mixing or perhaps analytical errors in the testing establishment.

The size of any solid sample submitted for chemical analysis is also important, as laboratories need sufficient material for the preparation and testing of duplicate samples and also to retain enough soil for any retesting which might be wanted later.

The minimum amount of material required will depend on the proposed testing programme. A sample size of 1 kg is recommended by Lord [34] for most testing. However, some tests may require more. Up to 10 kg will be required for a full combustibility analysis for instance. It is therefore essential that, at the design stage, an anticipated testing programme is formulated and

Table 3.12 Variations in chemical proof-testing of an apparently clean imported clay

Sample	Total sulphate content (mg/kg)	Total arsenic content (mg/kg)
1	363.0	7.4
2	123.4	8.2
3	6330.0	23.5
4	3995.0	17.4

(Samples were taken from the same 3000 m^3 stockpile and each was prepared by mixing and quartering 10 subsamples.)

the appropriate sample sizes selected for each material to be sampled. This will of course need to be refined as the investigation proceeds, and during the implementation process the investigator must keep in mind what tests are likely to be needed when taking samples.

In the heterogeneous material, the size of sample taken and its preparation prior to testing will have a significant effect on the results. Any sample of soil may contain a mixture of materials, some of which will contain more of the alien contaminants than others. If the sample is tested 'as received', as is necessary when testing for volatile substances, the analyst must remove a small quantity of the overall sample for analysis. This will give a high or low result depending on whether a more contaminated 'vein' within the sample is selected. If, however, the sample is dried, ground, mixed and quartered to obtain a sample for analysis, the result will be an average figure for the sample as a whole. Samples tested on an 'as received' basis will therefore have a wider spread of results. Similarly, if the size of the overall sample is initially small, it will be less representative of the material on site as a whole and give a wider spread of results compared with larger samples. Since, when assessing contamination levels, the primary concern is with the highest results, testing of smaller samples will tend to give higher individual concentrations, but with less frequency.

3.6 Exploratory techniques

There are several exploratory methods available to provide access to underground materials and to obtain samples.

3.6.1 Hand methods

Inspection and sampling of the surface layers can be undertaken simply and cheaply by hand. Manual excavation down to depths of about 0.5 m is straightforward and can be achieved quickly provided hard obstructions are not present. This permits good access for observation and disturbed samples are easily obtained.

Samples of soil can also be obtained using hand augers. A variety of types are available and in suitable conditions they may be driven up to 5 m in depth. However, progress is usually very slow and their use is normally restricted to fairly shallow depths. In addition, only small disturbed samples are produced.

3.6.2 Trial pits

Excavation of pits or trenches is perhaps the most widely used technique in the investigation of contaminated land. It is usually carried out by mechanical excavators, and pits can be extended up to 7 m in depth or even deeper, if benching of the excavation is carried out. The most appropriate type of plant will depend on the required maximum depth, materials to be excavated and access constraints. A wide range of excavators is now available and details are given in the *Construction Plant Annual* [35]. Mini excavators can be used where access constraints prevent the use of large plant. Wheeled back hoe excavators are capable for excavation down to 5 m and 360° slew excavators enable trial pits to be extended down to 7 m depending on the machine. The size and power of the plant chosen will also dictate the speed at which the excavations can be advanced. Breaker attachments may be necessary if hard slabs are present.

Trial pits have the advantage of exposing the ground for inspection. They are normally about 1–1.5m wide with the length dependent on the depth. They can easily be extended into trenches in order to expose more ground if required. Excavations deeper than 1.2 m should not be entered unless the sides are supported and the atmosphere checked for gas. Observations in pits deeper than 5 m become quite limited towards the base.

Large samples of soil are obtained and materials should be examined carefully on the surface. Where possible, visually different materials should be placed in separate stockpiles as this facilitates examination and sampling. Details of the soils encountered can be logged as excavation of the pit is progressed. However, this should be checked after the excavation is complete, since the extent of individual strata is difficult to judge until fully exposed. There may also be variations across the width of the pit or along its length that need to be noted. A levelling staff should be placed into the pit and readings noted at the surface and at points of interest down the pit side. Trial pits should always be photographed before backfilling and the use of flash will produce much clearer results. Photographs of the main materials encountered should also be taken in the stockpiles on the surface.

When backfilling the pit, the materials should, as far as practicable, be returned in the same order as excavated. This will limit the transfer of contaminants into otherwise clean ground.

Trial pits are the only method that permits examination of relatively large cross-sections of the ground. This enables a good description of the conditions and allows differentiation between continuous bands of material and small local pockets; it also allows selective sampling of different materials that occur at the same depth.

Relatively undisturbed samples can be obtained by driving sampling tubes into the side of the pit. This should be carried out before the excavation progresses below 1.2 m. At greater depths the sides of the pit must be supported and access becomes expensive and time-consuming.

Trial pits are relatively cheap and up to 20 holes per day can be excavated. The main disadvantage of trial pits is the disturbance that they create. A large number of loosely backfilled holes 7 m deep across a development site will clearly create problems. Similarly, investigations within an existing development will cause extensive damage at the surface, which will need to be reinstated.

3.6.3 *Light cable percussion boreholes*

Commonly known as shell and auger, this is the traditional method of investigation for geotechnical work in soils. It essentially involves repeatedly dropping a steel tube known as a 'clay cutter' into the ground and withdrawing it to remove soil that has been wedged inside. A casing is progressively driven into the hole that has been formed in order to support the sides and enable the depth of the excavation to be determined. As well as disturbed samples from the clay cutter, undisturbed samples can be obtained by driving a tube, usually 100 mm in diameter, into the base of the hole.

Boreholes can be advanced by this method down to a depth of 50 m within soils, but will not easily penetrate hard obstructions or rock. The technique enables accurate measurements of depth to the various strata encountered and yields good quality samples. However, it does not permit examination of the ground in-situ and it is not possible to differentiate between continuous strata and pockets of contaminated material larger than the borehole diameter.

Standpipes and piezometers for water monitoring and gas wells are easily installed by this method and multiple installations can be used to monitor strata at different depths in the same hole.

The equipment necessary is fairly lightweight and can be manoeuvred into difficult locations, although it cannot be operated where there is restricted head room. 'Low head room' rigs need a minimum height of 3 m. The technique is generally slower than trial pits, but is capable of greater depths. It has the advantage of creating much less disturbance of the ground or surface damage.

3.6.4 *Power auger drilling*

In the United Kingdom, flight augers are the most commonly available. The drilling tool comprises a cutting edge, above which a spiral flange is fitted. This is rotated from the surface by means of a rod or 'Kelly' and draws soil into the spiral as the tool descends into the ground. In short flight augers the spiral section is of limited length and must be withdrawn after a small number of revolutions. This enables disturbed samples from known depths to be obtained and therefore allows the succession of materials encountered in the ground to be described. However, samples are highly disturbed and the method is limited to a depth of about 6 m unless a very expensive piling rig is used. It is, however, quicker than shell and auger drilling and causes less disturbance than trial pits.

Continuous flight augers allow much deeper holes to be drilled, but with this equipment the soil is pushed upward to the surface along the continuous spiral. The depth to the various soils encountered cannot, therefore, be determined and considerable mixing usually occurs.

3.6.5 Dynamic probes and small diameter boreholes

This method simply involves driving a rod into the ground by percussion. It was originally developed as a means of rapid determination of soil strength by penetration testing. Most rigs therefore utilize a falling weight to drive the rod downwards. The equipment can be adapted to recover small diameter disturbed soil samples, either continuously or from specific depths, by replacing the rod with a small sampling tube. The equipment can also be used to drive gas monitoring probes directly into the ground without excavation. The use of dynamic probes to depths of 50 m is reported [36] but the rate of progress for sampling at such depths is probably unsatisfactory. An alternative technique is to drive the sample tube using hand-held hammers powered pneumatically or by hydraulics. In this case, a separate extraction tool is necessary to withdraw the probe and sample.

The equipment only produces very small samples, and the opportunity of selecting representative materials for testing is very limited. The use of tubes with discontinuous side walls (or 'windows') enables examination of samples before removal from the equipment, but because of the very small diameter necessary to facilitate driving, even very local pockets of material can appear as significant strata. The problems of testing small samples has already been discussed in section 3.5.10. This technique should not therefore be used as the primary investigation method, but is very useful to gain additional data about known strata or previously identified zones of the site.

3.6.6 Soil gas surveys

The presence of volatile contaminants or gas-producing material can be determined by sampling the soil atmosphere within the ground. The process usually involves driving a probe to a predetermined depth and drawing out a sample of the soil atmosphere. This can then be analysed for the presence of the vapours or gases anticipated. Analysis is usually conducted on site by portable meters and this permits the sampling strategy to be adjusted in reaction to results. Sampling is usually carried out on a grid or radial patterns and the measured values are then plotted on a plan. Vapour plumes can be illustrated by contouring.

Soil gas surveys can be carried out as a precursor to exploratory excavations in order to identify areas that warrant closer scrutiny. They can also be used to assist in the delineation of previously identified plumes of contamination. This technique can only be used to detect volatile contaminants and will not locate heavier substances. A survey of the oil terminal site in Figure 3.1 using this technique could fail to locate large areas of soil which actually were contaminated by heavy oils. The measurement of landfill gases is discussed in detail in Chapter 8.

3.7 Implementation

3.7.1 *Specification*

To ensure that the site investigation is conducted in the manner intended and the correct information recorded, the work should be carefully specified in advance. The specification should clearly define the methods of exploration to be used, the observations to be made and the sampling strategy to be adopted. It must also define the protocols governing the method of sampling, size of samples required and testing strategy. Until standardized sampling preparation and test procedures are produced, either nationally or internationally, these will need to be defined for each investigation. Investigators may be competent to specify these themselves or, alternatively, if an external testing agency is to be used, the analyst may be requested to propose the most appropriate procedures. In any event, a full description of sample preparation and test procedures must be prepared. Methods of sample preservation must also be specified. This will affect subsequent laboratory testing procedures and the analyst must therefore be aware of how the samples have been treated between excavation and delivery to the laboratory. This is of particular importance where the testing agency is a separate party to the sampler, since the laboratory results will be representative of the samples 'as received' and not necessarily 'as sampled'. The specification should also set out the way in which the data are to be reported, so that they can be readily understood and interpreted.

3.7.2 *Site work stage*

This is the point at which all the earlier preparation (data collection and analysis, definition of a site model, and choices of investigation and sampling strategies) either give the anticipated benefits or can prove to have been a partial or total waste of time.

The outcome which occurs depends to a great extent on the knowledge, experience and carefulness of the person put in charge of the investigation. If he or she is fully aware of where various contaminated and uncontaminated soils are expected and the specific aims of this particular investigation then success is more likely. However, if to save on costs, an inexperienced investigator is sent out without a full briefing, then a less happy outcome is predictable.

This is because no site model will ever be totally accurate, and – as more and more data become available – the investigator has continually to test and retest the accuracy of the conceptual model, and decide where variations to the planned investigation strategy are needed. For instance, if contamination from the oil terminal (Figure 3.1) were found to have extended a good distance into zones 5 and 8a (Figure 3.6), then an additional contamination zone would have to be allowed, and tested by excavating a number of unforeseen trial pits; this could then reveal the quite unexpected occurrence of permeable sandy gravels within the site's underlying boulder clay and so force the investigator to undertake yet further unpredicted investigation work.

This type of situation can be stressful and matters are complicated by the fact that the person in charge has other important duties to discharge, e.g.

- recording subsurface information in a precise manner (section 3.7.3);
- taking and preserving samples which are truly representative and then ensuring that these are properly preserved (section 3.7.4);
- ensuring that the health and safety of other site personnel are not put at risk (Chapter 13); and
- preventing unnecessary environmental damage.

This last point is worth further comment. Spreading contaminated soils from deeper excavations over the site surface only ensures that remediation difficulties are increased. Far better is to refill each trial pit in reverse order, so that the general sequences of strata are maintained. Equally, if pools of oily liquors were found to have collected locally on the uppermost surface of the example site's laminated clays, then deeper excavations should not be permitted, since this could allow the ponded liquors to escape into more permeable materials and spread a water pollution risk. In such conditions, it makes sense to suspend investigations until a pump and a waste liquid tanker can be brought on to site to remove the problem.

In any well-run site investigation of a contaminated site, the investigator put in charge of field explorations ought to be:

- experienced;
- fully briefed and totally familiar with the conceptual site model;
- provided with precise guidance on the samples which are needed and the sampling protocols which must be followed;
- competent enough to recognize where modifications to the site model are necessary and authorized to initiate such extra works; and
- constantly aware of the risks to site personnel and the wider environment and able to revise the exploration strategy when this could give rise to unacceptable risks.

3.7.3 Recording of data

Data obtained during the investigation must be accurately recorded, in a manner that can be subsequently understood. If this is not done, the value of the entire exercise will be seriously impaired. Records should clearly indicate the following items:

1. *Accurate location of all exploratory holes and surface levels.* These must be sufficient to allow each location to be re-established with some accuracy.

2. *Clear description of all material encountered.* A detailed log must describe the characteristics of the materials (appearance, colour, texture, smell) and their vertical and lateral distribution should be noted. Similar materials are easily matched on site and these similarities should be recorded. This grouping process is much more difficult if it has to be undertaken using only descriptions and laboratory samples. However, it must be recognized that visual appearance is not the only criterion and further subgrouping may occur at the interpretative stage. Conversely, variations in moisture content can radically alter the appearance of otherwise similar materials. The presence of anomalous materials should be specifically mentioned, since these more abnormal soils could be particular sources of unusual contamination.

3. *Record of water encountered.* The depth at which water is encountered, together with the containing strata, must be recorded. The description of the rate of flow into the excavation should also be given where possible and standing water levels should be recorded. In boreholes the recording of water strikes and levels should be as described in BS 5930 [37]. The appearance and smell associated with the water should also be recorded.

4. *Clear identification of materials sampled.* Each sample must be allotted a unique number. This should be related to the exploratory hole number from which the sample was taken. A clear description of the materials sampled should also be given, together with a note of the sample type and size. Identification by depth alone should be avoided since this can lead to confusion, particularly when the depth is at or close to a change of strata. Where a sample is taken of an isolated pocket of material, which is unrepresentative of the surrounding soil matrix, this should be noted. In this case, a sample of the soil matrix should also be taken from the same depth and clearly identified as such. The depth from which water samples are taken should be recorded, together with a note of their description and the strata with which they are associated.

5. *Details of in-situ testing.* The location of each test carried out must be logged. The precise test method used must be identified and all the results recorded, including any observations or irregularities noted.

6. *Instrumentation for monitoring.* When instruments are installed into the excavation, their precise location with reference to the descriptive log should be given. A full description of the installation should be provided, including the method of backfilling.

3.7.4 Sample preservation

Soil and water samples are chemically analysed only to establish the presence of contamination hazards. Thus it is important that whatever results a laboratory reports, they relate to those concentrations which occurred in the samples when they were collected. If losses of more volatile constituents (e.g. volatile organic compounds) or the oxidation of reduced substances (e.g. sulphides) have taken place between sampling and analysis, lower concentrations and smaller contaminant hazards will incorrectly be reported, and later risk assessments will be flawed.

There is good evidence [38] to show that poor practice in choosing sample containers and in storing samples can lead to such losses. 'Chain of custody' arrangements [19] are thus today usually required to ensure sample quality control and prevent deterioration. These list:

- the date of sampling;
- the sample description;
- the sample container type and size;
- the sample storage necessary;
- the analyses required; and
- the date of sample analysis.

Normally, a sample storage protocol (Box 3.3) will accompany the chain of custody records. More complex protocols are needed for more volatile contaminants and preservatives may sometimes have to be added to prevent deterioration [39]. This is particularly the case for water and liquid samples.

In UK publications, the best advice is still that given by Lord [34] who advises that:

- containers must be air- and water-tight;
- their materials must not react with or contaminate samples;
- glass, polypropylene or polyethylene containers are most suitable;
- stainless-steel tools should be used to transfer soils and waters to their containers; and
- that stabilization or necessary cooling of samples be confirmed by reference to a source, such as Neilsen [39].

Box 3.3 Sample storage protocol for solid materials excavated from an old domestic landfill
(anticipated to contain enhanced concentration of sulphates, chlorides and various metals)

Soil sample size	1 kg
Sample container	pre-cleaned glass screw top jar
Filling necessary	total volume of jar
Storage arrangements	in insulated cool box
Storage allowable on site	8 hours maximum
Receipt by analytical house	on the same day as sampling took place

---

The still too commonplace use of tied plastic bags to hold contaminated soil samples obviously is not satisfactory and is likely to give rise to inaccurate chemical analyses.

3.7.5 *Safety precautions*

Work on any site produces risks and hazards against which precautions should be taken and these are well documented elsewhere. The investigation of contaminated land introduces hazards for which special precautions are necessary. The main hazards are contact with toxic or corrosive substances, risk of fire or explosion, and dangers from unstable ground. Precautions needed, therefore, involve protective clothing and working procedures designed to prevent harmful events. The subject is more fully discussed in Chapter 13.

3.7.6 *Testing programmes*

The testing programme must be consistent with the overall strategy of the investigation. It should therefore be designed to identify the chemical conditions of the soil, water and atmosphere in each zone of the ground. In particular it must identify the severity and form of the contamination, significant variations within the zone and local anomalies.

The contaminants to be tested for consist of the four main groups discussed in section 3.5.1:

1. contaminants anticipated due to site history;
2. commonly occurring contaminants;
3. contaminants to which the site or planned use is likely to be particularly sensitive;
4. unexpected contaminants.

The first three will have been predetermined during the design stage, but analyses for unexpected contaminants will depend very heavily on the skill of the analyst, firstly in becoming aware of the presence of such substances and then identifying them. The testing programme, like the rest of the invest-igation, should follow a rational strategy in order to maximize the value of the test results and minimize the costs. Before specifying the testing programme, the record of exploratory work and in-situ testing should be examined. This may indicate the need to adjust the original subdivisions of the site to ensure that each zone represents materials of similar character-istics. However, assessment must not be based on visual appearance alone. A sample of apparently innocuous 'soil' tested during one investigation showed it to consist of about 40% of an explosive substance.

An initial testing schedule, which includes testing or representative samples from each segment, should be produced. Where sampling points have been distributed either randomly or in a grid or herringbone pattern within each zone, sufficient samples should be tested to provided the requi-site degree of confidence as discussed in section 3.5.3. Visually anomalous samples should not be included in this schedule, since they are likely to be unrepresentative of the segment as a whole. They should, however, be tested separately along with an adjacent sample of the 'normal' material for com-parative purposes.

Samples from different depths within each zone will require testing and account must be taken of the irregularities that the method of transportation and deposition may have caused. Contaminants may wash down easily through coarse granular materials and concentrate at less permeable strata. An investigation of a tar works site on Tyneside, which had been constructed on top of tipped gravel ballast, showed the gravels to be relatively clean to a depth of up to 8 m. A highly contaminated layer of tar liquors approximately 1 m thick was found to be located at the base of the gravel immediately over-lying a clay stratum. Similarly volatile materials may have evaporated in the upper layers, but still be present at greater depths. In disturbed or filled ground, contaminated materials may have been placed at any level. How-ever, if the ground is undisturbed, has relatively low permeability and con-tamination is seen to diminish with depth, then samples should be tested sequentially downwards until 'clean' soils are proven.

Where targeted sampling has been adopted, testing should be conducted progressively outwards and downwards from the source.

The testing itself may also be structured so that simple tests such as solvent extraction methods are initially undertaken. More detailed analysis for spe-cific contaminants may follow where appropriate. This itself may also be structured to check first for total concentration and only for specific forms where the previous tests indicate this is necessary. In order to identify the potential for leachate production, leaching (or solubility) tests may also be appropriate.

As many samples of water as possible within the resources of the investigation should be tested. Contamination that has become dissolved in the groundwater is potentially the most mobile and usually will have the highest priority for treatment. At the very least, the investigation must identify the quality of one sample of water taken from each zone, at each and every level at which it is encountered.

Laboratory testing of gases and vapours will usually supplement meter readings taken in-situ and should be carried out where it is necessary to know, with accuracy, the constituent gases and concentrations. The testing of landfill gases is discussed in Chapter 8.

3.8 Reporting and interpretation

3.8.1 *Presentation of data*

During the course of any investigation, a considerable amount of relevant data will have been assembled. In the case of a comprehensive investigation of a large site, this may amount to thousands of individual pieces of information. It is therefore essential that the data are presented in a logical manner that can, first, be understood and second, allow easy retrieval of individual items.

It is always prudent to distinguish between factual information and any interpretations which have been drawn, and – indeed – it is good practice to produce two separate reports; one a factual account of what was done and found, and the second an interpretation of the factual evidence. Site investigation reports often have a wider audience than the original client, e.g.

- planning authorities;
- water and waste regulatory authorities;
- consultants engaged by statutory bodies;
- funders and insurers; and
- future purchasers of the land.

It is very important that these very different readers should not be confused. Legal actions over inconsistencies and alleged misrepresentations in site investigation reports are increasingly common, as large financial investments can rely on the wording of site investigation reports. A prudent investigator will appreciate this and clearly separate factual information from his or her personal estimate of what this information signifies.

It is always helpful if a summary of the investigation is provided to give a fairly rapid overview without recourse to examination of the entire report. The overall format suggested, therefore, comprises an executive summary, factual report of data, interpretative analysis, conclusions and recommendations.

The factual section should describe what has been carried out during the various stages of the investigation and clearly identify the sources of information provided. This is of particular importance when reporting a preliminary investigation, since the information presented will be attributable to a variety of sources of differing ages and reliability.

The reporting of works conducted on site should comprise the following detail:

- description of work carried out;
- logs of exploratory holes, with accompanying photographs where available;
- details of samples taken;
- in-situ test results.

Factual details of laboratory testing should comprise a description of the preservation methods and analytical procedures. The quantitative results of the testing completed for soil, water and gases should be presented in tabular form.

The interpretative section of the report should present a description of the 'model' of the site, deduced from the investigation. This will normally be presented in the same manner as the description of the preliminary conceptual 'model' described in section 3.4.4, but refined to take account of the later stages of the investigation. The interpretation should describe the characteristics of each zone, the distribution, type and concentration of contaminants, and evaluate background conditions. Reference values for all contaminants tested should be proposed stating how they have been derived. These should normally represent contaminant levels at which a hazard could exist and for which an assessment of risk needs to be undertaken. The results of analyses undertaken should then be compared against these values.

Finally, conclusions and recommendations that satisfy the original objectives of the investigation should be provided. These objectives should, of course, also be defined.

3.8.2 Characterizing zones

Detailed examination of the records of exploratory excavation and sample description should be carried out in order to adjust the subdivision of the site. Interpretation of chemical test results should be conducted by reference to the individual zones, and results should be taken as representative of the whole of each zone and not necessarily the variations of contaminant levels within it.

It may be possible to judge that one part of the zone consistently shows different characteristics from the rest, but this deduction is only possible if there is a large number of results to compare. In a zone represented by only four samples, for instance, where two are found to be clean and two contaminated above a given reference value, it could not be confidently stated that half of the zone was free from contamination. The zone would, by definition, consist of materials with similar physical properties and exposure to contamination. Conclusions may be drawn about variations with depth only if it can be sensibly expected that contamination has advanced downwards from the surface in a progressive manner. Unless there is other supporting evidence, such as visual differences, the test results for all locations and depths should be considered together in order to classify the chemical condition of the zone as a whole.

Tests on visually anomalous materials should be excluded from this overall assessment, but should be reported separately as being present within the zone. This may be of particular importance if the anomalous material is substantially more contaminated than its containing matrix. On a former oil terminal site similar to that shown in Figure 3.1, large quantities of materials were classified as requiring removal off site for disposal. However, on excavation it was possible to remove selectively a relatively small proportion of badly contaminated 'veins' of soil, leaving the main body of material on site. Filled gravel pits in the south-east of England contained a mixture of different wastes. These were divided into groups such as clays, rubbles and ashes and it was found that only the ashes contained contaminants that were of concern. It was therefore a relatively simple matter to excavate the ashes selectively by visual identification, and only limited confirmatory testing of other materials was then necessary.

The material within any zone should therefore be described with reference to the physical and chemical characteristics of:

1. the main soil matrix;
2. separate, visually identifiable anomalous materials.

It may also be possible to indicate general variations within the zone either laterally or vertically, provided the reservations described earlier are satisfied.

The chemical results should be presented as the range of values measured and compared against reference values. The proportion of all test results in each zone that exceeds each reference value should be expressed. If there are individual exceptionally high results, additional investigation should be considered.

In some circumstances, the above approach will not be acceptable. If the hazard represented by the contamination is particularly acute, then the whole zone must be assumed potentially to contain the same hazard. The ashy fills tipped around a Royal Navy dockyard is an appropriate example. The dockyard association had meant that asbestos wastes occurred (usually as fibres too small for confident identification on site) and testing suggested that this problem was predominantly contained in the uppermost ash layers. However, suggestions of separating the ashes into asbestos-bearing and asbestos-free layers (and so dramatically reducing Health & Safety and off-site disposal costs) could not be entertained. It simply could not be argued that because one sample of deeper ash was free from asbestos fibres, all adjacent materials were equally non-hazardous.

Other examples could be where other substances are known to be present which can be lethal in low doses (e.g. explosive residues, radioactive materials and especially toxic chemicals).

3.8.3 *Assessment of targeted testing*

This concept relies on a knowledge of a contamination source at a particular location and the strategy is designed to establish the extent to which the contaminating substance has spread. Contaminant concentrations at various distances and depths should be examined to establish whether there is a relationship between the measured values and distance from the source. The method of translocation of contamination must be taken into account as described in section 3.5.4 but the reservations described in section 3.7.6 should also be considered. Reference has earlier been made to the probability that the culverted water course, on the example site's eastern flank, can be acting as a linear channelway for oily liquors spilled in the adjacent oil terminal.

3.8.4 *Assessment of water testing*

Water-borne contamination is potentially very mobile and its effect on the aquatic environment needs to be carefully assessed. It is therefore necessary to establish what water bodies exist within the site, where these are contaminated and where they are likely to migrate, either if left undisturbed, or as a consequence of disturbance. The effects of the contamination on river or other water courses and aquifer quality must be determined, since these will be the main criteria by which water contamination will be judged. Access to abstraction points will be of particular significance.

The results of the investigation should first be carefully assessed to establish the location, and likely extent of all individual bodies of water. These will include surface waters, perched and contained groundwaters, and aquifers. If possible, their direction of flow should also be determined. The water samples that have been taken in each of these bodies of water must then be identified and the test results examined to establish overall quality and variations across the site. The quality of surface and groundwaters both entering and leaving the site should be determined, as should discharges from site drainage. Where site waters are likely to enter adjacent rivers, such as the River Swift shown on the example site, the quality both upstream and downstream should be compared.

A full assessment of the hydrogeology of any particular site is usually beyond the scope of a general contaminated land investigation, but enough

information should be collected to indicate whether more detailed work is necessary.

The water quality within each zone of the site should also be assessed with reference to the contamination in the soil samples. This will provide a useful indication of the solubility of the contaminants within the soil, although it must be recognized that these will usually be in a relatively stable condition, which may be upset if the ground is subsequently disturbed.

3.8.5 *Report recommendations*

Reporting and interpretation is arguably the most important part of any site investigation, since it represents the culmination of all the work undertaken and will probably receive the most attention. Its content will depend on the objectives of the study, but it should in any event give a clear description of the conceptual model of the site.

The conclusion should identify both contaminated and uncontaminated zones as well as the quality and location of waters and gases. It should also identify implications of these with respect to planned uses and any possible environmental harm. Where there are areas of uncertainty or where information is limited, these should be stated. Recommendations should be provided in the report with respect to future work that is necessary and the treatment options that might be available.

3.9 Conclusions

Site investigation is necessary in areas of contaminated land to establish what contaminant hazards occur and whether pathways exist which could lead to future risks to a range of targets.

Because of the inevitable site-specific differences encountered, a logically planned multi-phase exploration process is more certain to reveal necessary information at an acceptable cost.

The initial assessment of available information and the derivation of a site model subdivided into zones is an especially powerful tool, which can bring meaningful benefits to the investigation design. If investigators are competent to update the site model continually as subsurface facts come to hand, then the final site investigation report is likely to adequately describe site conditions and the risks these might generate.

This phased approach really is necessary, since investigations can seldom expose more than a tiny fraction of the soils below land surface. With more than 90% of a site's soils not examined, contaminated land site investigation has to be seen as a risk assessment exercise which will be most accurate when investigators utilize all available data to target their investigation points effectively.

Execution of a site investigation on a logical and rational basis will enable the maximum value to the drawn from the information obtained and will lead both to economy of expenditure and confidence in the end result.

4
Reclamation options

4.1 Introduction

While an impressively large array of reclamation choices is today available[1], the selection of the most appropriate solution is governed far more by national perceptions of the problems which might arise (and by the differing legislation, regulations and soil quality criteria these have generated) than by technical or economic factors. Thus a solution found to be entirely acceptable in (say) Denmark can be rejected as quite inadequate only a few kilometres across the international boundary, into the neighbouring German regional state.

As political biases dominate the choice of the acceptable reclamation approaches to land reclamation, it has to be remembered that political shifts, such as the enactment of new legislation or the imposition of apparently minor regulations, can abruptly affect which reclamation methods are used. Likewise economic changes, such as a sudden rise in the sale value of land, could make hitherto unaffordable solutions cost-effective and so widen reclamation choices.

Thus a consideration of different national regulatory approaches is necessary to appreciate why the pattern of reclamation methods differs so much in various countries.

4.2 Reclamation choice in the United Kingdom

4.2.1 *Evolution of the UK's emphases on land contamination*

In the United Kingdom, the main emphasis has been to bring disused land back into productive use as quickly and cheaply as possible. Where land contamination would preclude or limit redevelopment, remediation has to be undertaken, but within the limits of that improvement necessary to permit a particular re-use.

This attitude stems directly from Central Government's policy and resource priorities and has been emphasized by such statements as [2]:

> The base objective is to enable contaminated land to be used safely and economically, to achieve this a balance has to be struck between the risks from contamination and the need to restore the land to beneficial use.

Wider environmental concerns have thus been of lesser priority than bringing land back into use. Given this guidance, a 'free-market' approach has developed in which:

- No legislation (until the passage of the 1995 Environment Act, whose practical application has yet to develop) specifically addressed the potential problems from contaminated land.
- No policies existed to enforce reclamation of contaminated land, until such time as its owner sought approval for some different land use. This, to some extent, may be altered when the 1995 Environment Act begins to be applied by local authorities.
- The limited soil quality criteria that exist are advisory in nature, and their application implies a large degree of professional judgement.
- Regulation of land use decisions is left to the planning system and is applied on the basis of local priorities.
- Land developers are responsible for treating contamination, although their proposals have to be acceptable to the local planning authority and to statutory consultees.

Of these, the most important is the lack of legislated soil quality standards, since this allows developers and regulatory bodies to agree reclamation solutions of varying thoroughness and to give importance to the ultimate costs of reclamations.

This has led to a much larger private sector reclamation industry than exists elsewhere, that is routinely capable of recycling old industrial sites at a surprisingly high rate. Thus, very many contaminated sites have been successfully returned to use, despite the lack of a specific policy or programme on land contamination.

Obviously this situation differs markedly from conditions in countries where more vigorous legislation and soil quality standards exist, where governmental funding for technology trials and reclamation is more abundant, and where the emphasis is on removing or reducing hazards to the wider environment.

Given its unique economic and legislative origins, the UK industry has thus come to emphasize three particular criteria that influence the choices of reclamation techniques:

(a) *Cost-effectiveness.* Contamination is usually not identified until a prospective redevelopment site is explored. Since such contamination could limit the land's use and value, remediation is undertaken to remove this obstacle. However the final end value of the reclaimed site usually imposes economic limits, which in turn encourage the use of simpler and cheaper remediation methods. Innovative or experimental techniques are seldom favoured, since land developers are disinclined to enter into research trials. Indeed the reverse is usually true in privately funded schemes, where a proportion of the agreed reclamation fee is commonly withheld until the specified end quality has been demonstrated. Any organization offering the use of an innovative technique thus has to price it competitively, and be sure that it will in fact function successfully on that particular site. This obviously is difficult to do until that new technique has been proved in the UK conditions.

(b) *Speed of reclamation.* The sooner a reclamation is completed, the quicker can redevelopment proceed and a return on the initial investment be obtained. Thus, techniques that take years to implement (such as some bioremediation approaches), or which require long-term effectiveness monitoring (as can be the case for many of the innovative stabilization methods) are unlikely to be adopted unless no other options are possible. In addition to this financial aspect, social pressure can tend to favour reclamations which are completed quickly, to reduce the duration of disruption and local inconvenience.

(c) *Flexibility.* Site specific variations are inevitable on contaminated sites, particularly where the land might have been used for a range of industrial processes over a century or more. These variations can impose limits on the use of many of the newer techniques (Table 4.1), but are of less importance to the broad spectrum engineering methods, which in most cases have to be included to remove floor slabs and buried foundations, and to improve ground bearing capacities. Thus it has been easier to avoid the need to prove whether site conditions will adversely affect one of the newer techniques, and fall back on the established and cheaper engineering solutions.

Table 4.1 Examples of site specific factors that can impair some newer reclamation techniques (after [3])

Reclamation technique	Adverse factors
(1) Bioremediation	Biotoxicity in the wastes
	Presence of some chlorinated organic materials can create carcinogenic vinyl chloride
	Will not degrade chlorinated hydrocarbons
(2) Incineration	Presence of mercuric, lead, bromide and reduced nitrogen compounds can create serious emission problems
	Presence of alkali metals and fluorides can damage refractory linings
	Presence of chlorinated hydrocarbons can create dioxins in the emissions
(3) Stabilization	Metal complexing agents (cyanides and ammonia) can make the stabilization process ineffective.
	Chromium (III) can be oxidized to the much more leachable chromium (VI) form
(4) Extraction/soil washing processes	Not effective on clayey soils

Obviously these reclamation criteria are not those that would have arisen had environmental concerns been the driving force for contaminated land reclamations. They are, however, entirely logical for the current UK emphases, and have resulted in a greater achievement of land recycling than has been possible elsewhere. The UK priority is to bring land back into re-use at no greater risk than that generally acceptable on a 'green field' site, and within this limited requirement the reclamation industry has been generally successful.

4.2.2 Challenges to current UK emphases

Six distinct challenges to current UK land reclamation attitudes have recently developed:

- scientific and technical criticisms
- the availability of newer techniques, from the United States and Europe
- changes in policy and legislation over waste disposal and groundwater protection
- changes in the type of land being reclaimed
- the appearance of foreign investors
- Increases in the disposal costs of contaminated soils [4].

Scientific criticism.   Scientific criticism of the limited range of established broad spectrum reclamation techniques (section 4.4) has been voiced for some years. Comments such as 'innovative techniques are relatively poorly developed in the U.K., particularly when compared to our European partners' [5] are widely encountered, but have had very little impact, since the established reclamation techniques appear to have been successful, and certainly are far cheaper than the innovative methods. If US costs [3], for example, have to be accepted (at between $500 000 and $1 000 000 per acre), this at a stroke would destroy the UK policy of recycling old industrial sites for beneficial re-use; the acceptable reclamation costs for private-sector participation in the recycling of contaminated areas of land tend to be about one-tenth of those found acceptable in the USA and seldom are in excess of £100 000 per hectare.

Appearance of innovative techniques.   A wide range of techniques that have been proved in the United States, The Netherlands and Germany has increasingly been offered within the United Kingdom in the last few years. While these are still too expensive to gain ready acceptance, their mere existence has emphasized that it is possible to reclaim contaminated land without freighting vast volumes of waste [6] to licensed tips, and that some reclamation problems that cannot be resolved by established techniques, can in fact be rectified, if the cost can be afforded.

To encourage the use of these newer methods, trials and demonstrations of soil washing, chemical stabilization, incineration and vitrification techniques have taken place in the UK, but without any apparent effect on the attitudes of commercial developers. The costs of these methods simply proved to be too high. Other demonstrations – of the vacuum extraction and bioremediation methods suitable for resolving hydrocarbon spills – have had much more impact, particularly on oil companies eager to clean up below leaking oil storage tanks and refinery plants. For smaller private-sector developments, however, even these proven remedial techniques have attracted little application. The reason appears to be essentially the length of time needed for treatment completion, which is far longer than excavating oil soaked soil and tipping this off-site.

Changes in legislative policy. It could have been anticipated that the strengthening of the UK's environmental legislation (i.e. the Environmental Protection Act 1990, the Water Resources Act 1991 and the Environment Act 1995) might, in itself, have encouraged greater use of these newer reclamation techniques, able to destroy many of the contaminants which occur in soils. These laws allow regulatory bodies the right to enter land, identify sources of environmental degradation and (if the landowner refuses to co-operate) initiate necessary remediation work and reclaim reasonable costs from the courts.

However, the words 'necessary' and 'reasonable' imply a real uncertainty that the courts will actually require a recalcitrant landowner to recompense the regulators fully, and that a significant financial risk could result if these powers were ever exercised. Because of this, negotiation and compromise have been preferred to legal challenges in most cases (as has been the pattern as far back as the 1936 Public Health Act which provided similar unused powers [7]) and this has inevitably resulted in the most economical of clean-up actions, and a lack of application of newer and more expensive techniques.

The sole impacts of this primary legislation have been the increased stringency governing waste disposal (which has made the use of off-site disposal to licensed landfills significantly more expensive) and the increasingly pro-active determination by the water regulatory authorities to protect the quality of surface and groundwater resources. The former National Rivers Authority [8] made clear its determination to bring legal prosecutions in cases where groundwater could be at risk from any contaminated site, and used its status as a statutory consultee, in the planning process controlling the re-use of land, to force a proper consideration of water quality protection.

While these evolutions in waste disposal and water resource protection regulations have certainly made land reclamations more thorough and expensive, they have yet to make it necessary to employ reclamation methods other than those which traditionally have been preferred [9].

Changes in the type of site being reclaimed. A point that is often overlooked is that the 'typical' UK contaminated site more resembles an uncontrolled waste tip than the less complicated sites described in the US or European published case studies.

Decades to centuries of often very different industrial uses have resulted usually in a thick surfacing of dissimilar waste products intermingled with generations of demolition debris, and complicated by areas of non-engineered tipping of variable refuse. Such sites became available for redevelopment with the decline of manufacturing industry from the 1960s. Since they are generally located on the outskirts of urbanized areas and served with existing roads and other facilities, their redevelopment has been attractive both to developers and to local planners.

Given the variability of materials, the differences in contamination type and concentrations, and the variations in ground permeability, these sites are less than suitable for the application of the generally narrow applications spectrum that typifies the newer reclamation methods. Thus, the UK adoption of broad spectrum engineering-based solutions has a practical basis, in addition to the economic criteria discussed earlier.

In the past few years, however, a quite different range of sites has undergone remediation. Typically these are owned by the larger oil and chemical companies and are often reclaimed not for development but to remove potential environmental hazards. In many cases, contamination from only a limited range of substances may exist (e.g. the petroleum and diesel spills on oil company production and tank farm sites) and the site's subsurface conditions may well be relatively uniform.

These sites have offered more suitable conditions to prove the effectiveness of newer techniques [10] and have allowed the first real introduction of US and European methods into the United Kingdom. Such non-redevelopmental reclamations are likely to increase, and so bring greater diversity into the UK reclamation industry.

Influx of foreign investors. An important recent factor for change has been the influx of overseas investors into the United Kingdom, mainly from the United States and Japan. These companies have tended to bring with them higher environmental expectations, a familiarity with more severe standards, and a greater concern for environmental liability. Thus an economic force now exists, which does not find acceptable many of the established UK reclamation approaches.

The introduction of a landfill levy. What could have had the greatest effect on the UK's contaminated land practices was the intention to extend the levy on the use of landfill disposal [4] (primarily imposed to give impetus to recycling and materials re-use) to contaminated soils. This, together with the existing landfill scarcities, could have been the final straw which forced UK practitioners to utilize more innovative, narrow spectrum reclamation techniques. However, lobbying by the construction industry has allowed the contaminated arisings from reclamation projects to remain exempt from these additional disposal costs.

Summary. The UK's contaminated land reclamation practices arose – and still remain in force – because of the need to obtain development land at affordable prices in an over-populated island. The Netherlands has a similar crying need for re-usable land, but has chosen to fund its reclamations from national taxation receipts, and then employ particularly expensive reclamation methods, because Dutch national perceptions are that land contamination is undoubtedly a serious risk to human health and the environment. In the UK, these Dutch certainties have never been accepted, and the British Government, by avoiding the restrictions inevitable if soil quality standards had been enacted, has found it acceptable to encourage the growth of a large commercial sector development industry, and so obtain the necessary land at lowest possible costs. Thus a relatively restricted range of broad-spectrum, engineering based, reclamation techniques (section 4.4) has come to dominate UK practice.

This, however, is gradually changing, particularly as the traditionally cheap solution of exporting vast quantities of contaminated soils to licensed landfills becomes less and less affordable. Some new (narrow spectrum reclamation techniques (section 4.5) have already become accepted as the appropriate solutions for hydrocarbon contamination of soils and waters, and others (particularly soil washing techniques which reduce necessary volumes of off-site disposals) can be expected to gain a greater usage in the near future. Greater attention is now being paid to establishing whether 'contaminated' soils really pose risks, since this allows treatment to be restricted to smaller volumes of material.

However, despite the various challenges to UK reclamation traditions, no major emphasis change is predictable, unless – as seems highly improbable – concerns such as those voiced by The Netherlands Government become proved. Until a clearer link between land contamination and the probable risks to people and the environment is established there seems no good reason to adopt the more expensive reclamation choices which dominate the USA's and The Netherlands' approaches.

4.3 Reclamation choices where health and environmental concerns dominate

4.3.1

In the USA, European needs to recycle old industrial land have yet to become enough to influence policy-makers, but public fears over the effects which chronic exposure to contaminants might generate are especially acute. Thus USA policy is driven by a wish to protect human health and, to a lesser extent, to reduce the environmental impairments which old industrial sites can cause. This national fear of health effects (particularly of cancers) has led to a very highly regulated land reclamation policy, and to the routine use of the widest range of narrow-spectrum and expensive reclamation techniques anywhere in the world. The selection process, leading to the identification of a preferred reclamation solution, is thus very distinct from that described for the UK.

4.3.2

The USA's selection of reclamation choices is governed by a comprehensive list of 'priority pollutants' and 'target compounds', which arose not from land contamination worries but from the widespread public acceptance that exposure to anthropogenic inputs must result in health impairment and possibly in early deaths. Given this belief, it is unsurprising that the USA's target compounds list gives much greater emphasis to organic chemicals (as these often are known to be carcinogenic in laboratory trials and so are likely to pose the same risks to exposed humans) and pays less note to the heavy metals on which European practice focuses.

These listed chemical substances arose after a 1976 law case which revealed the need for specific contamination standards before regulatory bodies could successfully prosecute polluters, and grew from the controls devised for discharges to water resources and for the standards for waste disposal (1972 Water Pollution Control Act Amendments and the 1976 Resource Conservation and Recovery Act). The initial 127 'priority pollutants' of the late 1970s have now been extended to a much more extensive 'target compound list'.

The existence of what are in fact very comprehensive soil quality standards, and the very low concentrations of remnant listed chemicals which are allowable in soils, has led to a risk-based ARAR (i.e. Applicable, Relevant and Appropriate Requirements) system being adopted for reclamation option selection.

4.3.3

The ARAR method calls for the existence and concentrations of a vast range of soil contaminants to the proved; has to assume that feasible pathways, which could lead to human beings becoming the targets of contamination exposure, are available; and requires that cross-media contamination possibilities (i.e. of air or water being affected by soil contaminants) are established. All this has to be completed to the satisfaction of regulatory bodies (state or federal) and so is a much more thorough risk assessment process than any developed elsewhere. This point is more easily appreciated when the small number of contaminants listed by the ICRCL (Appendix I) is compared to the vastly larger list of the US's target compounds.

The overall consequences are that site investigations are much more expensive than those outlined in Chapter 3, can easily cost up to $100 000 per hectare [11] and call for advanced techniques to map out areas of higher than acceptable soil contamination. This mapping process – to identify discrete sub areas which will require remedial action – is necessary simply because the costs of reclamation are so very significant.

4.3.4

When the sub-areas of a site, which must be remediated, are known and risk assessments have proved convincing to regulatory bodies, USA practice is to identify feasible reclamation choices from one or other of the large number of reclamation process databases which exist. The fact that such databases are available, and in such numbers (for example the Dialog® system itself has hundreds of process databases), will seem surprising to European readers, familiar with only a handful of reclamation process providers, within a whole country. However, the USA reclamation market is far more chemical engineering dominated than is the case elsewhere, and within one single reclamation method (say incineration to remove hydrocarbon contamination) there are dozens of slightly different and patented incineration systems, each with its unique operation limitations and advantages. Factors such as

- safety;
- reliability;
- possible reduction of contaminant toxicity, volume and mobility;
- necessity for process proof monitoring;
- treatment rate;
- treatment by-products (discharges/emissions) and the problems these pose;
- acceptability to regulators and local residents; and
- reclamation costs

can be identified and evaluated against each site's quite specific conditions.

Finally a list of those few acceptable reclamation choices is produced and given to the site owner. Some processes selected will carry especially high capital costs but will be completed in only a few weeks or months (e.g. incineration), while others (such as bioremediation) will be far cheaper but will call for much longer completion times. The site owner thus has to choose between length of disruption and the scale of financial outlay.

4.3.5

Reclamation practices in the USA are driven by factors entirely distinct from those in Great Britain or in other industrialized countries. Their costs tend to be much greater than those affordable elsewhere, and their adoption basically reflects the wealth of the United States, and the need for its industry to obtain statutory confirmation of adequate levels of site clean-up to protect landowners from future expensive litigation.

Whether or not the USA reclamation selection system is justified or necessary is a matter of personal opinion; what is undoubted is that it is necessary. A particular pattern of legislation, public opinion, and a national willingness to sue landowners who own land which could generate risks to human beings or the environment, makes it essential that such a land clean-up process is undertaken.

4.4 Engineering-based (broad spectrum) techniques

4.4.1 Excavation and disposal

Off-site disposal to a licensed tip has been by far the most widely used reclamation solution. The Department of the Environment in its evidence [6] to the House of Commons Environmental Select Committee estimated that between 1×10^6 and 5×10^6 m^3 of contaminated soil was disposed of each year in this way. These very large volumes indicate the traditionally low cost of tipping (often as low as £10 to £15/m^3) and the technical simplicity of the process.

However, this situation is changing. Suitably licensed tips are now scarcer and more expensive. As a result, some costs are now in excess of £30/m^3, and when transportation is included, together with the costs of importing clean fill to replace the excavated volumes, it is already obvious that off-site tipping has become as expensive as some of the newer remediation techniques, in at least a few parts of the United Kingdom. Also, the regulations governing the disposal of wastes have become more stringent. Thus it is predictable that off-site tipping must become a far less attractive option in the near future.

Tipping costs can, of course, be reduced if the site investigation has been properly conducted and has accurately identified the horizons and materials that carry the highest contamination levels. It is then often possible to visually identify these more contaminated materials and to separate them when a site is excavated. A typical example of this occurred on a disused iron works site, mantled by a variable surface capping of demolition rubble, slags, combustion ashes and general debris. While the total volume of these wastes was in excess of 15 000 m^3, only the combustion ashes proved to be highly contaminated. These ashes were easily separated when the site was necessarily excavated to improve its bearing capacity, and accounted for a total volume of less than 2000 m^3. As local and suitably licensed tips were available, transporting this relatively small volume of contaminated wastes off-site was by far the cheapest remediation option. However, site supervision and quality control had to be of a higher than normal standard (Chapter 13).

Methods, which demand no more than widely available civil engineering crushing and screening plant, and which can very markedly reduce the necessary volumes of off-site disposal (section 4.4.5) have already increased in popularity in those areas where landfill costs are higher. This trend seems very likely to accelerate.

The predictable decline in the amounts of off-site tipping has the significant advantage that it will remove the cheapest reclamation option and so make other newer techniques more cost-effective. It also has the wider benefit of reducing the past policy of simply relocating environmental problems, and of limiting the possible hazards when large volumes of contaminated soils are moved by road transport.

4.4.2 On-site encapsulation

As tipping costs have risen, developers have begun to accept on-site encapsulation as a more economic solution. Chapter 5 (section 5.5.3) indicates the approach that is advocated for the construction of environmentally secure on-site encapsulations.

In a recent case, land adjacent to an airport was to be re-used for long-stay car parking. Site investigations, however, showed that the area had been utilized for uncontrolled waste dumping for many years and, in particular, that the bottom sludges from the airport's own fuel tanks had been tipped there. The costs of removing these wastes to a licensed disposal facility, or of degrading the hydrocarbon residues, both proved too great for the project budget, and secure encapsulation was identified as the sole cost-effective answer. This proved to be possible largely because the sludges had weathered to the level at which their mobilities and water solubilities were of such trivial scale, that water resources pollution would not occur, and because regulators in the area saw no requirement for a waste disposal licence for the encapsulation. In other British regions, regulators are more likely to insist that a licence be taken out. If this is required, encapsulations would have to be managed in a manner similar to licensed landfills and substantial costs and potential future liabilities would result. These would inevitably make on-site encapsulations much less attractive to developers.

This would be unfortunate, in a technical sense, since it is ground disturbance which introduces oxygen and moisture, disturbs the equilibrium which has developed between contaminants and surrounding soils, and generates further leachates. After a few years, the more mobile components of contaminated materials have already leached away and the subsurface materials then pose much reduced potentials for risk. Wholesale ground disturbance generally will disrupt this situation and is often not the best environmental option.

The logic of regulatory demand for landfill licences on designed encapsulations is debatable, since rapidly constructed and well-designed encapsulations simply do not begin to resemble landfills, which invariably are built up over many years and seldom can be securely enough capped to minimize environmental degradation.

A waste disposal licence may also be required by some Waste Regulatory Authorities, or by the Environment Agency when is established. The logic of requiring a licence is that such encapsulations are depositories of unwanted 'wastes'. However, as Chapter 11 makes clear, if it can be demonstrated that the encapsulated soils were *not* excavated as unwanted materials, but for deliberate re-use in the core of a sound/noise screen mound, then the encapsulated materials would not legally be 'wastes' and a case for arguing any need for a waste licence would exist.

4.4.3 *Dilution of contamination*

This can be a cheap and simple process where contamination is uneven, or where only a thin capping of contaminated material overlies clean natural materials. Recently a site in central Scotland was treated in this way. The local contamination came from ashes, which averaged only 500 mm in thickness and were absent over large areas of the site. Below the ashes a clean lacustrine clay existed. The developer had to strip out the site, to remove the various concrete floors, slabs and foundations that surfaced almost its entire area, and found it convenient also to excavate the ash band together with a thickness of the underlying clean clay. These were mixed in a high speed rotary mixer and returned to the excavated areas for compaction, once site analyses had shown that the metallic contamination had been reduced to levels below the ICRCL guidelines for land to be used for domestic housing. The main environmental risk in this case was the possibility of creating air pollution, and allowing workers and nearby residents to inhale metal rich dusts. A stringent application of dust suppression methods was thus enforced. Similar results may be achieved by deep ploughing, though this is seldom possible as most sites have abundant subsurface obstructions.

The advantages of the dilution technique are simplicity, cheapness, and the need only for widely available equipment. Thus, this approach has grown in popularity, where access to waste disposal tips is especially difficult and expensive.

Two disadvantages are, however, apparent. The need for quality assurance is especially high, and systems have to exist to identify unacceptable materials and prevent their reincorporation into the site. The second difficulty would occur if the dilution process made any contaminants more readily leachable, and so increased groundwater pollution. Given the modern emphasis on protecting groundwater quality, this last point has to be regarded as a critical limitation.

Thus, the dilution method is likely to be best used only where essentially insoluble contaminants exist and where there is already a need to significantly improve a site's bearing capacity by excavation and recompaction.

4.4.4 *Clean covers*

Clean covers are considered in detail in Chapter 5 and are a particularly cheap option (at £15–30/m^2 of site surface). Their use is limited to those sites where surface level increases can be accepted, and they are best employed after site compaction and the removal of buried tanks, vats and pipelines has been completed. Clean covers are most appropriate for the heterogeneous sites with a history of various prior uses and contamination, and should not be used to attempt to contain gaseous or oily contamination.

Clean covers are often used as the clean surface above soils remediated by other methods (e.g. by dilution), and where the primary remediation has not fully attained desirable soil chemical concentrations. This 'gradation-with-depth' approach is particularly appropriate where the contamination is essentially from heavy metals and where the primary risk is to human health via soil ingestion or dust inhalation. Other countries (e.g. Denmark[12]) have accepted that this approach is an adequate solution and permit contaminant concentrations in soils to increase with depth.

4.4.5 *Soil dry screening*

When a site's soils necessarily have to be excavated to improve ground-bearing capacities and remove more troublesome layers and deposits, the opportunity exists to amend and improve the chemical nature of the excavated materials and possibly re-use most of the dug soils. A combination of a crushing plant (to reduce the sizes of coarser fragments) and a vibrating powered screen is usually the most effective in granular deposits, and tends to attract a cost (less than £6/m^3 in 1995) far below that for the more advanced wet soil washing systems developed in the USA and in Holland. (section 4.5.4).

The basis on which dry screening relies is that metallic contaminants (in particular) tend to be most concentrated in the finer particles of granular soils. Thus a prior sieving trial, on a large and typical soil sample can, in most cases, identify which particle size distinguishes 'clean' from 'contaminated' soil. When this is known, the vibrating power screen can be set to separate excavated soils at this particle size, and produce re-usable materials together with a separate (and hopefully much lower) volume of unusable contaminated fine soil.

Routine proof-testing sampling and chemical analyses are of course necessary (usually on a frequency of at least one bulk representative sample per $300\,m^3$ of 'clean' screenings) to confirm the adequacy of the particle separation process.

While cheap and rapid (up to $500\,m^3/day$ of output per power screen is routinely attainable), the method is subject to weather restrictions, as heavy rainfall can 'stick' finer (and more contaminated) particles on to the surfaces of coarser soil grains, and so render separation ineffective. Additionally dust suppression will usually be necessary in inhabited areas to prevent local residents being exposed to the risks of inhaling metal rich dusts.

One increasingly attractive variant of the dry screening method is to recover for resale some particular soil fractions. Ashes, which pose contamination risks because of their heavy metal and PAH contents and because of their ability to support subsurface smouldering and combustion if heated to $130\,°C$ or higher, are actually a valuable resource to a number of industries (building block manufacturers, steel producers who require additional carbon contents to improve their iron ore raw materials, and manufacturers of carbon electrodes). In periods of higher industrial and building activity, screened ashes can be sold for up to £8.00/tonne, and so markedly reduce the costs of reclaiming such sites as former railway yards.

Similarly crushed concrete and brick attracts a ready market and can be sold for £3.00 or more per m^3.

The decision to utilize a dry screening reclamation also offers the real advantage that particularly troublesome materials (such as landfill gas producing organic waste layers) are relatively easily removed in screening.

Dry screening – like wet soil washing – is, however, relatively ineffective in more clay-rich soils.

4.4.6 *Summary*

Engineering-based reclamation techniques have so far dominated the UK industry, largely because of their relative cheapness, speed of implementation and technical simplicity (Table 4.2). Very little evidence is available to indicate that reclamations based on these methods have been other than successful. However as the availability of landfill tip disposal for liquid and more soluble soil contaminants decreases (in response to cost rises and the unwillingness of landfill operators to accept wastes which would create difficulties with

Table 4.2 Salient features of various engineering techniques

	Excavation/disposal	Excavation/ encapsulation	Dilution/ reincorporation	Cover	Dry screening
Applicable to which contaminants	All	Most (except very mobile contaminants)	Relatively immobile contaminants	Contaminants in the liquid and solid phases (excluding oily contaminants)	Solid granular contaminant
Effectiveness?	High	High	High	High	High
Time necessary for reclamation	Short, if tip access available	Short, if construction programmed with the excavation	Short	Relatively short	Short
Cost indication?	Still the cheapest solution, but costs likely to escalate	Relatively cheap	Relatively cheap	Relatively cheap	Relatively cheap
Is quality assurance critical?	Quality assurance requirements are minor	Quality assurance requirements are particularly high	Quality assurance requirements are high	Quality assurance requirements are high	Yes
Are design requirements high?	No	Especially high and detailed	No	Yes	No
Limitations?	Restricted only by tip availability cost, and the legalities over waste disposal	Leaves contamination in situ, long-term performance has to be demonstrated	Can increase groundwater pollution if the diluted contaminants are leachable	Leaves contamination in situ, long-term performance has to be demonstrated	Limited by weather conditions and clay contents in soils

their own regulators) it is predictable that the engineering-based approaches must be complemented by an increase in the employment of narrow spectrum techniques. This is already occurring and can only be expected to increase.

4.5 Innovative (narrow spectrum) techniques

4.5.1 *Introduction*

The number of innovative reclamation methods is already large and is constantly being increased. For example, 63 innovative techniques were proposed for governmental evaluation in 1989 in The Netherlands alone [13]. It is not possible or useful to attempt to describe every possible technique, and attention is directed below only to those UK situations where a need is apparent for innovative reclamation methods.

4.5.2 *Remediation of oily contamination*

Corrective action at leaking underground oil storage tanks has become one of the most prominent procedures in the highly developed US waste management industry. As a result, a wide range of remedial techniques for the treatment of ground soaked with the lighter oils, and for the recovery of free oil product from groundwater tables has been proved in field applications. Since the United Kingdom has a similar need to treat contaminants leaking from oil tanks – and also because in the wider range of contaminated land, oily contamination is a prevalent problem not easily resolved by any of the engineering-based techniques – the US experience is thus of interest.

Treatment of oil soaked ground. Eight distinct techniques have been evaluated and ranked by Haiges and co-workers [14]. While these rankings (Table 4.3) are of value, given their derivation from a large number of field trials, note has to be taken of the different soil quality and environmental standards that apply in the United States and the economic differences between the two countries. Any future UK evaluations of these same technologies may well give different rankings.

Bioremediation currently is the most commonly available of these newer techniques in the United Kingdom and essentially mirrors the natural degradation of organic material to water and carbon dioxide. Use can be made either of naturally occurring microbial populations, or of specifically tailored micro-organisms. In either case, nutrients and oxygen are required and the remediation can be carried out in situ (although this can cause the clogging of soil pore channel ways), or in mounds of excavated soil.

Land farming is a simpler variation of the bioremediation process. The contaminated material is mixed into the local topsoil and biodegraded.

Table 4.3 Ranking of techniques for the treatment of soils contaminated with light oils (after [14])

Technique	Technical feasibility	Achievable treatment levels	Adverse impacts	Costs	Time of treatment	Overall ranking[a]
Bioremediation	3	5	1	4	7	**1**
Soil washing	6	2	4	5	2	**2**
Soil flushing	4	4	3	8	4	**3**
Land farming	5	3	2	3	5	**4**
Vacuum extraction	2	6	5	2	6	**5**
Passive venting	1	8	6	1	8	**6**
Thermal destruction	7	1	8	7	1	**7**
Stabilization	8	7	7	6	3	**8**

[a] 1 indicates best, 8 indicates worst.

The natural degradation capacity of the topsoil can of course be increased by introducing additional micro-organisms and nutrients.

Soil flushing involves abstracting local groundwater, mixing it with a sur-factant and nutrients, and spraying the liquid on the contaminated soil. The infiltrating fluids wash out absorbed oily contamination, which then is bio-degraded in situ.

Soil washing is a similar process, which calls for the addition of water, solvents and/or surfactants into the excavated soil. Separation of the mixture into solid and liquid phases concentrates the oily contamination in the liquid. Typically, several stages of washing are necessary to achieve the required contaminant reduction. Coal washing, which is widely used in UK reclamations, is a simpler form of soil washing.

Vacuum extraction and passive venting rely instead on the diffusion of more volatile constituents to reduce the soil contamination. The natural diffusion rates are increased in the vacuum extraction method by imposing a vacuum at suitably located extraction wells. Where excavation of the contaminated soils is possible, natural venting by exposure to sun and wind can be very effective.

Thermal destruction is taken by Haiges to include all the various incineration methods (section 4.5.3).

Chemical stabilization includes the various solidification and stabilization processes (4.5.4) that do not destroy the oily contamination but fix it in physical and/or chemical bonds.

Table 4.3 shows that bioremediation is Haiges' preferred option, despite it being almost the slowest technique and giving lower treatment levels than most competing methods. This preference reflects US standards on environmental emissions and need not reflect any likely UK emphasis. Vacuum extraction, which has successfully been used on one UK site [10] is distinctly cheaper, but does entail air pollution risks unless complex air cleaners and filters can be included. Thermal destruction methods rate poorly in Haiges' rankings because of their costs and the difficulties in ensuring acceptable emission standards. However, this judgement may be dated, since the newer oxygen enhanced incinerators [3] are able to double treatment rates despite their reduced capital costs, although the technical complexity of the incinerator units is inevitably increased.

Removal of free oil product from water tables. A range of pumped extraction techniques is available, together with two long established engineering techniques (Table 4.4).

As the problem of floating free product is commonly encountered on such sites as old timber treatment plants and coke works, there is a real need for effective remediation techniques in the United Kingdom. Haiges' weightings (Table 4.4) favour the use of the dual pump system, in which a deeper pump (in the groundwater) creates a cone of depression into which the free floating product migrates to be collected by a skimmer pump set at the oil/water interface. In UK conditions, trials have indicated that the simpler surface pump system, which collects both oil and water, can be effective and cheap, particularly since a range of oil separation processes can be included as required (i.e. filters, presses, thin layer biological filters, etc.).

An odd aspect of Haiges' ranking is the low rating given to interceptor trenches and subsurface barriers. UK experience is that bentonite and similar cut-off works can be very effective in the short term (Chapter 6) and that floating oils can be easily skimmed off once the groundwater flow is constrained. Thus doubt exists over Haiges' judgement of these techniques.

Table 4.4 Ranking of techniques to remove floating free light oil product (after [14])

Technique	Technical feasibility	Achievable treatment levels	Adverse impacts	Costs	Time of treatment	Overall ranking[a]
Dual pump	4	2	1	4	2	**1**
Surface pump	1	6	4	2	6	**2**
Vacuum enhanced recovery	6	1	3	6	1	**3**
Scavenger filter pump	3	3	1	4	3	**4**
Single skimmer pump	1	4	6	1	4	**5**
Cyclic pump system	5	5	5	3	5	**6**
Interceptor trench	7	7	7	7	7	**7**
Subsurface barrier	8	8	8	8	8	**8**

[a] **1** indicates best, **8** indicates worst.

In summary, a range of proven techniques is available for the remediation of soil soaked with fuel oils and for the removal of floating free product. Some are already available commercially in the United Kingdom and the others are likely to follow if market demand increases as the need to protect groundwater quality is enforced.

4.5.3 *Removal of cyanides and heavier hydrocarbon contamination*

The cyanides and semi-solid tarry wastes that typify many gasworks sites can be rectified by a range of thermal techniques. Most currently available methods involve the excavation, sorting and crushing (to a maximum 40 mm particle size) of waste soils, which are then fed into rotary kiln thermal units. The soil is preheated, and moved to the main heater unit where combustion occurs. The cleansed soil is then moved to a mixer for cooling and moistening, and the flue gases are passed through an afterburner with additional fuel and oxygen. Finally, flue gases are cleaned through various dust collector and filter units to give the required emission standards.

Very high achievement rates in reducing cyanides and polychromatic hydrocarbons are routinely reported, and it is fair to accept that the incineration processes are relatively insensitive to material variations and contamination concentrations. However, heavy metals and most inorganic contaminants remain unaffected. Thus a contaminated ash has to be disposed of, and this (should volumes be high) could be a particularly adverse factor.

The main concerns over the use of incinerator methods are of air pollution, although most providers do claim to achieve EU air emission standards, and treatment costs (about £70/tonne). Current trials of oxygen-enhanced incinerators, however, indicate that cost reductions are likely and that reduced air emission problems will be achieved, since the use of oxygen decreases the total volume of gases involved in the combustion process. Concerns have been voiced, however, over the sensitivity of these techniques to variations in the rate and size of materials entering the burner units.

Trials are also advanced in the United States on the vitrification of wastes, including those contaminated by heavy metals. This solution has in fact been used to treat the asbestos wastes uncovered in the reclamation of the Faslane site, but the energy demands and costs (in excess of £100/tonne) would seem to preclude vitrification from any widespread UK use. The impetus behind vitrification is that it does give a permanent solution, without such environmental problems as the leaching of contaminants over a long period of time, and that the product can be shaped into blocks for uses such as the armouring of coastal breakwaters.

Trials of low temperature fusion to either destroy or encapsulate contaminants in melted asphalt are currently in progress in the United Kingdom. These offer a permanent solution for low ignition point compounds, and a

potentially secure encapsulation for substances of higher melting and ignition point. The likely advantages are low energy and remediation costs, the
use of standard asphalt boilers and a minimal air pollution risk.

Incineration and fusion/vitrification techniques have yet to make a significant impact on UK practice because of treatment costs and the concern
over emission hazards. However, if the experimental oxygen-enhanced incinerators and the low temperature asphalt fusion approaches live up to their
initial promise this situation could change.

4.5.4 *Treatment of metal contaminated sites*

Two approaches, soil washing and stabilization, are likely to be appropriate
for metal contaminated sites.

The soil washing concept rests on the fact that contamination is usually
concentrated in a particular soil fraction, often in the finer material. The
technique thus involves passing excavated soil through various sieves and
scrubbers (using water or oxidizing chemicals) to concentrate the contamination. The concentrate is then taken as a sludge to a hazardous waste tip. A
basic problem in the United Kingdom would be the prevalent clay rich
soils, which could give rise to an excessive sludge problem and, consequently,
unacceptably high tipping charges. Variations including froth flotation have
been successful and will remove low density hydrocarbons and cyanide in
addition to the heavy metals. Costs, however, are far too high (at about
£100/tonne) for current UK market conditions.

Potentially more usable is the chemical stabilization approach. In this, a
variety of cement, lime, thermoplastic and soluble silicate reagents have
been used to fix the contamination in low permeability matrices, which hopefully will isolate the contamination permanently. All the existing processes
involve excavation, sorting, mixing and injecting the reagents. Usually the
product can be returned to the site and compacted to a high density (600–
1800 kN/m^2 bearing capacities). The method can stabilize hydrocarbons as
well as heavy metals.

Important doubts over the long-term stabilization of the contamination
have been voiced for some years [15] and even some of the more newly introduced stabilization processes have failed to achieve the standards required in
the US toxicity characteristic leaching procedure [16]. Thus, the permanence
of a stabilization solution always has to be subject to question.

One variation of this process in the United Kingdom [17] was offered for
some years. This uses a patented hydrophobized calcium oxide reagent,
which preferentially combines with any liquid contaminants before it reacts
with soil moisture. The calcium hydroxide thus formed retains the water
repelling ability, unless the hydrophobic calcium hydroxide agglomerates
are physically broken down, and an absorption of atmospheric carbon dioxide gradually converts the calcium hydroxide to a surface capping of calcium

carbonate. Thus, this process has a degree of self-healing ability, even if excavated at some later date. The method has a record of success on German sites and while still rather expensive (at £50–90/tonne) is comparable in costs to bioremediation, although much faster to implement. Given its ability to deal with lighter hydrocarbons and heavy metals, it could be attractive in some situations, particularly if its long-term durability were confirmed. Market take-up, however, did not develop and the technique appears no longer to be offered.

4.6 Summary

Changes in the UK approach to contaminated land reclamation appear certain to increase costs, and so make the use of some newer remediation techniques more attractive. The extent to which this will occur will depend on how stringent the applications of policy and legislation become.

Some established reclamation techniques inevitably will become less widely used, though none is likely to fall into total disuse. Until UK experience with the newer treatment methods is more widely available, land developers will have to accept the achievement records attained in the United States or elsewhere in Europe. This does entail risks, since UK conditions are not always easily comparable to those elsewhere, and it will be prudent to examine all achievement claims carefully.

The most likely situation will be that a particular mix of established and innovative treatment methods will prove cost-effective. This is likely to complicate site control, and so force the need for more vigorous quality assurance.

5
Clean cover technology

5.1 Introduction

In its simplest form, a clean cover need consist of no more than a thickness of a clean material, laid over a contaminated site to separate the proposed re-use from whatever contamination still exists at depth. Covers are cheaper than other reclamation options, and fit easily with the usual need to remove buried foundations, patches of more extreme contamination and improve ground-bearing capacities. As their installation calls only for the equipment and experience already widely available in the construction industry, very many clean covers (e.g. [1–3]) have been installed over the past 15 years.

However, covers are no more than containment systems, whose failure can have serious consequences for the planned re-use of a piece of land. The apparent simplicity of covers often seems to have obscured this critical point, and it is quite rare to find clean cover designs fully justified, to have design lives detailed, and to be given specified efficiencies and factors of safety. Consideration of the predictable failure modes that can affect covers is also often absent.

Because of this apparently casual approach, concern has grown that clean covers need not represent the best practicable environmental option in many cases, and that some covers might indeed offer only temporary relief from contamination migration [4].

This concern is reasonable, and can be seen as a growing maturity in what is still an evolving industry. The use, so far, of only a restricted portfolio of engineering reclamation solutions has been a limitation, and the introduction of the newer thermal, chemical and microbial clean-up technologies has to be welcomed. These newer methods can offer superior solutions, particularly to situations where near surface oils, fluid tars and organic pollutants exist (Chapter 4).

The more extreme criticisms that are sometimes voiced, and which suggest that 'the effectiveness of all engineering solutions will inevitably decline with time' are, however, unreasonable and go far beyond any factual comment. Such comments ignore achievable engineering quality and the fact that almost no examples of cover failure are known [5].

As detailed in the following section, clean covers can be quantifiably designed to provide whatever design lives and factors of safety are required. Their cost-effectiveness and construction simplicity advantages are well worth retaining for a large proportion of contaminated sites, and it seems likely that they will increasingly be used, perhaps in conjunction with newer techniques, chosen to destroy or remove particularly hazardous contaminants.

The essential caveat in this belief is, of course, that appropriate design care will always be included, and that any suggestion of a casual approach to designing a clean cover will be avoided.

5.2 Basic design decisions

The wide variation in the details of those clean covers that have been published, has suggested that no underlying rules exist for cover design [6]. This, in fact, is not the case. Site specific variations and the range of different end-quality requirements will quite properly lead to variations in the final cover details (sections 5.4 and 5.5), but a logical design methodology has to have been followed if a particular clean cover is to provably satisfy the necessary end-quality needs.

The design methodology can be posed as simple questions:

- What does a particular cover have to do?
- How long does it have to remain effective?
- What materials can be included in the cover?
- How can the design properties of these materials be defined?
- How is the design quantified?
- Have possible failure modes been checked and potential failure pathways been closed?
- How quickly can failure occur?
- Does the client clearly understand the design basis and any possible liabilities this could present?

It is worth considering each of these in some detail.

5.2.1 *What does a cover have to do?*

Various end-use requirements can exist. In some cases, all that is wanted is a clean site surface. If this is the case, the design process is simple and calls for very little quantification. However, such covers are only appropriate for sites where contaminant mobility is extremely limited (section 5.3).

In other cases, the emphasis may be to prevent any upward migration of contamination in rising soil water, during extreme drought periods (section 5.4). Such capillary break layer covers are more complex to design and call for a detailed knowledge of the fluid transmitting properties of the cover materials, as well as those of the unsaturated layers that underlie the cover.

Preventing or minimizing the downward movement of contaminants, dissolved in infiltrating rainfall, might also be necessary in situations where groundwater pollution is to be avoided. The design of such covers requires a knowledge of the same material properties as those intended to combat upward migration of contaminants and utilizes a similar design methodology (section 5.5).

Finally, covers usually have to support some surface vegetation, and a normal requirement is to ensure, as far as possible, that plant roots do not move down into the contaminated layers (section 5.6). The soil layer into which vegetation is planted should not be seen as an element in the engineered clean cover, but as a separate layer, designed to meet specific needs.

In many cases, a single clean cover will, of course, be required to fulfil several of the above requirements.

Deciding on the functions that have to be achieved is the first and critical stage of design, and calls for a detailed consideration of the site investigation and risk analysis data (Chapter 3) against the planned re-use of the land.

5.2.2 How long does a clean cover have to remain effective?

On first sight, choosing an appropriate design life for a clean cover appears difficult. Covers are obviously affected by climatic factors (droughts, high rainfall rates, wind erosion, etc.) whose severities will differ from year to year. For example, a cover mainly intended to combat the upward movement of contaminated soil moisture in droughts, will experience less severe droughts quite frequently, but will be tested by extreme droughts perhaps only once or twice in a century.

It could, therefore, be argued that the climatic event chosen to typify the design life should be related to the planned duration of the land's re-use. Typically [7], reclamations for domestic housing re-use have been designed for 100 years of use, and against the drought effects that will occur on a once-in-a-century return period, while reclamations for industrial re-use have been designed against less severe events with shorter return periods (of perhaps 50 years). The assumption behind this belief is that as a cover's design life is increased, so its thickness and costs become significantly greater.

Experience with clean cover reclamation design, however, indicates that this level of subtlety is unnecessary. The height to which a flow of contaminated soil water can be lifted in a more normal dry summer is in fact only a small amount less than that which will occur in a once-in-a-century drought in which surface soils are so desiccated that a soil suction equivalent to 1000 cm of water head is developed, and persists for 100 days. Equally, it is known that if desiccation becomes so extreme that all plants wilt and die (at

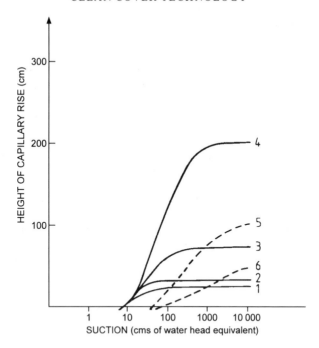

Figure 5.1   Height of capillary soil moisture rise with increasing surface suction (groundwater at 2 m depths. Materials 1–4, sands and sandy clays; materials 5 and 6, heavy marine clays. After Bloemen [12]

soil suctions of 15 000 cm of equivalent water head), the height to which a particular flow rate of contaminated soil moisture can rise is only a little higher than that which occurred in the once-in-a-century drought (Figure 5.1). This latter level of desiccation has, of course, never occurred in recent history.

Thus, it will be appropriate to design capillary break layer covers against a once-in-a-century drought event. This gives the advantage that any change in the planned land use, from a relatively insensitive purpose to a more sensitive re-use, will not force a re-evaluation of the cover's effectiveness.

A similar argument applies to those clean covers intended to minimize or reduce rainfall infiltration and subsequent groundwater pollution and it is reasonable to design such covers to cope with the wettest conditions likely to occur in a century.

5.2.3   *What materials can be included in a cover?*

The fact that coarse granular materials, with a minimal content of silt and clay sized particles, are less capable of permitting upward movement of contaminated soil water and more effective in diverting the downward movement of infiltration carrying washed down contaminants, is now well recognized.

Thus, the general rule is that the coarser the materials in a clean cover, the thinner (and usually the cheaper) can that cover be to perform its design requirements.

However, this is not intended to suggest that covers composed of silty or clayey materials cannot be equally effective (section 5.4). These covers will have to be thicker to achieve the same results, but that need not be an adverse feature, if locally available and cheap materials are to hand. Indeed, the above use of the words 'clayey' and 'granular' is overly simplistic, as it is the combination of the particular suction and hydraulic conductivity properties that determines whether a cover material will be more or less suitable.

Some consulting engineers seem to have preferred to avoid the costs and complexity of the clean cover design process, by adopting standard covers, specified perhaps as 'washed gravel, with no particle larger than 6.3 mm diameter, laid to 300–500 mm thicknesses, and covered with a geotextile blanket' (to prevent siltation), and then 'capped with a rolled clay layer' (to reduce rainfall infiltration). Such covers may well be effective (although this is not certain, since important variables, i.e. the depth to the design groundwater level, the contamination concentrations in that groundwater and the infiltration properties of the cover, have been ignored), but could be overly expensive, if their use forces the rejection of cheaper, locally available cover materials, and requires the importation of large volumes of washed gravels.

A final and obvious point is to ensure that all cover materials are themselves clean. Use is often made of materials from road schemes, adjacent reclamations, or from the crushing of demolished concrete foundations. Thus the possibility of importing extra contamination has to be considered. One recently encountered example (where a capillary break layer cover had been laid over tarry and copper/zinc contaminated fills and then covered by between 1 m and 4 m of imported fills and soils, to produce the required topographic levels) proved to be particularly ill-designed. To supply the volumes of the uppermost fill layer, use had been made of soils from redundant railway embankment, which actually proved to be almost as contaminated as the layers below the capillary break layer which had originally forced the reclamation of this site. Routine sampling and analysis of all imported materials is an essential precaution, as is the rejection of any material that is more contaminated than the ICRCL threshold trigger concentrations, devised for domestic gardens and allotments [8]. As well as the contaminants listed in this ICRCL document, it is usual to require analyses for mineral oils, PCBs and asbestos.

5.2.4 *Detailing the necessary material properties*

Once the possible sources of materials for use in a clean cover have been identified, it is necessary to obtain (except for the very simple covers discussed in section 5.3) two material property curves, i.e.

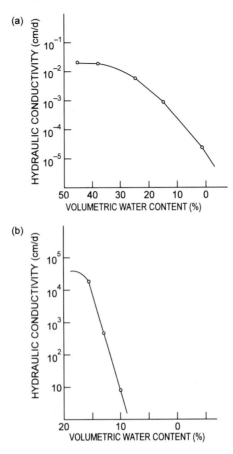

Figure 5.2   Hydraulic conductivity/moisture content relationships. (a) Stiff clay; (b) Crushed
concrete

    1 the hydraulic conductivity variation as a material's moisture content
    changes (Figure 5.2);
    2 the variation in soil suction with the state of material wetness (Figure
    5.3)

not only for the cover materials themselves, but also for the unsaturated
materials that will exist below the cover (Figure 5.4).

    These properties are essential for covers that are intended to minimize
either the upward movement of contaminants in rising soil water, or those
that are to reduce any downward transference of contamination in infiltrating rainfall.

    Soil suction can be seen as the driving force that encourages fluids to
move, either upwards or downwards, while the hydraulic conductivity
controls the rate of such movement. These two properties vary in different

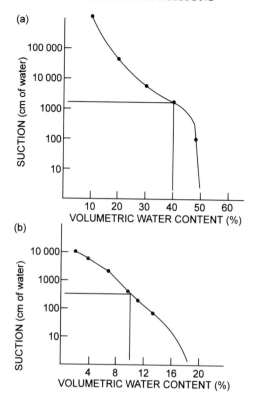

Figure 5.3 Suction/moisture content relationships. (a) Stiff Clay; (b) crushed concrete.

directions, as a materials saturation state changes, and are related in the appropriate versions of the Darcy flow formula, i.e.

$$V = K(\Psi) \left(\frac{d\Psi}{dz} - 1 \right) \text{ for upward fluid movement} \qquad (5.1)$$

$$q = K(\Psi) \left(\frac{d\Psi}{dz} + 1 \right) \text{ for downward fluid movement} \qquad (5.2)$$

where V and q are the flow rates (fluxes) in cm^3/day through each cm^2 of soil, at a chosen position in or above the cover, $K(\Psi)$ is the value of the material's hydraulic conductivity that relates to the moisture content at the position of interest, Ψ is the value of soil suction that relates to the material's moisture content at the position of interest and z is the height above the design groundwater table of the position of interest.

Since a material's degree of saturation will change from one level to another in both the clean cover and in the underlying unsaturated materials,

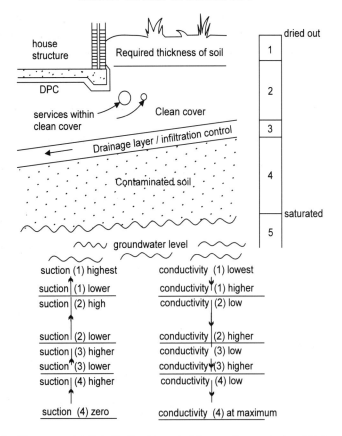

Figure 5.4 Clean cover details indicating the material properties required and their pattern of variation in a prolonged drought.

the conductivity and suction values will differ at each level and it is necessary to define the full range of property variation, from the completely saturated to the near totally dry states for each layer.

Sharrock [9] has defined simple laboratory tests that allow these curves to be measured at their intended compactions. The emphasis on ensuring that all tested materials are at their design compactions is an important point. Under-compaction will reduce the possible soil suction effect and increase that of the hydraulic conductivity, over-compaction will often give the reverse effects.

If direct measurement is not possible, various predictive methods (e.g. [10, 11]) do exist, which permit the hydraulic conductivity and soil suction curves to be derived from particle size analysis data, moisture saturation and saturated permeability values. Care should be taken when using these predictive methods, since none is fully accurate for all material types.

Without access to accurate and repeatable material property curves, appropriate for the planned compaction of a site and its clean cover, it is not possible precisely to design the more complex clean covers.

5.2.5 *Quantifying the cover design*

The assistance of an accurate mathematic model is usually necessary to quantify more complex cover designs (sections 5.4 and 5.5), particularly if a variety of possible cover arrangements is being considered.

Bloemen's model [12] is easily used for those covers intended to combat upward migration of contaminated fluids and the required computer program is fully detailed in the reference. Anders [13] evaluated the predictive accuracy of the Bloemen model against soil column studies of clean covers, and found it to be sensitive and accurate, although prone to slightly over-predict the concentration of contaminant uplift in particular imposed droughts. A graphical output from the Bloemen model is easily included and is particularly useful for a rapid evaluation of different possible arrangements of the layers in a cover (Figure 5.5).

For covers where preventing the downward movement of contamination carried by infiltrating rainfall is important, Bhuiyan's model is perhaps the most convenient [14]. Al Saeedi [15] has proved that this accurately reproduces the actual progress of a wetting front through a cover. A simpler manual computation is also possible for those cases where only the final position of the downward moving wetting front is required (section 5.5).

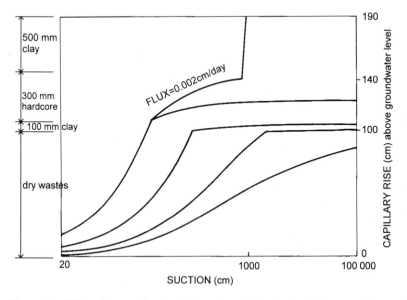

Figure 5.5 Effect of a granular break layer in reducing the possible upward flow of soil moisture.

5.2.6 *Identifying possible failure mechanisms*

Because the contamination that was of concern when a site was investigated still exists below a clean cover, considerable care has to be given to identifying and providing protection against possible failures of the cover. The more likely failure mechanisms can include:

(a) Siltation of the pore voids in granular covers
(b) Desiccation cracking of clayey covers
(c) Chemical attack on a cover's materials
(d) Settlement
(e) Erosion

(a) Siltation will clog the pores of a granular material and, where the cover is to act as a capillary break layer, will reduce its hydraulic conductivity properties and increase its suction potential. Thus, the predicted upward contaminant migration in a design drought may well be far too small (Table 5.1). Similarly, siltation of the granular material in a cover designed to minimize groundwater contamination will significantly reduce the cover's effectiveness. Avoiding siltation is not difficult, if either a designed filter layer or an appropriate geotextile blanket is included above the granular band. Filter layers are probably more robust than geotextile blankets and can be cheaper on those sites where modern crushing plants are in use, and are able to produce precisely limited granular materials.

(b) Desiccation of clayey materials has been very apparent in the sequence of recent dry summers, and cracks to 600 mm depths have been not uncommon. Such cracking can be included in the designs of both the main types of clean cover, but is an additional complication that preferably should be avoided. Predicting the cracking susceptibility of

Table 5.1 Siltation effects of a clean cover (shown in Figure 5.5)

Level of interest in the cover	Concentration of contaminant (mg/kg) added in design drought			
	Sulphate	*Boron*	*Zinc*	*Free cyanide*
(a) Unsilted clean cover				
Base of hardcore layer	1100	11.00	5.5	5.5
Top of hardcore layer	8.8	0.09	0.04	0.04
(b) Silted clean cover				
Base of hardcore layer	1380	13.8	6.9	6.9
Top of hardcore layer	31	0.31	0.2	0.2

Notes (1) Siltation from the overlying clay converted the hardcore to a gravelly silty sand. Presiltation state of the hardcore layer as shown in Table 5.5(b).
(2) Groundwater table in both cases 1.0 m below the base of the clean cover.

clayey materials has been studied by many workers [16, 17] and it is apparent that cracking potential can be related to easily measured properties (clay type and percentage, plasticity index and bulk density). However, it is preferable to quantify the particular cracking potential by placing the clay (at its design compaction) in large diameter moulds (500 mm or greater) and exposing these to heat lamps. This will identify the precise moisture content at which the particular clayey material will just commence cracking. Use of Bloemen's model (run against the design drought's severity and length, with the design position of the groundwater table, and with the soil layers that are intended to lie above the clean cover to support plants) will then identify whether the clayey material could desiccate to the point at which cracks appear. If this occurs, the solution is to alter the type or thickness of soil above the clay cover, until it can be stated with confidence that the cover cannot dry out to its cracking point.

(c) Chemical attack on the materials used in a clean cover is a not improbable situation. Oil and tar vapours are known to deflocculate some clays, and the use of dolomitic or limestone gravel over a site of very low pH does imply a degree of acid attack on the cover will occur. Avoiding such risks should be a priority when considering which materials can be used in a particular clean cover.

(d) Settlement should not occur on any well engineered reclamation if normal compaction quality control has been exercised. However, the risk has to be considered, since clean covers are not only relatively thin, but also have very limited abilities to compensate for more than trivial settlements. Avoiding this risk calls for careful attention to the quality of the site compaction before the clean cover is installed (Chapter 10).

(e) Erosion of a clean cover has so far not been reported for any UK reclamation and would be an unusual event, given the subdued surface topographies applied to most reclaimed UK sites. However, for sites that include landscaped mounds (section 5.5.3), risks might be appreciable. These would call for a deliberate channelling of any storm run-off and the inclusion of soil reinforcing fabrics in these channel ways. Erosion where a clean cover fronts on a large water body is a much more predictable event, and one that should be counteracted by established erosion protection methods.

In addition to these general failure mechanisms, there will be site specific factors that could be of concern. The commonest of these seems to arise from the location of buried services. If these are installed within the clean cover, later maintenance and repair should not pose a hazard. Equally if the services are within an identifiable service reservation and set in an oversized trench, backfilled with clean cover materials, little difficulty should arise.

However, deeper sewers that have to be laid within the contaminated materials, below the clean cover, can pose a very real difficulty. Later excavations to repair broken services could disrupt the site's clean cover. Particular care should be taken to avoid this occurrence.

Sometimes the site construction activities themselves can adversely affect a cover. In one particular case, a cover was laid over the whole site early in the reclamation (to prevent dust blow risks and to allow construction to go ahead without special health and safety precautions), but a large volume of surplus concrete foundation material had then to be crushed. Since this could not be done off-site, the crushing plant was located on the cover and was in operation for several months. The result of this extra loading and vibration was that the granular clean cover became over-compacted and larger particles were crushed. This so altered the cover's design properties that a large area proved inadequate and had to be excavated and replaced.

Plant root migration is often cited as potentially disruptive to the engineered layers in clean covers. This, however, is a grossly overstated claim since the soil densities, at which plant root penetration is precluded, are well established [18] (Table 5.2) and these preventive densities can easily be achieved as cover layers are laid and compacted. If this is ensured, then other biological disruptions (rodent burrowers, warm tunnels, etc.) are also prevented. Plant root penetration and animal burrows occur only if advantage can be gained by going deeper; ensuring that cover layers are too dense for penetration and that no additional food sources exist in deeper layers takes away any possibility of advantage and the feared penetration does not occur.

Concerns over biological disruption of covers is reasonable only where cover layers are loosely laid and where a new higher land surface offers drier conditions to animals such as rabbits. While these problems are not difficult to avoid, it is prudent for designers to bear the possibility in mind and check that the design is proof against such degradations.

Table 5.2   Mechanical resistance to root extension

Soil density (g/cm^3)	Effect on root extensions
1.37	Root growth begins to be affected
1.37 to 1.77	Root growth diminishes linearly
1.74 to 1.83	Root growth entirely ceases
1.55 (in clay soils)	Root growth severely impeded
1.83 (in granular soils)	Root growth severely impeded

Note: properly compacted soils should exhibit densities in excess of 1.7 g/cm^3.
Source: after [18].

5.2.7 The time before any failure becomes apparent

Clean covers are unlikely to display any failure until many years after their installation. Even if the extreme event against which a particular cover was designed were to occur within the first few years, the processes of moisture and contaminant movements are relatively slow and take many months to become apparent. In most cases, much less severe climatic events will occur in a cover's first few years, and no severe testing of the cover will take place.

Thus, an inadequately designed cover will tend to appear effective for some years, and the most rapid failure so far encountered took 7 years to become obvious, despite that particular cover being nothing more than 150–300 mm thicknesses of the local glacial clays, laid over acidic and cyanide rich gas works wastes, which had a near surface and highly contaminated groundwater table. No design basis for this particular cover is discernible and yet it appeared adequate for quite a long time.

The uncertainty imposed by this order of delay indicates that careful design is essential. Accelerated testing of covers is possible [19] and can be useful, but it usually will be more effective to monitor a cover for the first two or so years of life, and to record the variation of soil moisture content at chosen levels, the position of the groundwater table, and the soil moisture contamination concentration change at particular levels of interest (Figure 5.6). If an accurate mathematical model [12, 14] is available, this can be

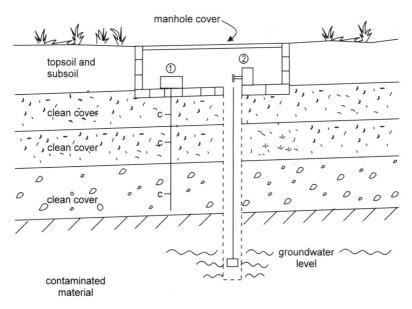

Figure 5.6 Monitoring clean cover performance. Soil moisture cells (c) set in vertical array and connected to an electronic readout (1). Groundwater level monitoring (2) using a punched tape recorder.

used to confirm that the measured data fit the model predictions and that the original design was in fact well founded.

The monitoring necessary is neither extensive nor expensive, requires only equipment that is commercially available, and can easily be undertaken. It is a matter of concern that very few clean covers are in fact monitored. This is in marked contrast to the routine confirmatory monitoring that is seen as necessary for most other engineering works. Without confirmatory monitoring it is difficult to see how any quality assurance can be factually claimed for a clean cover.

5.2.8 *Ensuring that the client appreciates the design methodology and any liabilities this may imply*

As environmental legislation becomes stricter, the liabilities on a landowner or occupier will become more onerous (Chapter 10). Clean covers leave contamination in place, albeit at depth and in such a condition that this should not affect the planned re-use of the land. However, if this remnant contamination affects adjoining land or water, the client might be faced with liabilities at some future date. Care should thus be taken to ensure that the client fully understands the design basis and any possible liabilities this could present.

5.3 Simple covers

The simplest of covers are those intended to separate very immobile contamination from whatever re-use is planned for the site. Such covers need consist of no more than a thickness of clean material, adequate to prevent direct contact with the buried contamination, and compacted to give the required bearing capacities and settlement characteristics.

A suitable situation for such a cover could be over a site where, below a surfacing of metallic contaminated combustion ashes, a stiff clay exists without any groundwater table or contaminated water. The hazards in such a situation would be to human health (ingestion or inhalation of metallic contamination) and to migrating plant roots, which could feasibly abstract phytotoxic contaminants.

Covering the site to a depth great enough to ensure that no future excavations would be likely to penetrate into the buried ashy layers, would remove the hazard to health. The chosen cover thickness would be a matter of judgement but would presumably be in excess of 0.6 m, plus the soil thickness needed for the planned vegetation. If the separation cover is laid to a high density, this will also discourage casual excavations, and if a self-cementing layer (such as freshly crushed concrete laid to a thickness of at least 300 mm) is included, then excavations without the assistance of powered plant will be near impossible. Manual excavation through 600 mm, or more, of highly compacted cover is a much more difficult matter than is sometimes believed. Removing the plant phytotoxicity hazard would consist of no more than providing enough soil depths, in garden and open space areas, to make plant root migration out of the soil layer unlikely. If a particularly fertile and water retaining soil were chosen, there would be a minimal probability of any movement of roots into the less hospitable ashy bands. Quite obviously, such a cover would only be suitable if contamination mobility were trivial.

For cases in which more mobile contamination does exist, but at significant depths below a site's surface, a similar approach can be used. For example, if highly contaminated groundwater occurs at depths greater than 3 m, then upward migration of contaminated soil water is unlikely to be measurable, even after extreme hot droughts. A consideration of equation (5.1) indicates why this should be so. As the depth (z) to the contaminated groundwater increases, so the possible height to which capillary rise of contaminated soil water can occur, will decline. Beyond 3 m, below the original site surface, this is likely to be of trivial amount. In any such case, it will, however, be prudent to check that the site is not underlain by especially sensitive materials that will promote abnormally great heights of upward contamination. The most sensitive materials, in this context, will be sandy clays with appreciably high effective conductivity and suction values (section 5.4).

Simple covers can include the works needed to prevent or minimize ongoing pollution of groundwater or surface water, by controlling the rainfall infiltration that will be needed to leach out the more soluble contaminants (section 5.5).

Simple covers have been widely employed without any known problems, but are only suitable for conditions of very limited contamination mobility, where the mobile contamination exists only at significantly great depths, or where on-going site management is adequate to prevent future excavations into to buried contamination.

5.4 Covers intended to combat upward migration of contamination

5.4.1 Designing a capillary break layer cover

The stages in designing a capillary break layer cover are outlined in Table 5.3. This reveals that the essential data required are:

(a) the conductivity–moisture content and suction–moisture content properties of the proposed cover materials (Figures 5.2 and 5.3) and of those site materials that lie above the groundwater table's level, during the design drought conditions;

(b) the chosen drought severity (section 5.2.2);

Table 5.3   Stages in the design of a clean cover

Identify the site contamination that could adversely affect its planned re-use

Is this contamination of a type against which a clean cover will be effective (section 5.7)?

If so, precisely what does the clean cover have to achieve?

- Is a simple separation cover all that is wanted? (section 5.3)
  → Proceed with simple cover design

- Is protection of the site's surface the requirement?
  → Proceed with capillary break layer design

- Is protection/improvement of the groundwater quality the requirement?
  → Proceed with infiltration control design

In either case, determine the likely location of the groundwater level at the critical design condition (i.e. drought conditions for capillary break layer, high water level for infiltration control cover)

In either case, is it necessary to limit the possible fluctuations of the groundwater by installing drainage works?

Determine the contamination state of the groundwater (section 5.4.1)

Identify the thickness of unsaturated existing materials that will lie above the design groundwater level

Obtain the suction and hydraulic conductivity properties of those materials. Note where site compaction is to be improved, ensure that the properties are those at the required final densities. Ensure that the drying or wetting curves are obtained as is appropriate

Identify the locally available materials available in the required quantities for use in the clean cover

Check that these materials are clean and unlikely to be attacked by contamination on the site

For materials which satisfy the above requirement, obtain their appropriate suction and hydraulic conductivity properties

Utilize an appropriate mathematical model to determine the likely upward migration of contamination in a design drought (section 5.4)	Utilize an appropriate mathematical model or the simplified manual method (section 5.5) to determine the infiltration that can occur
Consider possible cover arrangements and thicknesses, until the required diminution of possible contamination rise is achieved	Consider possible cover arrangements and thickness until the most effective infiltration control is achieved
Check that the topographic levels imposed by the selected cover are acceptable	Check that the topographic levels imposed by the selected cover are acceptable
Check the chosen cover against all likely failure modes	Check the chosen cover against all likely failure modes
Identify necessary quality controls for the cover construction activities	Identify necessary quality controls for the cover construction activities

Check the requirements for any soil layer needed to support vegetation.
Check that this will not compromise the clean cover's effectiveness

(c) the depth to the groundwater table in the design drought;

(d) the worst contamination concentrations in the local groundwater.

Selecting the design level for the groundwater table is usually not difficult. In some cases, the control of groundwater levels by adjacent rivers or by the sea will indicate what can be anticipated in the design conditions. In other cases, enough groundwater monitoring in the area might allow the design level to be predicted. In either case it should be noted that the shallower the depth to the design groundwater table, the greater will be any contaminated upward migration, thus a conservative design choice of as high a groundwater level as is reasonable should be adopted. Care should also be taken to determine if tidal effects on the groundwater are likely, and if so to use a groundwater level raised by the influence of high tides.

Selecting the worst likely contamination concentration of the local groundwater is a less precise matter. Groundwater quality tends to be less well defined than are the concentrations of the solid contaminants and, as Chapter 3 indicates, the confidence possible from site investigation data is often much less than is sometimes assumed. Thus, a conservative approach is usually essential, which entails examining every groundwater analysis, determining if these are likely to typify the site, or if it is likely that more contaminated waters might underlie areas that have not been explored. Usually such an examination suggests that higher water contamination levels could occur in unexplored areas of the site, and this in turn suggests that a particularly contaminated groundwater quality, which includes at least the highest measured concentration of each contaminant, should be used (Table 5.4) Once the four necessary items of data are available, as many clean cover variations as are required, can be evaluated.

5.4.2 Example

An example of the design of such a cover is that installed over the site of a former metal refinery. Site exploration had revealed a near surface (0.5–0.8 m depth) contaminated groundwater, which was affected by the water depth in the adjacent main river. From river low flow records, it proved possible to predict that the groundwater table would exist at depths of 1.0 m or more, during an extreme design drought. Thus, the conservative choice of a 1.0 m depth to the design groundwater level was made.

Groundwater contamination concentrations proved to be very variable, and obviously related to the contamination from specific production units that had existed on the site. Since the density of the site investigation works had been inadequate to give a 95% confidence that the likely worst contaminated water had in fact been encountered, the groundwater quality detailed in Table 5.4 was utilized.

Table 5.4 Design groundwater contamination quality

	Selected design concentration (mg/l)	Highest known concentration (mg/l) in any sampled groundwater
pH	6.9	6.9
Sulphate (as SO$_3$)	2500.00	632.00
Phenols	100.00	72.20
Copper	161.00	116.20
Lead	116.30	84.00
Chromium	99.70	72.00
Total PAH	27.90	20.12
Cadmium	6.00	4.30
Arsenic	4.70	3.40

The site itself was mantled by up to 5.5 m thicknesses of decomposed ashes and mine tailings, which now behave as silty, sandy clays, with meaningfully high suction and hydraulic conductivity values. Thus, the potential for upward movement of contaminated soil water is high if the site were exposed to surface desiccation.

Sources of clean cover materials in the necessary quantities were limited, and finally were reduced to a choice of two:

(a) a stiff local clay (Figures 5.2(a) and 5.3(a)) whose properties are summarized in Table 5.5(a);
(b) crushed concrete and brick derived from the demolition of the refinery buildings (Figures 5.2(b) and 5.3(b) and Table 5.5(b)).

Table 5.5 Properties of the materials considered for use in the clean cover

(a) Clay material

Liquid limit	75
Plastic limit	36
Saturated permeability	0.018 cm/day
Particle size distribution	–

Sieve size (mm)	600	200	60	20	6	2
% passing	100	99	98	84	74	64

As laid moisture content 40% (volumetric)
Hydraulic conductivity and suction properties shown in Figures 5.2(a) and 5.3(a).
Dry density (as laid) 1330 kg/mg^3

(b) Crushed concrete material

Saturated permeability	66 000 cm/day
Particle size distribution	–

Sieve size (mm)	37.5	20	10	6.3	2.0	600 μm	63 μm
% passing	100	69.5	33.4	17.4	3.8	1.8	0.3

As laid moisture content 10% volumetric
Hydraulic conductivity and suction properties shown in Figures 5.2(b) and 5.3(b)
Dry density (as laid) 2170 kg/m^3

The prime requirement for this cover was to ensure that the site's new higher surface would not be adversely affected by upward migration of fluid contaminants.

If the site were not covered, its surfacing of silty sandy clay materials would (in a 100-day-long design drought whose severity were such as to desiccate the surface to produce a suction equivalent to 1000 cm of water head) permit very significant upward movement of contaminated soil water. For the phenolic contamination, whose concentration is 100 mg/l in the groundwater, this would lead to the addition of an extra 42 mg/kg of phenols to the site's surface materials.

The planned re-use of the site called for the worst contamination concentrations to be no more than 1% of the levels in the design groundwater, and the level at which this reduced contamination could be allowed was specified as the top of clean cover (i.e. below the soil layers and inert fills brought in to provide the required final topography).

Using the crushed concrete material, a cover of 25 cm thickness produced the required reduction. With the stiff clay cover, a rather greater thickness of 100 cm was necessary to give the same effect.

In each case, the results were obtained by identifying the various fluxes that would rise to different heights above the design groundwater level (see Figure 5.5). The fluxes were graphed against the heights to which they rose and the various dv/dh ratios established. Equation (5.3) gives the amount of contaminant that can rise to whatever level in or above the clean cover is seen as important.

$$\text{Contaminant addition} = \frac{dv}{dh} \times d \times \frac{\rho w}{\rho s} \times c \ (\text{mg/kg}) \qquad (5.3)$$

where d is the design drought duration (days), ρw is the groundwater density (kg/m^3), ρs is the compacted density of the cover (kg/m^3) and c is the concentration (mg/l) of a particular contaminant in the design groundwater. For any other contaminant, it is only necessary to scale the above listed phenolic contaminant additions in the groundwater contamination ratios, as is obvious from equation (5.3).

The selection of the top of the clean cover as the level of interest for contaminant addition came entirely from the planned re-use of the site and an identification of the most sensitive targets for contaminant attack and hazard. If any other level of interest or any different allowable contamination addition had been chosen, the same process would have been followed.

This example confirms the view expressed earlier that non-granular covers can be as effective as granular covers, though as rather thicker layers. Granular covers do, however, have the real advantage that minor increases in thickness give sizeable improvements in cover efficiency. If, for example, an allowable contaminant concentration of half that chosen above had been selected, the granular cover thickness would have had to be increased by some 7%. For the cohesive material cover, however, a thickness increase of 47% would have been necessary to give the same effect. Thus, if the crushed concrete cover had been laid as a slightly thicker (27 cm instead of the planned 25 cm) band than required, the cover's effectiveness would have been 100% increased.

5.5 Covers designed to minimize groundwater pollution

5.5.1 Introduction

The emphasis that a clean cover's primary function is to prevent the new and higher surface of a site becoming contaminated is now seen as too limited. Although most contaminated sites are underlain by polluted groundwater (due to past spillages and to the on-going leaching out of more soluble contaminants by infiltrating rainfall), reducing this pollution locally can be important. This is particularly so where the groundwater flows to surface water bodies or to groundwater abstraction points. Thus, covers intended to control rainfall infiltration have become more widely used.

The greatest experience with infiltration control is in the design and construction of clay caps for landfill sites. The priority in these is to reduce infiltration and so minimize the generation of polluting leachates. The UK guidance on this is contained in *Waste Management Paper, No. 26* [20] which recommends:

- a cover of topsoil/subsoil thick enough to protect the clay capping below from desiccation and cracking this cover must also be of a thickness suitable to support whatever vegetation is anticipated;
- a rolled clay cap of about 1 m in thickness and laid to a permeability of 10^{-9} m/s.

It could be argued that this guidance is appropriate for a clean cover intended to minimize groundwater pollution, but this view ignores the ineffectiveness of many landfill cappings. Knox [21] has summarized the infiltration results available from a number of monitored landfill caps (Table 5.6), and has shown that high infiltration rates can occur in some cases. It thus is more appropriate to examine the causes and mechanisms of rainfall infiltration, to identify where solutions are possible.

5.5.2 Rainfall infiltration and cover design

Infiltration is a complex process affected by a number of variables, including:

- rainfall duration and intensity;
- absorptive capacity of the surface soils;
- the prior state of the surface soil's wetness;

Table 5.6 Infiltration data from experimental studies and completed landfills (modified from [21])

Site	Capping	Infiltration (mm/year)
Undrained sites		
1(a)	900 mm topsoil/subsoil	>84
2	750 mm topsoil/subsoil	11→66
	500 mm clay	
	200–500 mm weathered shale	
3	1 m topsoil	200
Drained sites		
1(b)	900 mm topsoil/subsoil	<20
	600 mm Gault Clay	
4(a)	Soil	5–7 (on land slopes of 4%)
	300 mm drainage layer	
	500 mm clay	
5	Topsoil 700 mm	56–89
	Drainage layer 300 mm	
	Clay 500 mm	
Sites with steeper land forms		
4(b)	Soil	1–2 (on 20% land slopes)
	300 mm drainage layer	
	500 mm clay	
Sites without a surface soil layer		
6	300–600 mm boulder clay	20
7	> 3 m silty clay	90–140
8	1 m Keuper Marl	40→86
9	1 m Oxford Clay	49
10	500 mm weathered shale	30–120

- the permeability of the surface soil;
- the topographic slopes;
- ambient temperature and season.

It is extremely difficult to quantify these over other than a relatively long time span. If an annual quantification is acceptable the task becomes more achievable, and it is obvious that two quite separate forces encourage rain to enter a soil and then move down the profile.

The first of those is the surface soil's suction, which (Figure 5.3) is highest when the soil is dried out and pulls in moisture. Obviously, as the infiltration continues, the surface material becomes wetter, its suction falls, and so the widely known decline in infiltration rate with the duration of rainfall occurs. As the suction reduces, the second force (the gravity potential) comes to dominate the process, and within a relatively short time the progress of the wetting front down a soil profile becomes controlled by that soil's saturated hydraulic conductivity value. This allows the speed of advance of the wetting front (Figure 5.7) to be evaluated.

The rate of rainfall entry into a soil will be reduced if the surface layers have as low an absorptive capacity as possible. Thus, the use of porous, granular, or high organic content soils above a clay cap will inevitably increase

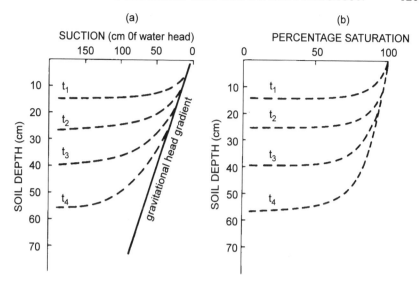

Figure 5.7   Progress of wetting front at times, t_1, t_2, t_3 and t_4 in a uniform soil. (a) Suction/
depth relationships; (b) water content profiles. After Hillel [24].

the ultimate amount of infiltration. This indicates that the usual practice of
laying topsoil and subsoil over a clay cap is bound to increase the infiltration
(Table 5.6), but omitting these soil layers would expose the clay cap to erosion and desiccation, and would also limit the vegetation cover that is usually
required. Thus, an acceptable compromise has to be made.

Increasing the topographic slope of the cover will, of course, minimize the
rate of rainfall entry (Table 5.6, site 4(b)) by encouraging rainfall to move off
as surface run-off.

The initial progress of infiltrating rainfall through a clay cap is thus controlled by its surface condition and type, and if a clean cover of the *Waste
Management Paper No. 26* type was installed, all that would be achieved
would be the delay until the wetting front had moved down into the materials
below the cap. Thus, adequate infiltration and leachate formation control
would not occur.

If, however, a coarser layer with large, open, air-filled pores is laid below
the clay cap (Figure 5.8), infiltrating moisture moves down in the wetting
front until the interface of the finely porous clay and the underlying coarsely
porous granular layer is reached. At this stage, downward progress of the
wetting front is halted for a time, since moisture cannot move out of a material that still has a high suction into one with a lower suction. If the materials
detailed in Figure 5.3 and Table 5.5 had been used, the crushed concrete laid
at a 10% moisture content would have a suction of 500 cm, and infiltration
down from the stiff clay would not take place until the clay's suction had
fallen to at least this value.

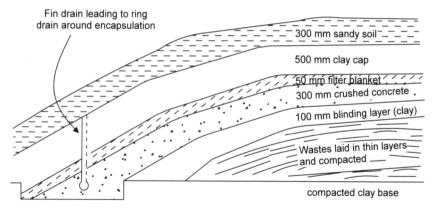

Figure 5.8 Encapsulation design.

Halting the wetting front would not affect infiltration into the site's surface, and so the overlying clay would become wetter and its suction would decline.

At the 500 cm suction level, the stiff clay's moisture content would be close to its saturated level. In practice, an even greater moisture content in the clay would be necessary to move moisture into the underlying crushed concrete. This is because the crushed concrete's pore voids are still air-filled and this air has to be forced out before water entry can take place.

Thus (with the materials shown in Figure 5.3), the stiff clay would be almost totally water saturated and the crushed concrete would become entirely water saturated when moisture finally crosses the material interface.

It could be said that the above argument is flawed, since the clay band (layer 5) below the coarse granular layer (Table 5.7) will be relatively dry, and thus have a very significant suction ability. This, it could be argued, would pull moisture down from the saturated granular layer, and so allow the wetting front to continue its downward progress to leach any soluble materials from below the cover.

This, however, ignores the facts that:

(a) the speed of outflow in a saturated granular material is extremely rapid, while the progress of a wetting front in a clay is very slow and equivalent to its saturated hydraulic conductivity (i.e. in this case to a mere 0.18 mm per day);

(b) the suction ability of the basal clay layer will decline as it manages to abstract some moisture from the overlying coarse granular band.

The net result should be a near total interruption of the downward movement of the rainfall infiltration.

Table 5.7 Functions of the layers in the encapsulation capping

Layer	Functions
No. 1. Top- and subsoil (300 mm)	To support grass cover; to minimize inflitration
No. 2. Clay (500 mm)	To act as soil moisture storage reservoir of high suction capacity material to be compacted to 10% air void ratio and maximum practicable compaction (Table 5.5(a))
No. 3. Filter blanket (50 mm)	To prevent siltation of the crushed concrete layer
No. 4. Crushed concrete (300 mm)	To underdrain the overlying clay by virtue of low saturated moisture content and high saturated permeability
	Material compacted to maximum practicable compaction (Table 5.5(b))
No. 5 Blinding layer (100 mm)	To cap the contaminated material and prevent airborne contamination; to underseal the crushed concrete layer
	Material to be compacted to 10% air void ratio and maximum practicable compaction

However, this will only be achieved if suitable drainage outflow facilities are provided for the coarse granular band, and that usually will imply that the cover as a whole will have been laid to at least a 10% topographic slope.

As the coarse granular band is the critical layer in achieving the required control over infiltration, it is valid to consider where this should be placed within the cover profile. Arguments are possible that the drainage layer should be laid below the surface soils and above the clay cap (Table 5.6, site 4(a)), but it seems preferable to locate the drainage layer below the bulk of the clay capping, since this ensures that the moisture storage capacity of the clay is utilized (Table 5.8), and that a maximum inflow to the coarse granular band takes place.

Table 5.8 Moisture storage capacity of the capping layers

Layer	Initial volumetric moisture content (%)	Saturated volumetric moisture content (%)	Equivalent rainfall infiltration to achieve saturated moisture content (mm)
1 Sandy topsoil (300 mm)	30	50	60 ⎫
2 Clay (500 mm)	40	50	50 ⎪ 187
3 Filter blanket (50 mm)	10	32	11 ⎬
4 Crushed concrete (300 mm)	10	32	66 ⎭
Outflow to ring drain comes into operation			
5 Blinding layer (100 mm)	40	50	10

5.5.3 Encapsulation covers

Covers of the type outlined in Figure 5.8 are particularly appropriate for those cases where a developer requires a provably clean site, and so has the contaminated surfacing scraped off and deposited in a capped mound.

In such cases, it is relatively easy to define the functions required of each layer in the cover (Table 5.7) and to demonstrate the infiltration storage that will occur (Table 5.8). With relatively steep topographic slopes to encourage run-off and with care taken to select a topsoil cover that will not encourage excessive water entry, infiltration is likely to be no more than a maximum of 5% of the winter rainfall value (infiltration from summer rainfall will normally be minimal given the water consumption by plant cover and the higher evaporation rates), and the clay layer (layer 2) will provide several years' water storage, before the outflow drainage from the granular band commences.

A thorough compaction of the contaminated wastes that are to be encapsulated, will ensure that the surface area available for leaching will be minimal.

While little or no infiltration is likely to move down to the encapsulated wastes, these encapsulated contaminated materials were of concern when they occurred on the site's surface. When concentrated in an encapsulation, they have to be regarded as potentially more hazardous, and particular care has to be taken to minimize the effects even of the negligible rainfall infiltration that might reach them. If tarry deposits are present on the site and are air-dried and oxidized prior to encapsulation, it is possible to compact such materials to highways quality, and so reduce any leaching effects [22] to trivial proportions (Table 5.9).

Table 5.9   Leachate from oxidized tarry wastes (W. Midlands site)

Parameter	When wastes finely ground (mg/l)	When wastes compacted to highways standards (as would be the case in an encapsulation cell) (mg/l)
Naphthalene	84–24	< 1
Acenaphthylene	12–10	< 1
Acenaphthylene	7–2	7–2
Fluorene	19–7	2–< 1
Anthracene	15–7	< 1
Phenanthrene	20–5	3– < 1
Fluoranthene	18–5	3– < 1
Pyrene	27–6	4– < 1
Benzo anthracene	43–7	< 1
Chrysene	16–7	< 1
Benzo [b] fluoranthene	7	3– < 1
Benzo [k] fluoranthene	< 10	
Benzo [a] pyrene	< 10	< 1
Indeno [1, 2, 3 – cd] pyrene	< 30	< 1
Dibenzo [a, h] anthracene	< 30	< 1
Benzo [g, h, i] perylene	< 30	< 1
Toluene extract	46–96	6–1
Phenol	420–610	1.9–< 1
Sulphates	80–312	30–95

Encapsulation covers thus necessarily include the works to minimize rainfall infiltration, and the treatment to reduce contaminant leachability. They have to be evaluated against potential failure mechanisms (section 5.2.6), and their efficiencies as capillary break layers should be confirmed, to guard against any upward movement of contaminations, in the unlikely event that contaminated leachate were to collect in the encapsulated wastes.

It has been argued that covers of this type can be compromised by the presence of high water capacity channel ways (e.g. rabbit burrows, decayed tree roots, etc.). Various researchers (e.g. [23]) have shown, however, that water will not enter such air-blocked macropores until the soil around these is almost totally saturated. Macropore flow is only significant in circumstances where rainfall intensity exceeds the surface material's infiltration capacity and surface run-off into these voids can occur. This is usually easily prevented by minimizing surface ponding and channelling run-off into defined channel ways. Additionally, if proper note is taken of the advantages gainable from adequate compaction of clean cover layers (section 5.2.6), the existence of biological channel ways can be precluded.

When designing such covers, it is necessary to use material curves that accurately represent the changes in suction and hydraulic conductivity properties as the material's moisture contents increase, since this will be the actual condition that will occur. The analogous material properties needed for a capillary break layer design (section 5.4) should, in contrast, be obtained from tests carried out as the materials are progressively dried out. For many materials, the wetting and drying curves will differ to a significant extent.

While no use of artificial materials is included in Figure 5.8, some practitioners would prefer to install a high-density polyethylene liner in addition to the clay blinding layer, above the compacted wastes, as this will provide added assurance. Use of such a liner in place of the clay layer, however, would be open to criticism, since damage to liners during construction is likely to result in leakage holes.

For an encapsulation to be predictably effective, several conditions have to be achieved, i.e.

- local groundwaters should never be able to rise into the core of the encapsulation;
- settlement caused by inadequate soil/waste compaction should be precluded;
- a logically defensible choice of soil-capping materials is essential to divert rainfall infiltration; and
- the outlet perimeter chain which collects infiltration must be able to drain freely to a stream or a sewer.

5.6 Soil cappings

In most cases, it is necessary to allow for a vegetation cover above at least parts of a reclaimed site. This obviously necessitates the provision of an appropriate type and thickness of soil capping. As stressed above (section 5.2.1), such cappings should be regarded not as part of the clean cover, but as separate works, designed to meet specific and defined horticultural requirements (Table 5.10). Chapter 9 gives appropriate guidance on rooting depths for various plant species and on the soil characteristics that are required. All that need be added is that care should be taken to ensure that the soil cappings do not compromise the underlying clean cover, by allowing either siltation or desiccation. The migration of plant roots into an engineered cover is only likely if the plants have not been provided with an appropriate type and thickness of soil for their root development.

Table 5.10   Desirable characteristics for garden soils

Bulk density	$< 1.5\,g/cm^3$
Stoniness	$< 30\%$ by volume
Maximum stone size	$< 50\,mm$
Water retention	Similar to a loam or clay loam (see Chapter 9)
Organic content	about 10%
Nutrient content	Equivalent to a good quality topsoil
Laid thickness	300 mm (grass, etc.)
(minimum)	500 mm (shrub areas)
	1 m (fruit and specimen trees)
pH	5 to 8
Contaminant levels	Below those cited by ICRCL 1987 Guidelines

5.7 Appropriate and non-appropriate applications of clean covers

Clean cover design philosophy is based on the proven principles of soil moisture movements. Thus, the use of clean covers is most appropriate when the contamination mobility is mainly in a liquid phase. When gaseous contamination is a particular hazard, clean covers can be used, although only on sites where the gas hazard is trivial, and slotted gas collection pipes can be set in a coarse granular cover, and then vented to the atmosphere. On sites with more severe gas hazards (i.e. where measurable outflow rates occur, or where high gas pressures exist), clean covers will not be a successful solution.

Equally, it will be inappropriate to expect any clean cover to deal effectively with organic vapours (from oil soaked ground or from free product floating on the groundwater) or with layers of semi-liquid tars, which can move up through the cover as the load on the site's surface is increased. Technologies are now commercially available to resolve such contamination problems (Chapter 4) and these should be applied before any clean cover is installed to cope with any remnant and non-organic contamination.

Clean cover technology is an example of the wide spectrum reclamation technologies, and so is particularly useful on those diversely contaminated UK sites that have had several prior industrial uses. If a site, however, is affected only by a contaminant from a point pollution source, it will be more appropriate to select one of the newer technologies that can destroy or immobilize that particular contaminant.

6
In-ground barriers

6.1 Introduction

Adjacent to any centre of population anywhere in the world there is likely to be a landfill site. Many of these sites will have been used because they were in a convenient position or simply because they were available (for example, holes left by gravel, chalk, clay extraction, etc.). Inevitably some of these sites will be in areas where the hydrogeology is such that they pose a threat to groundwater resources and thus remedial action is necessary. Furthermore, developments in landfilling practice, such as daily cover layers and greater compaction of the waste, have tended to reduce the vertical permeability within the landfills. This has promoted horizontal migration of landfill gas, which may need to be controlled. Ironically many carefully operated landfill sites today pose a greater hazard from landfill gas migration than those where the material was deposited more loosely and the gas could escape vertically. This serves to demonstrate that there has always been a crucial lack of information on the behaviour of contaminated sites. Currently, new specifications for landfills are being developed and yet very few landfills have been exhumed to see how man-made barrier materials or even native clays perform in situ.

Landfill sites are now being selected much more carefully and liner systems installed if they do not exist naturally. Thus, it is to be hoped that the number of problem landfills will reduce with time although this does presuppose that the current design procedures will pass the test of time. Whether or not the number of problem sites is actually reducing, the absolute number of problem sites is still large and this has stimulated a substantial amount of work on remedial treatment.

Fundamentally, existing technology allows only two treatments for problem landfills: (a) excavation and re-burial in safe(r) storage; or (b) installation of containment systems. In general, excavation and re-burial will be extremely costly and generate substantial secondary pollution for all but the smallest landfills. Thus techniques are necessary to contain pollutants in situ. In principle, three types of barrier may be needed at any contaminated site: (a) vertical barriers to form a cut-off wall around the site; (b) horizontal barriers to provide a low permeability base (although for small sites there may be potential for inclined interpenetrating walls formed by a mix-in-place process [1]; (c) cover layers to control the infiltration of rainwater. Cover layers are addressed in Chapter 5. In practice, it is often found that there is a natural base layer of low permeability material beneath the contamination and thus all that is needed is a vertical wall that connects with this layer.

If no natural low permeability layer is present, a base may be installed by grout injection or by jet grouting and clearly developments in this area may considerably extend the use of barriers systems generally.

It should be noted that although a base layer is normally a prerequisite before a vertical barrier can be effective, there are some situations where it is not required.

Figure 6.1(a–d) shows four barrier layouts that have been used in the United Kingdom. Figure 6.1(a) shows a standard barrier taken down to an aquiclude. This type of barrier will be suitable for liquid control, although for contaminated groundwater it will be necessary to consider the durability of the wall material. Figure 6.1(b) shows a standard wall with a vertical membrane. This type of wall may be suitable for gas and the more aggressive groundwaters (although the durability of the membrane will now have to be considered). Figure 6.1(c) shows a partially penetrating wall for gas control. The design groundwater level will need to be carefully assessed (it may be seasonal) as will the maximum gas pressure. The wall need not extend below the membrane unless leachate is also to be contained (the membrane

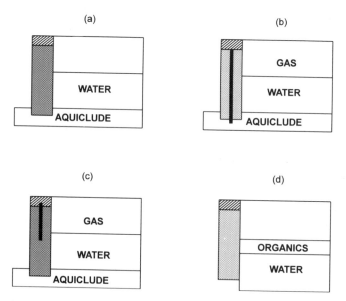

Figure 6.1 Types of cut-off wall. (a) Standard cut-off, (b) membrane barrier, (c) partially penetrating gas barrier with leachate barrier, (d) partially penetrating light organics barrier.

is necessary for gas control, see section 6.3). When using a partially penetrating membrane, careful attention must be given to the integrity of the wall at the toe of the membrane. Figure 6.1(d) shows a partially penetrating wall without a membrane. This might be used for the control of light non-aqueous liquids in a region where there is no aquiclude within a reasonable depth; for example for oil spill control at a tank farm founded on permeable soils.

Barriers may also be required as part of the clean-up strategy for contaminated industrial sites or as preventive measures at new sites. At the present time in the United Kingdom, most barriers are used at landfills and particularly for the control of gas migration. However, it is to be expected that they will find increasing use at contaminated sites in general as controls on groundwater pollution are progressively tightened.

6.2 The requirements for a cut-off

The contamination at landfills and other sites need not be limited to aqueous solutions. A barrier may have to retain not only aggressive chemical solutions but also light and dense (relative to water) non-aqueous phase liquids (NAPLS) and gases. For all of these the cut-off should represent a sufficient barrier that the rate of escape of pollutants (or transit time through the barrier for degradable materials) is such that it will not adversely affect the surroundings. Thus the barrier performance ought to be related to the local geology, hydrogeology, the nature of the contained materials and the land use adjacent to the site.

For the control of aqueous liquids there is some international consensus that clay liners for landfills should have a permeability equivalent to 1 m of 10^{-9} m/s material or less (i.e. a permittivity of 10^{-9} s^{-1}; permittivity is the permeability divided by the thickness of the barrier and thus an indicator of the overall hydraulic resistance). Although 1 m of 10^{-9} m/s material seems to be generally accepted, there can be no confidence that this will remain the standard. For example the proposed EC Directive on the Landfill of Waste [2] requires that the base and sides of a landfill should have a permeability equivalent to not less than 3 m of 10^{-9} m/s material, i.e. a permittivity of 3.3×10^{-10} s^{-1}. However, for slurry trench cut-offs, which are often thinner than 1 m (typically the thickness may be 0.6–0.80 m depending on depth, deeper walls being thicker in order to be able to accommodate a thicker excavator arm), the specified permeability is not usually reduced to compensate for the reduced thickness and thus permittivities rather over 10^{-9} s^{-1} seem to be permitted for such vertical barriers.

In fact there can be little case for requiring permeabilities less than 10^{-9} m/s, not because this represents the level at which all escape is insignificant but rather because it is about the level at which permeation ceases to be the most important vector for escape. As the permeability drops below 10^{-9} m/s, other transport mechanisms become progressively more important (see section 6.4).

It will be interesting to see how the permeability/thickness debate develops. It is to be hoped that the actual engineering of landfills will be addressed rather than the simplistic parameters of permeability and thickness. In addition, the overall environmental balance needs to be considered. The lining of a landfill with 3 m of clay may leave a substantial hole somewhere else, a hole which itself may represent substantial environmental damage.

6.3 Gas permeability

For the control of gases, the *Waste Management Paper 27* [3] says 'reworked natural clay or calcium bentonite linings are probably the most suitable commonly available materials for gas barriers. They should be laid and compacted to achieve a maximum water permeability of 10^{-9} m/s.' Thus, gas barriers are to be designed to achieve a specified water permeability although it is noted that 'the effectiveness of liners in preventing gas migration is not yet fully understood and their permeabilities to gas have yet to be determined. It would seem likely that they would be more permeable to gas by several orders of magnitude compared to leachate.' Indeed they will be! With the current state of knowledge, to design gas control measures on water permeability is unsatisfactory. For clay, bentonite–cement and most mineral-based materials, the gas permeability will be very sensitive to the moisture content and there can be no correlation between the gas and water permeabilities.

The proposed EC Directive on the Landfill of Waste [2] gives no guidance on the formation of gas barriers and merely requires that 'appropriate measures are taken to control the accumulation and migration of landfill gas' and for biologically active landfill sites receiving or having received more than 10 000 metric tons per year of waste that 'landfill gas shall be collected and properly treated and preferably used'.

The gas permeability of mineral materials is complex. If the material is fully saturated, there can be no bulk gas flow and methane, etc. can move only in solution by diffusion through or advection with the water. (Gases could be produced downstream of a barrier by degradation of a leachate that has passed through the barrier. The liquid permeability must be designed to be sufficiently low to ensure that the gas generation rate is minimal.)

However, if the material dries out it may shrink, crack and so become highly gas permeable. In the partially saturated state, unless the gas forms a continuous phase within the material, then the flow may be effectively blocked unless the pressure is sufficient to force the water from the pore space and establish a flow path. The necessary pressure may be large and thus the partially saturated material may still behave as an effective barrier. However, a small further loss of water (from the blocked partially saturated state) may open up flow paths of quite significant permeability. For example a 1 m thickness of good quality structural concrete when air-dried might show a gas flow rate of 50 000 ml/m^2 per day per atm of pressure differential (compared with perhaps 50 ml for a membrane). Clearly, the permitted gas flow rate through a membrane must be related to the venting arrangements, etc.; i.e. it must be designed to match the site conditions. Appropriate design procedures need to be developed.

In many situations there is probably sufficient moisture in the ground to ensure that most fine grained materials such as clay or a bentonite–cement are of low permeability to gas. However, droughts do occur and these have led to the shrinkage of clays to significant depths. The author therefore prefers to regard the in situ performance of mineral materials such as clay and bentonite–cements as unproven for gas control. Unless or until it is shown to be satisfactory, a membrane should always be included when designing a gas barrier unless the full depth that may be subject to drying is adequately protected by vent trenches (although these will exacerbate drying and thus the problems become circular). For exposed barriers such as clay capping layers, some gas permeability must be expected in dry seasons.

If a high density polyethylene (HDPE) membrane is used, the gas permeability might be of the order of 50 ml/m^2 per day per atm of differential pressure for methane, or 200 ml/m^2 per day per atm for carbon dioxide [4]. These are very low figures and indeed it is not strictly correct to consider them as permeabilities. Furthermore, it does not follow that, simply because they can be achieved with an intact membrane, such low values are actually required. In practice, there may be some additional leakage at joints between membrane sections and possibly at the toe of the wall where the membrane links with the formation. Clearly, when designing membrane systems, careful attention must be given to the joints and mechanical protection of the membrane. There is minimal design data on what represents an acceptable gas flux through a barrier.

6.4 Potentials tending to cause flow

The following are some of the potentials that ought to be considered when assessing the magnitude and direction of flow through a barrier. A fuller treatment is given by Mitchell [5].

(a) *Hydrostatic pressure* due to difference in liquid or gas pressures. The flow may be into or out from the contaminated area depending on the relative pressures. Furthermore different phases may move in different directions. For example, if the leachate level in a landfill is below that of the surrounding groundwater there will be a tendency for water flow into the landfill, whereas gas generated within the landfill will tend to move outwards.

(b) *Chemical potentials* tending to cause diffusion. In general, the chemical flux will tend to be outward as concentrations of chemical species should be higher inside the contaminated area than outside.

(c) *Osmosis*. In general, this will promote inward flow of water as osmotic potentials within a contaminated area generally will be higher than outside.

(d) *Electrical potentials*. These may be generated by the difference in chemical conditions between the contamination and the surrounding soil. For example, the environment within a landfill may be significantly more reducing than outside. Electrical potentials may induce electro-osmosis (movement of the water), electrolysis (movement of ions) or electrophoresis (movement of colloidal material). The effect of any electrical potential will depend on the charge of the migrating species (e.g. in the reducing environment of a landfill, cation migration rates may be reduced and anions accelerated, although the local charge balance must be maintained).

(e) *Temperature differentials* may drive flow. Such differentials may result from the generation of heat by microbiological activity within a landfill or the microbial oxidation of methane escaping from a landfill.

(f) *Gas flow* may also be influenced by changes in atmospheric pressure and, more particularly, the occlusion of flow paths as a result of rainfall or other changes in soil moisture content.

Thus, there are many different potentials and, as shown above, they may act in contrary directions and influence the water and ions differently. Furthermore, in a multibarrier system, each component of the barrier will have a different response to the potentials. Fluid could accumulate between layers or there could be a net removal. For example, for a waste underlain by a membrane-compacted clay-membrane sandwich, the clay layer could be desiccated by osmotic flow.

At the present time, permeability seems to be regarded as the fundamental parameter for barrier design. Clearly this is simplistic and the other driving forces must be considered. In addition, the chemical and electrical as well as the physical boundary conditions need to be established.

6.5 Types of vertical barrier

Vertical barriers are now regularly used for the containment of pollutants. In contrast, very few horizontal barriers have been installed except after bulk excavation. Vertical barriers may be broadly classified into three groups: (i) driven barriers, (ii) injected barriers; and (iii) cut and fill barriers.

6.5.1 *Driven barriers*

These barriers may be formed by driving steel or concrete elements into the ground. The fundamental requirements are that the elements are durable and can be joined to form a suitably impermeable wall. Driven systems may have considerable potential in situations where barriers are required to both control pollution migration and provide some mechanical support. Furthermore, they have the advantage that there is no need for excavation or the disposal of possibly contaminated arisings. With the rapidly developing demand for barrier systems the essential simplicity of driven systems suggests that new developments are to be expected in the coming years.

6.5.2 *Injected barriers*

Grout injection has been regularly used to form seepage barriers under earth dams, etc. To obtain a low permeability barrier, it is necessary to use several stages of grouting at reducing centres and therefore costs are relatively high. A permeability of order 10^{-7} m/s is often regarded as reasonable for grouting operations but the actual value will be very dependent on the soil type, heterogeneity, etc. By its nature, the injection process may lead to a grout curtain that contains a number of ungrouted flow paths. These may allow relatively rapid erosion of the adjacent grout material, especially if there are aggressive chemicals present in water. Grouting is therefore not a first choice for pollution control barriers and it is unlikely to find major application except in circumstances where other procedures are impracticable. However, it may find considerable application for repairs to barrier systems and also for the local encapsulation of toxic wastes.

6.5.3 *Cut-and-fill type barriers*

For convenience, all processes where soil is excavated and replaced by a cut-off material may be termed cut-and-fill procedures.

Secant piling. Secant piling is often used in civil engineering to provide structural cut-off walls. The development of systems using soft primary piles and structural secondaries has allowed significant cost savings in situations where a full wall of structural piles is not required [6]. For pollution control, there must be concern about the joints between piles. In general, secant pile walls are unlikely to be competitive as barriers, unless there is a need for structural strength as well as impermeability.

Secant pile type walls can also be formed by mix-in-place techniques, for example using counter-rotating augers. Developments in this area are to be expected but the process may be restricted to a rather narrow range of soil types.

Jet grouting. Jet grouting may be used to form secant piles or thin walls. The process may involve cutting with a high pressure air-shrouded jet followed by filling with cement grout, or direct cutting with a cementitious fluid, in which case the process becomes effectively a mix-in-place procedure. Jet grouting may find particular application in situations where there is limited surface access such as under a structure.

The vibrated beam wall. For this process an 'H' pile is vibrated into the ground and then extracted. During extraction, a cement-based grout is injected at the toe of the pile. The technique produces a relatively thin wall, which is not ideal for pollution control. However, a more secure system may be achieved if a double wall is installed with cross walls to form a system of independent cells. If pumped wells are installed in these cells and the internal groundwater level maintained below that of both the contaminated area and the surrounding soil, then there should be minimal potential for pollution migration and the well flows can be used to monitor the performance of the system.

Shallow cut-off walls. Shallow cut-offs may be formed by excavating a narrow slot with some form of trench cutter and inserting a geomembrane. Clearly, the excavation must be of sufficient depth to reach an aquiclude and the membrane must be sealed to it. An alternative way of forming a shallow cut-off is to excavate a wide trench, and backfill with compacted clay. The trench will need to have battered sides to avoid any risk of collapse during the compaction works. In general most cut-off walls must be taken down to below the groundwater and this may lead to instability of the trench and/ or difficulties with compaction of the backfill. While it may be possible to control the groundwater by pumping, it may be more economic to pursue a slurry trench solution.

6.6 The slurry trench process

Slurry trench cut-offs are currently the most widely used form of remedial barrier system and therefore the procedure will be discussed in detail. It would seem that the first slurry trench wall was built at Terminal Island, California in 1948, although Veder had conceived the idea of a contiguous bored pile wall formed under slurry in 1938. However, trials were not carried out until the 1940s and structural walls not achieved until the 1950s [7]. Many variants of the process are now available and it is one of the most versatile techniques for the formation of vertical barriers.

If a trench is excavated in the ground to a depth of more than about 2 m it is likely to collapse. However, if filled with an appropriate fluid the trench can be kept open and excavated to almost any depth without collapse. For this purpose the fluid must have two fundamental characteristics: (i) it must exert sufficient hydrostatic pressure to maintain trench stability; and (ii) it must not drain away into the ground to an unacceptable extent. These requirements are typically met with a bentonite–clay slurry, a bentonite–cement slurry or a polymer slurry.

Once a trench has been formed, the impermeable element may be installed. This may be concrete (although the costs tend to limit the use of concrete to structural walls), a clay–cement–aggregate plastic concrete, a soil–bentonite mix, a geomembrane or even specially shaped glass panels [8]. A further variant is to excavate the trench under a bentonite–cement slurry, which is designed to remain fluid during the excavation phase and to set when left in trench to form a material with permeability and strength properties similar to those of a stiff clay. These slurries are known as self-hardening slurries and are the most widely used form of cut-off in the United Kingdom. In the United States, soil–bentonite cut-offs are often used for groundwater control but it appears that membranes are preferred for pollution control. In Germany, a chemically modified high density soil system has been developed and offers an interesting additional wall type.

6.7 Types of slurry trench cut-offs

There is now a significant number of different types of slurry trench cut-off and, as demands for pollution control increase, further developments must be expected. Currently the principal types of cut-off wall are as follows.

6.7.1 Soil–active clay cut-offs

These are formed by excavation under an active clay slurry (almost invariably bentonite) and backfilling the trench with a blended soil–active clay mix, which forms the low permeability element. Design procedures for soil–bentonite backfills are given by D'Appolonia [9].

The backfill in a soil–active clay wall must be at a relatively high moisture content so that it will self-compact when placed in the trench. Unfortunately, high water content clay systems will always be vulnerable to chemical attack, as any change in the chemical environment may lead to substantial changes in the interactions between clay particles. This may cause significant shrinkage and so have a dramatic effect on the permeability. For good chemical resistance, dense non-swelling clay systems are required but these cannot be installed by the conventional slurry trench process.

6.7.2 Clay–cement cut-offs

Clay–cement cut-offs are now regularly employed for groundwater and pollution control. With UK materials, typical mix proportions are illustrated in

Table 6.1. The use of bentonite contents above 40 kg appears to give a marked improvement in permeability, although with a slight increase in strength (this suggests that there is an interaction between the cement and the bentonite). The 40 kg figure is not absolute and will vary with the source of the bentonite and the nature of the cement. Above 55 kg, the slurry may become so thick as to be unmixable and unusable in the trench. Increasing the cement content significantly increases the strength but also reduces permeability (it may have rather little effect on the fluid viscosity). The best design compromise appears to be to use the maximum possible bentonite content and the minimum cement content. The cement may be a mixture of ordinary Portland cement and replacement materials such as ground granulated blast-furnace slag or pulverized fuel ash.

Table 6.1 Clay–cement cut-offs

Material	Quantity (kg)
Bentonite	40–55
Cement	90–200
Water	1000

Clay–cement slurries are usually used as self-hardening slurries and simply left in the trench to set at the end of excavation. If the excavation rate is likely to be particularly slow (for example for a very deep wall), it may be necessary to excavate under a bentonite slurry and replace this with a bentonite–cement slurry or a backfill mix. Normally it will be necessary to install such walls as a series of panels, panels 1, 3, 5 being excavated and filled in sequence, and the secondary panels 2, 4, 6 installed once the primaries have set.

6.7.3 Clay–cement–aggregate cut-offs

For deep walls, or if for some reason a more rigid wall is required, the slurry may be replaced with a clay–cement–aggregate plastic concrete prepared by blending the displaced slurry with a suitably graded aggregate (and cement if there was no cement in the original excavation slurry) to form a backfill mix.

For clay–cement–aggregate mixes, the bentonite and cement concentrations of the slurry phase (excluding the aggregate) generally will be towards the lower end of the ranges listed in Table 6.1. Despite this, the mixes tend to be much stronger and stiffer than simple bentonite–cement systems and the strain at failure may be low especially if the aggregate particles are in grain-to-grain contact (this may be difficult to avoid unless the slurry is specially thickened so that the aggregate particles can be held in suspension prior to set). For pollution control, the reduced content of clay and cement and the potential for cracking at the slurry–aggregate interfaces tends to militate against their use except for systems using specially designed aggregate gradings.

The addition of aggregate will reduce the volume of clay–cement slurry required to form the wall and, as already noted, the slurry may be of lower clay and cement content than that for a self-hardening slurry wall. Despite these savings, backfill walls are likely to be more expensive than self-hardening walls because of the extra materials handling.

6.7.4 *Cut-offs with membranes*

After excavation, and before the slurry sets, a membrane may be lowered into the trench. The use of such membranes may be appropriate where high levels of pollution exist or very aggressive chemicals are present, as membranes can be obtained to resist a wide spectrum of chemicals. For gas migration control, membranes may be essential. Most slurry-based cut-off materials have a relatively high water content. If this water is lost they may become gas permeable and thus the potential for such materials for gas control must be regarded as not proven.

The major problems with membranes relate to the sealing of the membrane to the base of the excavation and the joints between membrane panels. The base seal has been attempted by mounting the membrane on an installation frame and hammering the whole assembly a predetermined distance into the base layer. This operation has to be undertaken with some care to avoid damaging the membrane and the inter-panel joints. An alternative is to over-excavate the trench and use a longer membrane in order to extend any flow path around the toe of the membrane.

A number of different joint systems are available to form the inter-panel joints but all require considerable care in use. A membrane panel may be over 10 m high by 5 m wide and mounted on a heavy metal frame. The panel joints therefore need to be robust and simple to use. Details of a number of jointing systems are given by Krause [10]. Typically a bentonite–cement trench will be of the order of 600 mm wide. The insertion of a membrane will substantially disturb the behaviour of the slurry in the trench and the membrane may act as a sliding plane and encourage cracking. Furthermore, the membrane may take up a meandering path both horizontally and vertically within the trench. This can be advantageous as it will provide some flexibility if the membrane is stretched as a result of ground movements. However, it will mean that the bentonite–cement material may be divided into isolated units that are not continuous along either side or throughout the depth of the trench. If a membrane is used, therefore, it should be treated as the primary impermeable element, and the bentonite–cement regarded as fundamentally providing only support and mechanical protection.

6.7.5 *High-density walls*

These are relatively new cut-off materials developed specifically for polluted ground. The hardened material is hydrophobic and the water permeability may be very substantially lower than with conventional soil–bentonite or bentonite–cement walls. Indeed, it is held that the permeability is so low that diffusion is the most important transport process. The chemical resistance to aqueous pollutants may be better than for conventional slurry walls. Material costs are higher and a special excavation plant may be necessary because of the high density of the slurry that follows from the high solids content. Details of the system are given by Hass and Hitze [11].

6.7.6 *Drainage walls*

Pollution migration (liquid or gaseous) may be controlled by drainage ditches. These ditches must extend to a sufficient depth to intercept all the pollutants. Deep drainage walls may be formed by excavating a trench under a degradable polymer slurry. On completion of excavation, the trench is backfilled with a gravel drainage material and the slurry broken down with an oxidizing agent or left to degrade naturally. In principle, the drainage wall system could be useful for the formation of gas venting trenches. However, such trenches need not extend below the water table and, unless this is exceptionally deep, direct excavation will be cheaper than a slurry trench procedure. There is potential for driven drainage wall systems especially as there would then be no need to excavate and dispose of possibly contaminated soil.

6.8 Slurry preparation

Good mixing and accurate batching of the mix components are fundamental to the performance of all slurries Figure 6.2 shows a schematic diagram of a typical plant layout for bentonite–cement slurry preparation. The first step is to prepare a slurry of bentonite and water. This slurry is then pumped to a storage tank where it should be left to hydrate for at least 4 h and preferably 24 h. If the bentonite is not properly hydrated prior to the addition of the cement, the resulting bentonite–cement slurry may bleed excessively.

Before any hydrated bentonite is drawn off for use, the contents of the storage tank(s) should be homogenized by recirculation. This is necessary because some of the bentonite solids may settle before the slurry has developed sufficient gel to prevent sedimentation. Without recirculation, the first slurry drawn off could be excessively concentrated and the last almost entirely water.

From the hydration tank the slurry is pumped to a second mixer where cement (and, if appropriate, cement replacement materials such as slag or pulverized fuel ash) is added. When cement and clay slurry are mixed

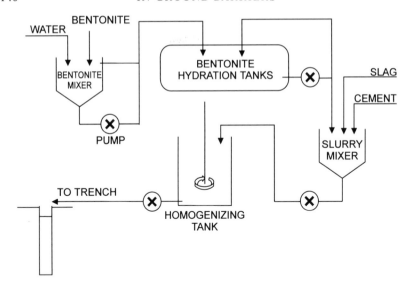

Figure 6.2 Typical plant layout for bentonite–cement slurry preparation.

there is a mutual flocculation, which leads to a flash stiffening that normally lasts only a few moments provided the slurry is kept moving. The mixing plant must be capable of handling this flash stiffening without stalling or loss of circulation.

From this mixer, the bentonite–cement slurry is pumped to a homogenizing/holding tank, which should have a capacity of the order of three batches from the bentonite–cement mixer. The slurry is then pumped to the trench. The pumping distance often will be in excess of 1 km. Centrifugal pumps are often sufficient but occasionally it may be necessary to use positive displacement pumps.

The type of mixing plant used can have significant effects on the fluid and set properties of a slurry. Thus, whenever possible, trial mixes should be prepared in the actual site mixers as it is difficult to simulate site mixers in the laboratory.

6.9 Requirements for the excavation slurry

The fundamental requirements for the excavation slurry are that:

(a) It exerts sufficient hydrostatic pressure to stabilize the excavation.
(b) It controls the loss of slurry to the ground by penetration and filtration.
(c) It should not bleed, i.e. the solids should not settle to leave free water at the surface.

(d) If a self-hardening mix is used, the setting time should be compatible with the excavation procedure, and typically at least 8 h and the slurry should not entrain excessive amounts of spoil as a result of stiffening during the excavation period.

(e) If a backfill mix is to be used, the slurry is sufficiently fluid to be fully displaced by the backfill.

6.9.1 *Excavation stabilization*

For bentonite slurries, satisfactory stabilization of the excavation is normally achieved if the slurry level is about 1.5 m above the groundwater level although with some soils and slurries a difference of less than this may be sufficient. Occasionally, and particularly in fine non-cohesive soils, a greater difference in level may be necessary. When clay–cement excavation slurries are used, the density of the slurry will be greater than that of a simple clay or polymer slurry and this will increase the stabilizing pressure. However, this extra pressure may be offset by the increased hydrostatic pressure of the groundwater due to fluid lost by filtration from the slurry.

6.9.2 *Slurry loss*

Fluid may be lost by filtration, with the deposition of solids to form a filter cake on the soil surfaces, or by bulk penetration of slurry into voids in the soil. Penetration may be reduced by increasing the gel strength of the slurry. The penetration, L, of the slurry into the soil may be estimated from the formula

$$L = 0.15DP/t$$

where P is the hydrostatic pressure difference between the slurry and the groundwater, t is the gel strength and D is the effective diameter of the soil particles (which may be approximated by the D_{10} of the soil). Thus, for a slurry of gel strength $10 \, N/m^2$ in a soil of effective particle diameter 10 mm, penetration might be of order 1.5 m/m of head difference between slurry and groundwater levels. If penetration is a problem it may be necessary to increase the clay concentration.

Filtration loss may be controlled by the addition of a fluid loss control agent. For pure bentonite or polymer slurries, fluid loss will seldom be a problem. With clay–cement slurries, the filter loss will be much higher than for pure bentonite slurries (very often by an order of magnitude or more). Occasionally this has led to problems of excessive reduction in slurry level in the trench prior to set. Despite this, fluid loss control agents have been used rather rarely in civil engineering (they are regularly used in oil well drilling muds). The reasons for this are probably partly economic and partly because most specifications prohibit the use of admixtures unless there is a proven track record of their use.

6.9.3 *Bleeding and settlement*

Bentonite–cement slurries may show some bleed, i.e. settlement of the solids to leave clear water at the surface. Bleed water represents a loss of useful cut-off volume. Severe bleeding suggests an improperly designed/prepared mix and, as a first step, the bentonite concentration, mixing procedure and, if necessary, source should be checked. Severe bleeding may lead to instability of the trench as the bleed water drains into the ground, thus producing a rapid drop in the fluid level in the trench and hence in the stabilizing pressure.

With backfill mixes containing aggregate there may be a tendency for the solids to settle prior to set of the slurry. This will leave a layer of thin slurry at the top of the trench. The formation of such a layer implies some instability in the mix but it may be difficult to prevent all settlement. The thin slurry layer should be removed during the capping operation. If excessive settlement occurs, this must be countered by modification of the mix design either by thickening the interstitial slurry so that the aggregate can be held in suspension, or by increasing the aggregate content to ensure grain-to-grain contact. Both of these measures may adversely affect the workability of the mix.

6.9.4 *Setting time*

The required setting time will depend on the method of excavation and the depth of the wall. In practice, set is rather difficult to define for these soft materials. The practical requirements are that excess slurry should drain freely from the grab during the excavation phase and that the integrity of the wall should not be compromised by the inevitable disturbance that occurs at day joints. Set time may be increased by the use of admixtures such as lignosulphonates. However, admixtures must be used with great care as most retarders increase bleed, and overdosing may kill all set.

6.9.5 *Displacement*

If used, a backfill mix must fully displace the excavation slurry. This requirement should not present any problem if non-setting bentonite or polymer excavation slurries are used. With clay–cement excavation slurries, stiffening will occur due to setting and thus backfilling should follow excavation as soon as practicable and certainly within the day of excavation.

6.10 Compatibility of materials

6.10.1 *Clays and water*

The predominant clay used in clay–cement slurries is bentonite. Indeed, unless bentonite (or more rarely attapulgite) is used it is very difficult to get satisfactory bleed control at a reasonable solids concentration.

Bentonite is a natural material, and supplies from different sources can show substantial differences in gel strength, viscosity, bleed control, etc. In general, it is necessary to carry out trial mixes with the actual bentonite and cement to be used in any application to ensure compatibility, low bleed and satisfactory permeability, etc. of the hardened material.

In the United Kingdom, it is normal practice to use a sodium exchanged calcium bentonite as the base for the excavation slurry. Such bentonite is quarried as a calcium bentonite and processed with sodium carbonate to produce a sodium bentonite. Generally it is not necessary to use a natural sodium bentonite such as Wyoming bentonite. Because of the sodium carbonate treatment, the pH of ion exchanged bentonite slurries tends to be higher than that of natural sodium bentonites, which may be near neutral. In the United Kingdom, most ion exchanged bentonites have a pH of about 10.5, although elsewhere pH values of 9.5 are not unusual.

In designing the slurry it is essential to check the behaviour of the bentonite and its compatibility with the proposed mix water and cement material. An outline procedure is as follows:

(a) Prepare a slurry of the bentonite in distilled water using concentrations in the range 3–7 g bentonite/100 g water. Ideally the laboratory mixing time and mixer shear rate should be matched to that of the site mixer. High shear mixers should not be used for laboratory trials unless they will be used on site. The mixed slurries should be poured into 1000 ml measuring cylinders and left undisturbed for 24 h. After this time the slurries should be inspected for bleed (separation of free water and settlement of solids). If there is any bleed, the slurry should be remixed and left quiescent for a further 24 h. If there is significant bleed at this stage, it is unlikely that the bentonite will be suitable for use in a slurry.

(b) If the bentonite has passed the test with distilled water, prepare a slurry of the bentonite using the mix water from the site. Again inspect the slurry for bleed at 24 h and, if necessary, at 48 h (after remixing). If there is bleed, it is likely that the site water contains dissolved ions at excessive concentrations (dissolved ions, particularly magnesium, can inhibit the swelling of the clay). It is possible that unsuitable water can be treated or dispersants used in the slurry preparation. However, often it is cheaper to find an alternative supply of water.

(c) Prepare slurries of 24 h hydrated bentonite and the cementitious material. When the bentonite and cement are first mixed, there is a rapid and substantial thickening of the slurry. The slurry will thin after a few moments of further mixing provided that the mixer is of sufficient power to maintain circulation within the mix vessel. (If full-scale mixers are used for the trials, it may be prudent to add the cement slowly as the thickening and subsequent thinning will occur after only a very little cement has been added; if the mixer stalls with a full batch of cement it may cause considerable problems.) The mix should be checked for bleed over a period of 24 h again using a 1000 ml measuring cylinder. Mixes that show more than 20 ml of bleed water at any time may be unsuitable. Generally the amount of bleed can be reduced by increasing the solids content and particularly the bentonite content. However, increasing the solids content may produce a very viscous slurry and an undesirably strong set material. If a low bleed cannot be obtained at a reasonable solids content the bentonite is probably unsuitable.

Bleed has been selected as the indicator of mix compatibility in the above procedures as it is the simplest and often the most informative parameter for a fluid slurry. Filter loss, viscosity, gel strength tests, etc. also ought to be carried out at the design stage to check that the fluid is suitable for use on site. However, considerable experience is required to interpret the results of these tests. Bleed can give a more unequivocal indication of materials compatibility. It should be noted that the above is a general procedure. Step (a) should not be necessary unless there is particular concern about the quality of the bentonite. Step (b) should not be necessary if UK mains water is to be used to prepare the mix.

Clearly, all the raw materials should also meet appropriate local standards. Thus, the bentonite is often required to meet the Oil Companies Materials Association Specification [12]. However, this is not always a sufficient specification for cut-off slurries. Similarly the cement and slag (or pulverized fuel ash if used) should be to appropriate standards to ensure consistency of the product.

6.10.2 *Cements and cement replacement materials*

The nature of the cement can have a profound influence on the properties of the slurry. In particular it has been found that slag cements give much more impermeable materials. In the United Kingdom, slag cements are not generally available and so it is normal practice to use a blend of ordinary Portland cement and ground granulated blast-furnace slag. The optimum percentage of slag in the mix is usually in the range 60–80% but may vary with the source of the slag. Below 60% there is little improvement in permeability and above 80% the material may be excessively strong and brittle. The use of slag enables permeabilities of the order of 10^{-9} m/s to be achieved at total cementitious contents (cement plus slag) of the order of 90–150 kg/m^3 of slurry. Without slag, such permeabilities may not be possible even at cement contents of 250 kg/m^3.

Pulverized fuel ash (PFA), if used as a cement replacement, has rather little effect on the properties of bentonite–cement slurries except to improve sulphate resistance. At replacement at levels over about 30%, the resulting slurry may be rather weak. However, it may be added, not as a replacement, but in addition to the cement to increase the fines content. High solids content mixes will show better resistance to drying than their low solids counterparts. However, long-term strengths may be rather high and the mixes thus show poor plasticity (see sections 6.13.2, 6.14.2). Proprietary hydration aids may be used to improve the performance of PFA.

A particular feature of PFA is that it will contain some unburnt carbon. The adsorption of organics from groundwater is well correlated with the organic carbon content of soils. While coal may be a rather refractory form of carbon, it is to be expected that PFA will offer better organic absorption and retardation than pure clay, cement or slag.

6.10.3 *Mix water*

As already noted, in the United Kingdom, water of drinking quality is usually satisfactory for the preparation of slurries. If there is any concern about the quality of the water, the trial mixes should be prepared. Chemical analysis can provide some guidelines on the acceptability of a water. However, it is difficult to predict how individual clays and clay–cement systems will perform. Of the common ions found in water, problems may be encountered for bentonite slurries at over 50 mg/l of magnesium, or 250 mg/l calcium, or 500 mg/l of sodium or potassium. These figures should not be regarded as safe limits and if concentrations approach these values, hydration trials should be carried out.

Attapulgite has been used in situations where fresh water could not be used [13]. However, the design procedures for attapulgite–cement–salt water slurries are quite different from those for conventional bentonite–cement slurries and the salts may have unexpected effects on the cement hydration [14].

6.11 Sampling slurries

Samples for testing the fluid or hardened properties of cut-off slurries must be taken from the trench or the mixers during the cut-off wall construction. Coring the wall after the slurry has hardened may damage the wall, and the resulting samples are generally so cracked/remoulded as to be unrepresentative. A number of sampling systems are available and the only requirements for site operation are simplicity, reliability and the ability to perform at depth in a trench (the hydrostatic pressure can jam some valve mechanisms). The sampler should be of reasonable volume as significant quantities of slurry are needed for tests (the Marsh cone requires at least 1500 ml of slurry, while samples for testing hardened properties may require considerably more). The properties of the slurry may vary slightly with depth in the trench due to the settlement of solids. It is therefore standard practice to test samples from at least two depths and, for deep walls, from three depths (the top 1 m, the middle and the bottom 1 m).

6.11.1 *Sample containers, storage and handling*

It is most important that all containers are compatible with the slurry. For example, aluminium components must not be used with alkaline systems such as converted sodium bentonites or cement as they will react to liberate hydrogen gas. Plastic tubes are usually satisfactory. Sample tubes should be wiped with mould release oil prior to casting to ease de-moulding. When preparing samples for testing in the laboratory, it may be necessary to trim a significant amount of material from the ends of the samples. Thus, sample tubes should be at least twice as long as the required length of the test specimen.

When preparing clay–cement samples, care should be taken to avoid trapping air. After filling, the tubes should be capped and stored upright until set. Thereafter they should be stored under water in a curing tank until required for test. Bentonite–cement slurries are sensitive to drying and, even in a nominally sealed (waxed and capped) tube, some drying will occur. Storage under water is the only sure procedure.

Great care should be taken when transporting samples from the site to the test laboratory. They must not be dropped or subjected to other impact. Ideally the tubes should be packed in wet sand during transit, for protection against drying and mechanical damage. Samples should not be de-moulded until required for test.

Samples of cut-off wall slurries will be relatively fragile and should be treated as sensitive clays. Before samples are extruded, the ends of the tubes should be checked for burrs, etc. to ensure that the samples can be extruded cleanly. Specimens must not be sub-sampled, for example, to produce three 38 mm diameter samples from a 100 mm sample, as this will lead to unacceptable damage.

6.12 Testing slurries

Quality control on site should be to confirm that the slurry has been batched correctly and that its properties are as designed. Experience shows that control of batching and mixing is fundamental. Checking fluid properties such as density and viscosity will only identify rather gross errors. Thus, the monitoring of these properties is not sufficient as a quality control procedure.

6.12.1 *Density*

For slurries, the most usual instrument for density measurement is the mud balance. This is an instrument similar to a steelyard except that the scale pan is replaced by a cup. The instrument thus consists of a cup rigidly fixed to a scale arm with a sliding counterweight or rider. In use, the whole unit is mounted on a fulcrum and the rider adjusted until the instrument is balanced. Specific gravity can then be read from an engraved scale. The balance was developed for the oil industry and the range of the instrument is rather wider than is strictly necessary for civil engineering work (typically 0.72–2.88), whereas construction slurries will usually have specific gravities in the range 1.0 to perhaps 1.4. The smallest scale division is 0.01; with care the instrument can be read to 0.005, although the repeatability between readings is seldom better than 0.01. A resolution of ±0.005 corresponds to a resolution of order 15 kg/m^3 for bentonite or cement in a slurry. Thus, the instrument is not suitable for site control of mixes as greater precision is needed (for example it would detect only rather gross errors in batching a typical bentonite–cement slurry with 45 kg/m^3 bentonite, 90 kg/m^3 slag and 45 kg/m^3 cement).

6.12.2 *Rheological measurements*

The most common instrument used for measuring actual viscosities (rather than ranking slurries for example by flow time) is the Fann viscometer (sometimes referred to as a rheometer). This is a co-axial cylinders viscometer specially designed for testing slurries. Two versions of the instrument are generally available: (i) an electrically driven instrument; and (ii) a hand cranked instrument. All versions of the instrument can be operated at 600 and 300 rpm and have a handwheel so that the bob can be slowly rotated for gel strength measurements. Some also have additional speeds of 200, 100, 6 and 3 rpm. For all versions of the instrument there is a central bob connected to a torque measuring system and outer rotating sleeve. The gap between bob and sleeve is only 0.59 mm and thus it is necessary to screen all spoil-contaminated slurries before testing.

The instrument is not entirely satisfactory for testing cement-based systems. In particular, the gel strength readings tend to be very operator dependent and are generally of such poor repeatability as to make their measurement irrelevant.

The viscosity readings tend to be more repeatable but are very dependent on the age of the sample (due to the hydration of the cement). Thus, it is very difficult to demonstrate repeatability of behaviour for a cementitious system with the Fann viscometer. However, the instrument can be useful in identifying gross variations in mix proportions, etc. but for this a much cheaper test, such as the Marsh funnel flow time could be equally effective.

The Fann viscometer is an expensive instrument and must be used with care by a trained operator if reliable results are to be obtained. It is best suited to use in the laboratory for investigation of mix designs, etc. where the detailed rheological information that can be obtained from it may be of great value.

6.12.3 *The Marsh funnel*

For general compliance testing on site, the Marsh funnel is more convenient, although the results cannot be converted to actual rheological parameters such as viscosity or gel strength.

It should be noted that the Marsh funnel is just one of a wide variety of different flow cones in common use. It is therefore important to specify the type of cone (dimensions, volume of slurry used, volume discharged and time for flow of water) when reporting results. The Marsh funnel should be filled with 1500 ml of slurry and the discharge may be 1000 ml or 946 ml (1 US quart). When reporting results, the flow quantity as well as the flow time should be reported.

For clean water at 21 °C (70 °F) the times should be as follows:

- For 946 ml, 25.5–26.5 s
- For 1000 ml, 27.5–28.5 s

The Marsh funnel is suitable for testing most bentonite and bentonite–cement slurries. However, for some thick slurries, the flow time may be very extended or the flow may stop before the required volume has been discharged. It does not follow that these slurries are necessarily unsuitable for use in a trench, merely that a different cone must be used to test them.

6.12.4 *Other parameters*

Other parameters that may be specified for an excavation slurry include pH, filter loss and bleeding.

For UK or Mediterranean bentonites, prior to the addition of cement, the pH should be of the order of 9.0–10.5 and results outside this range should be investigated. The pH should be consistent between batches and again any variation should be investigated. The pH for slurries containing cement is almost always over 12.0 and gives little useful information about the slurry.

Filter loss is always high for cement-based systems and again gives rather little useful information unless the application requires the use of a fluid loss control agent.

Bleeding represents a loss of useful volume of the cut-off material and thus is important. Typically, bleeding will be measured with a 1000 ml measuring cylinder, and a limit of 2% loss of volume in 24 h is often specified. Some specifications require that the measurement of bleed is carried out in a volumetric flask with a narrow neck. The intention is to concentrate the bleed from a large volume of slurry into a narrow column and so improve the accuracy of reading. Unfortunately, this procedure does not work as the bleed water will not migrate up the neck of the flask but remains on the upper surface of the bulb. Thus, the procedure gives a false low reading.

6.13 Testing hardened properties

Tests on hardened clay–cement materials will normally be carried out in a soils laboratory or, more rarely, a concrete laboratory. However, the materials are rather different from soils or concrete. It is important that laboratory staff are familiar with them. Very often, apparently unsatisfactory results have been traced to unsuitable sample preparation/testing procedures that have been borrowed from other disciplines without sufficient consideration of the nature of the material.

For cut-off wall slurries, tests that are often specified include: (i) unconfined compressive strength; (ii) confined drained stress–strain behaviour; and (iii) permeability.

6.13.1 *Unconfined compression tests*

For bentonite–cement slurries, the samples will normally be in the form of cylinders. The sample should be trimmed so that the ends are smooth, flat and parallel and then mounted in a test frame such as a triaxial load frame. The sample should not be enclosed in a membrane but should be tested as soon as possible after de-moulding to avoid drying. The loading rate is often not especially important. About 0.5% strain/min may be convenient as it gives a reasonably short test time and is not so fast that it is difficult to record the peak stress. Stress–strain plots from the test may be of interest but are seldom formally required as, under unconfined conditions, the strain at failure will generally be rather small. For example, the strain at failure is typically in the range 0.2–2%. It is only under confined drained conditions that cut-off materials show failure strains of the order of 5% or greater. The unconfined compressive strength test should be regarded as a quick and cheap test of the repeatability of the material, rather like the cube strength test for concrete. It does not show the stress–strain behaviour that the slurry will exhibit under in situ confined conditions.

6.13.2 *Confined drained triaxial testing*

To investigate the behaviour of the material under the confined conditions of a trench, etc., it is necessary to carry out tests under confined drained triaxial conditions. The sample is set up with top and bottom drainage and allowed to equilibrate under the applied cell pressure (which should be matched to the in situ confining stress) for an appropriate time, which will be at least 12 h. It is then subjected to a steadily increasing strain as in a standard drained triaxial test. To ensure satisfactory drainage, a slow strain rate must be used; 1% strain per hour is often satisfactory for bentonite–cement slurries, although this is quite fast compared with the rates used for clays.

The stress–strain behaviour will be sensitive to the effective confining stress. The following behaviour is typical:

(a) Effective confining stress > 0.5 × unconfined compressive strength. Brittle type failure at strain of 0.2–2% with a stress–strain curve very little different from that for unconfined compression save that the post-peak behaviour may be less brittle.

(b) Effective confining stress ≈ unconfined compressive strength. The initial part of the loading curve is comparable to (a). This is followed by plastic type behaviour, with the strain at failure increased significantly, typically to > 5%.

(c) Effective confining stress > unconfined compressive strength. Strain hardening observed in post-peak behaviour, and possibly some increase in peak strength. Initial stages of loading curve similar to (a).

Thus, when designing cut-off materials, it is important that the in situ confining stresses are considered and the mix proportions selected accordingly.

6.13.3 Permeability tests

For cut-off wall materials, permeability tests must be carried out in a triaxial cell. Samples for permeability testing should not be sealed into test cells with wax, etc. This invariably leads to false, high results.

The procedure is therefore to use a standard triaxial cell fitted with top and bottom drainage. An effective confining pressure of at least 100 kPa should be used to ensure satisfactory sealing of the membrane to the sample. The sample may be back-pressured to promote saturation if there has been any drying. Generally a test gradient of 10–20 is satisfactory. The sample should be allowed to equilibrate for at least 12 h (or longer if possible) before flow measurements are made (there may be very little actual consolidation). Measurements should then be made over a period of at least 24 h and preferably 2 days or more. It should be noted that there is usually a significant reduction in permeability with time (see Figure 6.3). This appears to occur irrespective of the age of the sample at the start of the test and appears to be a function of the volume of water permeated through the sample. For critical tests it may be appropriate to maintain permeation for some weeks although this does make for an expensive test.

The permeant from many slurries is markedly alkaline and so aluminium or other alkali sensitive materials should not be used in the test equipment exposed to the slurry or permeant (see section 6.11.1).

6.14 Specifications for slurry trench cut-offs

Typically, a specification for a cut-off wall will include the following components:

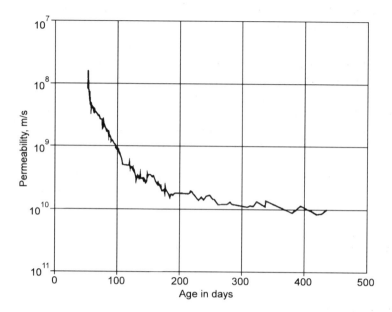

Figure 6.3 Loss of permeability of cut-off wall materials with time.

(a) Specifications for materials as supplied
(b) Specifications of fluid properties
(c) Specifications for mechanical performance of the wall
(d) Durability requirements
(e) Quality control on site during construction.

As the properties of most cut-off slurries develop rather slowly, it is important that the specification separates the on-site quality control testing from the laboratory tests required to confirm the hardened properties. For example, a 90 day strength or permeability is useless as a quality control measure for a job that may be finished before the first of the 90 day results is available. Very often, site control specifications will be concerned with fluid properties whereas hardened properties will be of concern for the mechanical performance and durability aspects of the specification.

6.14.1 *Mechanical performance: permeability*

In geotechnical engineering, for example for earth dams, a cut-off is often required to have a permeability of less than 10^{-8}m/s. As early cut-off walls were required for hydraulic control, most cut-off wall specifications required this level of permeability.

More recently the specifications for clay liners have come to be adopted for cut-off walls and a permeability of 10^{-9} m/s is usually specified; lower values are to be expected if the requirements of the EC Directive are to be applied (see section 6.2).

For a cut-off wall, a permeability of 10^{-8} m/s should give sufficient hydraulic control (i.e. allow minimal water flow) unless there is an unusually high gradient across it. Thus, a permeability $\leq 10^{-9}$ m/s is unnecessary for control, although it may give better durability for the wall material. It is significantly more difficult to form a cut-off of permeability 10^{-9} than 10^{-8} m/s, particularly if this permeability is to be achieved at 28 days (the permeability of bentonite–cements drops substantially with time under permeation). Indeed, with current mix designs, a 28 day permeability of 10^{-9} m/s cannot be reliably achieved without sacrificing strain at failure (see sections 6.13.2, 6.14.2).

Structural concrete, despite years of experience, is generally specified only by strength with perhaps minor constraints on cement content and water/cement ratio. Clay–cement materials are much less well understood and yet it is regarded as entirely reasonable to specify strength, permeability and strain at failure. Furthermore, the specified strength may vary between jobs by a factor of more than 30.

6.14.2 *Stress–strain behaviour*

A typical feature of many specifications is that the cut-off wall material should show a strain at failure of not less than 5%. This seems to be an arbitrary figure and it is rare that any justification is offered.

When re-moulded between the fingers, bentonite–cement materials behave as soft clays. However, the materials are cement-based and thus some brittle behaviour must be expected. Investigation has shown that large failure strains (plastic behaviour) cannot be obtained unless the material is subject to a confining stress of at least 50% of the unconfined compressive strength (see section 6.13.2). In situ, the cut-off material will be subjected to imposed stresses from the surrounding ground. Clearly, the large strain at failure is specified to ensure that if there are ground movements, the wall will deform plastically and will not fail as a cut-off. The stresses imposed on the wall are not easily identified and thus for many years it was the practice to specify a maximum strength for the material to ensure plastic behaviour at as low a confining stress as possible. Recently, engineers have refused to accept the concept of a maximum strength and it has become normal practice to specify a minimum strength. It is a popular misconception that strength is always an indicator of quality.

Thus, the effective confining stress is very important. If plastic behaviour is required at low confining stresses, for example near the surface of a cut-off wall, then a low strength material is required unless it can be shown that any ground movements will develop stresses in the wall sufficient to ensure plastic behaviour. The lack of plasticity at low confining stresses is an important reason for the use of a capping layer of clay on top of a cut-off wall (see section 6.15.1)

6.14.3 *Durability*

Bentonite–cement cut-off walls have now been used to contain pollutants for many years and it seems that there have been no reported failures. Laboratory tests show that the material is resistant to many chemicals provided that it is subjected to a confining stress (as will occur in situ due to the lateral pressure of the adjacent ground except in the near surface region). This provides a further reason for the use of a clay capping layer (see section 6.13.2).

A basic discussion of the effects of contained chemicals on cut-off walls materials is given by Jefferis [15]. However, it must be allowed that there is a lack of detailed information on the long-term performance of cut-off materials. Clearly, some interaction between contained chemicals and a cut-off material must be expected although it may not always be damaging and many reactions may lead to a reduction in wall permeability. When considering chemical attack, it must be expected that the general rules for durability of concrete will apply; i.e. the lower the permeability of the wall and the higher the reactive solids content, the better will be the durability. The addition of aggregate (particularly non-reactive aggregate) to form plastic concretes may reduce the durability of the wall under chemical attack. The reduced quantity of the bentonite–cement phase will be more rapidly removed to leave an open aggregate matrix. Without aggregate, the bentonite–cement material may compress continuously under the in situ stresses as attack occurs in order to maintain a homogeneous and relatively low permeability material. Such behaviour has been observed in the laboratory even when chemical attack has led to very large strains.

In addition to external chemical attack, the long-term internal stability of the material must be considered. Clearly, data on this are limited as the history of cut-off walls is still rather short. However, it can be said that there are many surviving examples of ancient construction materials that were mixes of lime, clay and pozzolanic materials. Thus, such mixes appear to have the potential to remain mutually compatible for generations without the formation of new and unstable mineral forms (although it must be allowed that there may have been many unstable mixes and that only the stable have survived).

An important feature of slurry trench cut-off walls is that they are easily renewable. Thus, the wall does not need to be entirely immune to pollutant attack. If the wall is designed to be a sacrificial element that degrades over many decades and is then replaced, the life cycle cost may prove to be lower than if a more complex barrier system is employed.

6.14.4 *The design compromise*

If a high strain at failure is required then a low strength material is necessary. However, a low permeability requires high clay and cement contents. Clay content is limited by the rheology of the slurry. High contents will lead to an unmanageably thick slurry (in contrast cement content has a rather more limited effect on the rheology unless very high concentrations are used). Durability also requires a high solids content and, perhaps particularly, a high clay content. Thus the design constraints are:

(a) Low clay content to give a fluid slurry
(b) Low cement content for high strain at failure
(c) High clay and cement content for low permeability
(d) High clay (and cement) content for durability.

Thus, the design of cut-off materials must be a compromise. When a permeability of only 10^{-8} m/s was required it was relatively easy to achieve a 5% strain at failure at low confining stresses. However, the requirement for a permeability of 10^{-9} m/s makes this much more difficult (unless an age at test of the order of 1 year rather than the usual 28 days is specified). If both the early age permeability and strain at failure are to be fundamental then new materials must be developed or membrane liners always used. Thus, the client must be prepared to pay substantially more. While this may be appropriate, it should be remembered that many existing walls designed to 10^{-8} m/s behave satisfactorily. In situ examination of these walls will provide much more useful design data than desk-based refinement of specifications. In particular, it would be most useful to know the actual permeability of these walls as it may be substantially lower than the design value.

6.15 Overtopping and capping

It is important to remember that a cut-off wall can be overtopped (there may be infiltration of rain, etc. even if the site is capped). It is therefore important that drainage is set to remove water from the contained region, which otherwise is effectively a pond. In principle the life of a cut-off wall may be substantially extended if there is always a tendency for inward flow of clean groundwater through the wall rather than a tendency for outward flow of contaminated water.

It seems that in future, the UK regulatory authorities will permit only a very limited build-up of water within a landfill and thus in general there will be the tendency for inward flow. However, it should be remembered that a low water level will substantially reduce the microbiological activity within the landfill and hence the rate of breakdown of organic matter.

The water removed by the drains may have to be treated. This will increase the running cost of the site but, if the cut-off and cover is effective, the quantities will be small. Leachate removed by drainage will contribute to a slow clean-up of the mobile and thus most hazardous components of the waste, which otherwise may remain as a permanent and unchanging risk.

6.15.1 *Capping*

A cap or some other form of cover is essential to prevent drying of most mineral cut-off materials (for a landfill this may be a part of the general cover layer or it may be a separate element). Bentonite–cement materials are particularly sensitive to drying and will crack severely if not covered. If the slurry is not protected, cracking may start almost immediately the slurry begins to stiffen. Cracks up to 1 m deep have been found in a wall left uncovered for a few weeks after construction. Figure 6.4 shows the detail of a sound (although seldom used) capping procedure for a bentonite–cement barrier:

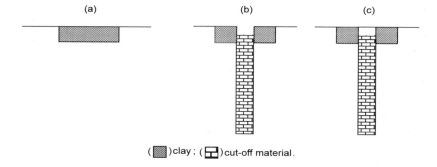

(▓)clay ; (⊞)cut-off material.

Figure 6.4 Capping procedure for a bentonite–cement barrier.

(a) Excavate a trench along the line of the wall to a width three times that of the cut-off wall. Backfill the trench in layers with compacted clay. The depth of the trench must be selected from consideration of the depth to the groundwater and the potential for drying during extreme weather conditions.
(b) After compaction of the clay, install the cut-off in the normal manner.
(c) As soon as any section of the cut-off has set, scrape off the surface layer of the cut-off material to a depth of 0.5 m and replace with compacted clay.

If a membrane is included in the barrier, a special capping procedure may need to be developed, in particular to link the membrane to any surface works.

6.16 Future developments

It is clear that the current level of activity with barrier systems will ensure that there is a continuing flow of new developments. Many of these may be fundamental improvements to existing systems. However, it is to be hoped that some will represent lateral approaches to the problem. Two examples of more novel approaches are outlined in the following sections.

6.16.1 *The bio-barrier*

In oil wells, bio-barriers have been developed to clog undesirable permeable regions. An outline of the procedure is as follows. A sample of the natural bacterial population in the formation is obtained, cultured and then starved. On starving, certain bacterial species may remain viable but reduce very substantially in size and develop an electrically neutral cell wall (which reduces attachability). These bacteria are termed ultra-micro bacteria. The formation to be blocked is permeated with a culture of the ultra-micro bacteria and a slow acting feed. The bacteria then slowly develop and expand, and so block the formation. At present the technique is limited in application to materials within a narrow band of permeabilities. It would be most elegant if a system could be developed that selectively clogged a leachate or gas migration path. More detailed discussions of microbiological effects in civil engineering are given in reference [16].

6.16.2 *Active barriers*

The fundamental aim of all barrier systems is to prevent all flow from the contaminated area, repository, etc. This is an ideal requirement and strictly unachievable in a passive system. However, it is possible that active systems may be developed where the flow is dynamically controlled. In many areas of science, if a no-flow condition is required then a guard ring procedure is used. In electrical measuring circuits guarding is well established. For heat experiments, adiabatic conditions may be achieved using a guard dynamically driven to match the sample temperature. Jefferis [17] has developed the guard ring concept for the elimination of edge leakage errors when measuring the permeability of concrete. Thus, the prevention of flow by use of an active guard driven to the sample potential (electrical, thermal, hydrostatic, etc.) is well established. Unfortunately, it is more difficult to develop a guard ring for a contaminated area site, for, as noted in section 6.4, many different potentials may influence flow through a barrier, and many different chemical species may be involved. Mitchell [5] has shown that the migration of cations may be retarded by applying a reverse potential for short periods during a permeation test. However, the procedure must of course accelerate the migration of the anions and it may have little benefit for organics. There may be some scope for the development of electrical sandwich layers designed to retard (or recover) anions and cations separately.

6.17 Conclusions

Vertical barriers are now widely accepted for the containment of gas leachate from contaminated sites, landfills, etc., although the actual number of sites that have been contained is still quite modest. However, the number that pose a threat to groundwater may be of the order of thousands. Thus the potential applications for cut-offs are enormous even if only a fraction of the problem sites are appropriate for such barriers.

To date there have been very few problems reported in the United Kingdom (all relating to installation rather than long-term performance). However, it is important to realize that applications are continually being extended, on occasion beyond the validity of the existing research base – an unsatisfactory situation, which has led to failures in many branches of civil engineering. A systematic investigation of the in situ behaviour of barriers is now urgently required.

There are now slurry trench waste containment walls that have been in the ground for 20 years (or perhaps more). It is important that some sections of these walls are exhumed to: (i) examine the mechanics of waste-wall interaction; (ii) establish whether reaction zones move uniformly through the wall or whether they localize or form fingers, etc.; (iii) investigate any reactions that may have occurred in the cut-off material; and (iv) quantify any change in permeability and pollutant migration mechanisms. Until a number of operational barriers have been exhumed, no guarantee of durability can be credible.

As a result of the wider knowledge and concern about toxic leachates, performance specifications for cut-off walls are becoming more rigorous and it would seem that new cut-off systems will have to be developed. However, it is illogical and potentially very wasteful to develop new systems before the in situ behaviour of the present systems has been properly investigated, particularly as no containment failures have been reported. No amount of laboratory testing or theoretical analysis can reproduce field conditions.

Specifications for barriers are a regular source of problems. Realistic common standards need to be established, addressing the compromises inherent in barrier design. Design procedures for gas barriers also need to be established. Most specifications for these barriers specify only a water permeability (generally 10^{-9} m/s). This is nonsense, as gas and water permeabilities (extrinsic or intrinsic) may be orders of magnitude different. Indeed the most elementary consideration of the microstructure of mineral sealing systems shows that there can be no unique relation between gas and water permeabilities.

7
Reclaiming potentially combustible sites

7.1 Introduction

The reclamation and re-use of derelict and contaminated land for housing and industrial use is increasing, and among the sites being reclaimed are many that contain large volumes of potentially combustible materials.

The high incidence of subterranean fires in the United Kingdom in recent years [1] is such that the hazards posed by potentially combustible materials on sites have to be treated as serious. While many of the reported subsurface fires have been on undeveloped sites, some have occurred on redeveloped land [2, 3] or on waste tips [4]. Typically these sites where fires have occurred have near surface layers containing materials of high calorific value. These are often colliery spoils, tars, wood and paper wastes, and domestic refuse layers, and can occur in such large volumes that their excavation and off-site removal is not practicable. The concern that such material could ignite is compounded by the known costs and complexities of dealing with such combustion incidents. Obviously, preventing combustion is a more attractive solution in such cases and this calls for a clear understanding of the combustion processes and the factors that influence it.

7.2 Combustion processes

7.2.1 *General combustion*

The general processes of combustion are well established. No substance can burn until its temperature is high enough to permit a sufficiently rapid reaction with oxygen. Visible combustion is normally preceded by a gradual rise in temperature, which initiates a slow ignition process. This is a complex process, affected by a variety of factors. Combustion, of course, is the consequence of the oxidation of materials, with the resultant evolution of heat and light. Flame, which features in some definitions, need not be present and is indeed usually absent in subterranean cases. Reactants inevitably are consumed in the combustion process, and thus steady state conditions are not possible. Instead, the temperature rises as the oxidation reaction continues, then peaks and finally decays, as the reaction fades out, and cooling

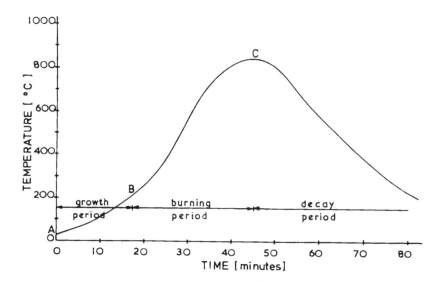

Figure 7.1 The temperature/time graph for a combustible material, e.g. coal. (A–B) growth period, temperature relatively low, heat transfer processes, production of gases and ignition; (B–C) burning period, flashover and fire develops, temperature rises sharply because heat is developing at a higher rate than can be dissipated; (C onwards) decay period as combustible material is burnt out and dissipation rate exceeds the rate of heat production through combustion.

and heat dissipation take over (Figure 7.1). This pattern suggests that the two important stages are:

- The ignition temperature (that at which the temperature build-up process just commences)
- The run-away combustion temperature (that at which the reaction with oxygen becomes very rapid and high temperature rises occur)

Essentially three conditions have to coexist before any material will burn:

(a) presence of a fuel source;
(b) availability of enough air or oxygen to initiate and support oxidation;
(c) presence of an ignition source, giving rise to whatever critical minimum temperature is appropriate for the particular fuel/air/inert fill concentrations.

This can be summarized graphically in the triangle of fire (Figure 7.2), which makes the point that if any one of these three factors is absent, then runaway combustion cannot occur.

This representation is of course simplistic, since other factors can be of importance. For example, the ignition temperature is to some extent dependent on the presence of catalysts or inhibitors, the dimensions of the sample,

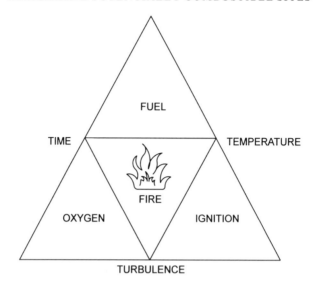

Figure 7.2   Triangle of fire.

the pressure on the fuel source, particle size and friability of the fuel source, the surface area open to reaction and the moisture content.

To ensure that combustion occurs, three conditions described as the 3 Ts, have to be met:

- Time
- Temperature
- Turbulence

Time is important simply because a long slow process of gradual temperature increase generally has to occur to permit runaway combustion. Very rapid temperature increases, such as those occurring in laboratory furnaces, need not mirror the actual reaction processes that occur on site and may give misleading indications of what will occur.

Temperature is critical in that it increases the oxidation reaction rate. A 10 °C (18 °F) temperature increase, for example, will double the rate of a chemical reaction. For any particular mixture of a burnable material and air with inert soil and fill, there will be a well defined temperature at which ignition will occur and reaction rates increase exponentially with temperature rise [5], while heat losses only increase linearly as a material temperature rises.

Turbulence is important as the control both on air and oxygen availability to the reaction, and as the mechanism that can cool and stop a reaction by removing heat.

Thus, the generality of combustion processes is well understood, and the normally observed course of events can be predicted (Table 7.1). This

Table 7.1 General events leading to runaway combustion for most organic materials

Temperature (°C)	Result
Above atmospheric	Water vapour is evolved and may be visible as a mist in cold humid weather
< 50	Slow absorption of oxygen
50–80	Oxidation accelerating
50–100[a]	Ignition and spontaneous combustion commences
About 80	A period of steady temperature probably associated with drying out of the material
100–120	Evolution of hydrogen and oxides of carbon begins
100–180	Interaction of oxygen and material accelerates rapidly
180–250	Thermal decomposition begins, evolution of tarry vapours, detectable by their odour and as bluish smoke; thermal runaway likely
250–350	The hottest zones within the material move against air flow, appearing as fires at the surface of the material

[a]At any temperature above 80 °C, direct gaseous oxidation generally proceeds at a faster rate in wet materials than in dry materials.

shows that a rise in temperature is the necessary first stage. This increases the rate of oxidation and so the temperature rises, which in turn tends to make the process self-accelerating. Finally, the process dies out as the oxidation rate declines when the reactivity of the fuel material declines. If the heating and heat dissipation rates are equivalent, no temperature increase will occur. If the dissipation rate is low, the fuel material will eventually become hot enough to ignite and, in most cases, combustion will ultimately take place.

An obvious practical consequence is that a material's liability to oxidize is less if it has been exposed on a site's surface for a long period and its more sensitive constituents have already suffered oxidation. Freshly excavated material will, however, pose a greater risk, as the more unstable constituents will still be present and largely unoxidized.

7.2.2 Spontaneous combustion

While the generalities of combustion are well known, the interactions of the various factors that control the detailed progress of the combustion process are much less well understood.

There have been detailed studies of the combustion of coals (although not on coals admixed with inert soils and fills), and data are available on the effects of coal and char particles sizes, coal particle porosities and coal composition. Other workers [6] have studied the effects of self-heating on non-coaly materials, but this research did not extend to the types of combustible fills that occur on derelict and contaminated sites. Thus, only indirect guidance is available for the prevention of subterranean combustion on reclamation sites, although some basic truths seem well established. For example:

- Particle size is important, since this controls the type of reaction that can occur. Smaller particles allow the rapid diffusion of the evolved gases away from the heated fuel particles. Large particles, however, can permit a more complex diffusion process.
- Reaction rates are initially governed by the amount of heat transferred by radiation, convection and conduction from the fuel particle surface.
- Air availability and the temperature of the air as it comes into contact with the fuel particles are known to be the controlling factors.

More practically biased work on coal storage facilities has shown that the carbon monoxide concentration within a coal heap is a good indicator of the onset of ignition, and a level of 25 mg/l has become accepted as a meaningful indicator that ignition has commenced [3]. More refined work on the gaseous precursors to actual ignition (e.g. the presence of hydrogen) has, however, met with little success [7], and so no prior prediction of likely hazardous conditions is yet possible.

Thus, the situation is that no substantial theoretical basis exists that can be used to predict whether combustible materials intermixed with soils, demolition rubble and other inert matter will pose a risk to a reclaimed site.

7.3 Combustion tests

7.3.1 *Introduction*

Given that no adequate theoretical foundation exists to allow subterranean fire risk prediction, it is necessary to determine if empirical tests are adequate for this purpose. A variety of possible tests exists and each is considered in the following sections.

7.3.2 *British Standard tests*

There is a British Standard (BS476) [8] on the fire tests appropriate for buildings and structures. This, however, considers the properties only of elements of building and structures, and is not relevant for any assessment of the underlying site materials.

7.3.3 *Direct combustion testing*

The simplest of the possible tests consists of no more or less than abstracting particles from a site sample, and heating these with a Bunsen burner to some 650 °C, for a time of 10–20 min. This gives very obvious data on a material's combustibility and is a useful initial screening test to reduce the number of samples that have to be tested more fully. The data obtained are largely subjective, e.g. 'smokes but does not ignite', 'high emission of volatiles and tars before any ignition occurred', 'material ceased to burn when Bunsen flame withdrawn', etc. The test is, however, not an adequate model of the actual conditions that will occur below a site's surface, since an abundant supply of free air is available and the ambient conditions permit the easy transportation of combustion gases and heat from the tested particles. Equally, the test cannot provide precise numerical data for analysis.

7.3.4 *Loss-on-ignition tests*

This approach has been seen as a useful indication of combustion potential and is commonly carried out on samples taken from contaminated sites. The test certainly provides data that can be used to estimate a material's calorific value [9]. Some authors have gone beyond this correct usage and have suggested that loss-on-ignition data give a cheap and simple indication of the organic content of a sample. This, however, is a false belief, since the measured weight loss includes the loss of free moisture, the loss of water of crystallization, and the weight losses caused by the breakdown of some inorganic materials (i.e. decarboxylation) [10]. Thus, even as a crude indicator of the amount of potentially combustible material present, the loss-on-ignition test cannot be recommended, and its regular use in the analysis of contaminated land samples is unjustified.

7.3.5 *Calorific value testing*

The calorific value (CV) is the heat given off when unit weight of a substance is completely burned. The general consensus is that the most accurate method of measuring calorific value is to burn a 1 g sample in a bomb calorimeter under an oxygen pressure of 3 MPa. With the bomb immersed in a water bath, the total heat evolved during the sample's burning can be determined precisely from the temperature rise in the surrounding water [11]. A measurement accuracy for fuels low in hydrogen (i.e. most British coals) of less than $\pm 0.1\%$ is claimed [12]. The test procedure is not difficult, testing facilities are widely available, and it has become accepted that calorific value data are useful and accurate.

In fact, this is not as certain as is assumed. The measured energy output will include a proportion due to the latent heat of moisture held in the sample, and an error of about 1.2 MJ/kg [13] is likely to arise from this. It also has to be noted that the sample size (1 g) is minute compared to the volumes of potentially combustible materials that occur on most derelict and contaminated sites, and this suggests that a very large number of CV determinations should be carried out to take due account of sample variability. In fact this is seldom done and a spurious accuracy is ascribed to perhaps one CV determination for a site area of 400–600 m^2. Apart from any doubts on the accuracy and repeatability of CV test results, it also has to be noted that the method was devised to measure the energy content of pure coals and not the combustion susceptibility of combustible materials intermixed with inert fills.

There is good evidence that CVs relate well to the rank of a particular coal where rank reflects the carbon, volatile and hydrogen content and the caking properties of different coals [14]. As carbon content increases, so CVs rise and, with carbon contents of 92% or less, calorific values and carbon contents relate very well. A typical British coal can be expected to have a CV of some 32 MJ/kg. When correlation with volatile matter is considered, it becomes obvious that an inverse relationship exists. The higher the hydrogen and volatile content, the lower is a coal's rank and the lower is its CV. This single point emphasizes the most obvious problem in using CV data to predict combustion risks. Low ranking coals are much more volatile rich, generally more reactive, more susceptible to self-heating, and so pose a much greater combustion risk. Yet these are the materials that have the lowest CVs!

Thus, the use of a particular calorific value (e.g. the level of 7000 kJ/kg advocated by the former Greater London Council) as an indicator of combustion risk is intrinsically flawed. Despite this, continued references to the risks posed by particular calorific value levels are made and Smith [15] has recently suggested that any values above 2 MJ/kg could pose a hazard. This would, of course, make most fertile garden soils potentially dangerous, and is, at best, misleading advice.

Apart from the bomb calorimeter tests, use has been made of simpler procedures to estimate CVs. The proximate analysis [9] and Ball's [10] loss-on-ignition tests for estimating organic matter, intermixed with non-calcareous soils, are examples of these approaches. Ball's method has the advantages of being designed for a mixture of materials and of not requiring hand picking of very small sub-samples from the site's samples. Additionally, both methods determine the volatile contents specifically and so have a better foundation for predicting the combustion risks.

A comparison of the CVs given by the bomb calorimeter, proximate analysis and Ball's method is given in Table 7.2. This makes obvious the discrepancies between the bomb calorimeter results and those from the other two methods. The table also indicates, on the accepted wisdom, that samples 3 and 4 could be expected to pose combustion risks. In fact, the use of the direct combustion test (section 7.3.3) proved that neither of these samples could be persuaded to burn, even in the ideal conditions of freely available air and imposed temperatures in excess of 600 °C.

Calorific value testing was designed to give the energy output for clean coals. Its use in this context is not in doubt, although the measurement accuracy and repeatability may be less than is often assumed. When used for subterranean fire risk evaluation, particularly where potentially combustible material is intermixed with inert materials, misleading results are inevitable, since the method is being misapplied.

Table 7.2 Calorific values determined by various methods (colliery spoil, former coal stock yard, NE England)

Method	Calorific value (kJ/kg)			
	Sample 1	Sample 2	Sample 3	Sample 4
Bomb calorimeter	1000	1000	22000	18000
Proximate analysis	4500	3000	7000	3750
Ball's method	3500	2750	6000	3750

7.3.6 Fire Research Station test method

The spontaneous ignition test developed by the Fire Research Station (FRS) [16–18] suffers from similar limitations to those of the calorific value test. Grain size and surface area are important factors in combustion, and therefore both methods are affected by the fact that the sample for testing needs to be ground to a fine particle size. The information that both tests provide is an indication of the total available energy rather than of susceptibility to combustion. The sample selected for grinding may also be non-representative of the original material, because larger fragments may have been removed.

In the FRS test, the sample material is ground to its most reactive particle size (less than 2 mm), placed in a wire mesh basket with sides ranging from 20 mm to 200 mm and suspended in a standard convected laboratory oven operating up to 250 °C. The furnace is then set at a known temperature and a thermocouple placed in the centre of the sample to monitor the temperature.

The test is continued until the sample temperature has reached its peak level and then fallen back to the set point temperature of the oven (Figure 7.3). Ignition is assessed from a series of tests performed at different oven temperatures and, from this, the material's behaviour under normal conditions is predicted. The FRS method fails to mirror site reality in that:

- the material is ground to a fine particle size to maximize the reactive surface area available;
- the test material is loosely packed in the basket and no assessment is attempted of the benefits that compaction could give;
- air is constantly circulating around the sample, in a way that does not mirror actual site conditions.

Thus the FRS method gives a useful indication of the ultimate combustibility in terms of the available energy of the material, but fails to provide any measure of the material's combustion susceptibility in the conditions that will actually occur on a reclaimed site.

7.3.7 Combustion potential test

If a test is to be effective then it should mirror as closely as possible the conditions under which it will occur on site. The advantage of the combustion

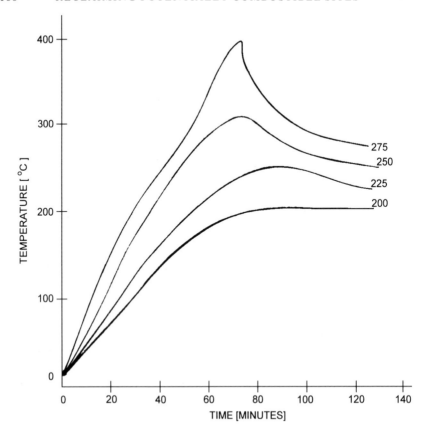

Figure 7.3   Typical result: soil admixed with coal waste material (Fire Research Station Test).

potential test described here is that little or no sample preparation is necessary. Additionally, the method of testing attempts to assess the two critical
factors that affect the potential for spontaneous combustion:

   (a) the gradual build-up of heat, which is retained and not dissipated;
   (b) the minimum supply of oxygen needed to allow combustion.

The test procedure adopted identifies these two properties by:

   • having the test sample in a central tube with a controllable air flow being
      passed through it (Figure 7.4);
   • surrounding the sample with a cylinder of an inert reference material;
   • gradually heating the two concentric cylinders with a wire wound elec-
      tric furnace up to a temperature of 600 °C.

The essential parameters for combustion to occur have been established and
are as follows:

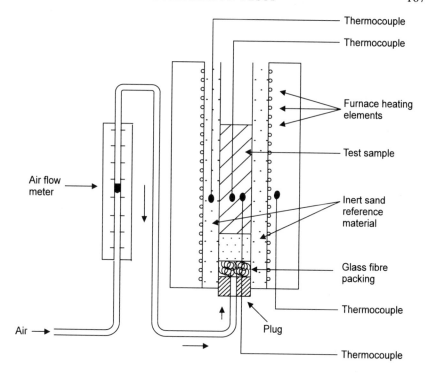

Figure 7.4 A schematic representation of the apparatus for the combustion potential test.

(a) there must be sufficient air or oxygen;
(b) the air or oxygen stream should be at the same temperature as the material before entering the bed;
(c) the heat evolved must be detected immediately by a rise in temperature and not dissipated;
(d) the temperature of the material should be uniform when the ignition point is reached;
(e) the two curves for the temperature–time of the material bed and the reference material should cross sharply to permit a precise determination of the ignition temperature.

From these and previous studies of coals and cokes it has been found that the sharpness of the cross-over point, and hence the precision of the ignition point, could be controlled by regulation of the heating rate. The same result can be obtained with samples composed of soil containing waste and contaminated materials.

In essence, the test employs the time/temperature relationship that is a feature of BS476 [8]. The apparatus and the procedure were adapted from earlier work by Sebastian and Mayers [19] on coke reactivity. In the arrangement used, the turbulence is kept constant and the temperature and time can be varied, although in each determination this is fixed by controlling the rate of heating of the furnace. The point of runaway combustion is identified when the temperature rise of the material is compared with that of the inert reference material (Figure 7.4). In the test, the temperature when the reaction proceeds at a higher rate than the heating rate of the arrangement is taken as the ignition point. This is the point at which the self-heating effect overtakes the external heating rate being supplied, and does not necessarily correspond with the sample starting to glow or with the first appearance of smoke or flame. The resultant temperature–time graph can be examined to establish if ignition and runaway combustion have taken place.

If the test sample is inert, the resultant graphs of the thermocouple outputs from the sample and the reference material will remain parallel (Figure 7.5). This separation of the two curves represents the thermal lag between the two materials and corresponds to the time taken for the heat to pass through the reference material to the sample.

Should the test sample be of a combustible material, the two curves will remain parallel initially but deviate when ignition commences, when the test sample curve's gradient will steepen. The curve will continue to rise and may eventually cross over the reference sample curve if combustion occurs (Figure 7.6).

By using a modified arrangement of the thermocouples it is possible to ascertain the point of ignition and, by making allowance for the thermal

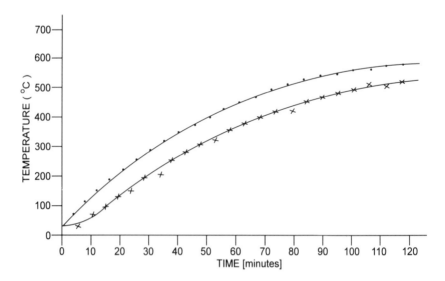

Figure 7.5   Test result for an inert sample. Air flow 2.5 l/min. (●) reference material; (×) shale plus coaling matter.

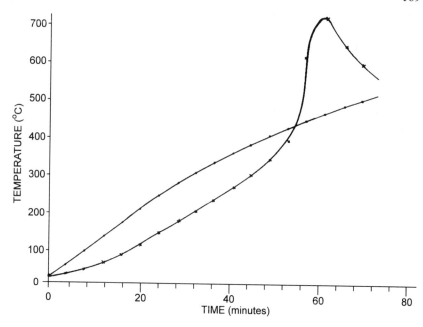

Figure 7.6 Test results for a reactive sample. Air flow 2.5 l/min. (●) reference material; (×) soil with coal admix.

lag, it is also possible to estimate the cross-over point, which corresponds to the point of thermal runaway (critical temperature for combustion). Thus, the method affords a means of identifying the point of onset of oxidation (ignition) and the point at which thermal runaway (combustion) occurs (Tables 7.3, 7.4, 7.5 and 7.6). The test procedure generally takes less than 2 h to complete and the analysis is simple to carry out. The value of the test is that combustion susceptibility is directly measured. The lower the temperature at which the test sample's temperature gradient increases, the more susceptible the material.

Table 7.3 Critical air flow to cause runaway combustion (colliery site, NE England)

Air flow (l/min)	Ignition (°C) (smouldering)	Combustion (°C) (runaway)
0	165	—
1.5	160	—
2.5	145	—
5.0	120	—
10.0	120	—
15.0	120	200
20.0	120	200

Table 7.4 Ignition and critical air flow to cause runaway combustion for materials with different calorific values (colliery spoil, former coal stock yard, NE England)

Air flow (l/min)	Temperature (°C)[a]							
	Sample 1		Sample 2		Sample 3		Sample 4	
	Ignt.	Comb.	Ignt.	Comb.	Ignt.	Comb.	Ignt.	Comb.
0	145	–	140	–	160	–	165	–
1.5	115	210	130	215	130	215	160	–
2.5	115	240	125	210	105	230	120	235
5.0	120	250	115	240	125	215	120	220
10.0	115	240	105	225	120	215	135	215
15.0							120	200
20.0							130	200

[a] Ignt., ignition point; Comb., thermal runaway.

Table 7.5 Ignition and combustion test (colliery spoil heap, NE England)

Sample type	Calorific value (kJ/kg) average	Temperature (°C)[a]				Risk factor
		Air flow 0.5 l/min		Air flow 2.5 l/min		
		Ignt.	Comb.	Ignt.	Comb.	
BH1 shale, clay, coal	4800 6800[b]	205	220	115	205	9.7
BH2 shale, sand, coal	4900 8400[b]	225	–	105	220	6.3
BH3 shale, clay, coal	4100 4700[b]	220	250	120	235	9.9
BH6 shale, gravel, coal	5700 9500[b]	190	–	115	180	7.2
BH7 shale, mudstone, coal	4200 6400[b]	205	245	95	200	7.2
BH8 shale, clay, coal	4400 7500[b]	215	–	90	185	0.3

[a] Ignt., ignition point; Comb., thermal runaway.
[b] Highest value.

Table 7.6 Ignition and critical air flow (former railway sidings)

Sample	Temperature (°C)[a]					
	Air flow 0 l/min		Air flow 0.5 l/min		Air flow 2.5 l/min	
	Ignt.	Comb.	Ignt.	Comb.	Ignt.	Comb.
TP4[b]	–	–	135	–	130	230
No. 21	–	–	125	–	110	200
No. 27	130	–	125	290	100	160

[a] Ignt., ignition point; Comb., thermal runaway.
[b] CV = 10 000 kJ/kg.

A second and very practical aspect of the test is that the air flow through the sample can be altered. If the test sample is run at different air flows it is possible to establish the critical air flow at which runaway combustion commences (Table 7.3). This can be compared with the worst possible situation anticipated on site and an assessment of the risk made (see section 7.3.8 and Table 7.5).

7.3.8 Air permeability test

Geotechnical research into soil compaction has revealed that at optimum compaction, the air permeability of a soil material is extremely low. The Proctor test [20] can be used to determine a material's dry density value at its optimum compaction level. This density value can then be reproduced in the combustion material and the achievable air flow through the sample at that density can be measured.

This essential supplement to the combustion potential test requires a sample of the material compacted to the optimum density in a cylindrical mould. The bottom is sealed and connected to a variable compressed air supply via a flow meter, with a mercury pressure manometer (or suitable pressure measuring device) in parallel. The arrangement is shown in Figure 7.7 and is similar to that used by Nagata [21] to determine the air permeability of undisturbed soils. The air flow is set to a suitable value and the pressure measured. This procedure is repeated for a range of values. The top of the mould is taken to be at atmospheric pressure and the pressure gradient can therefore be determined. From the measurements, the corrected air flow, and hence the velocity of the air flow through the compacted material, are calculated. A plot of inlet head pressure against velocity of air flow is constructed, from which the air flow velocity at any inlet head pressure can be established (Figure 7.8). The maximum likely velocity of air flow through the material is estimated and compared with the critical air flow velocity for combustion determined from the combustion studies. The results can then be evaluated in two discrete but similar ways. One is based on CP3 (Chapter V, Part 2, 1972, Wind Loads) [22]. This predicts a maximum value for pressure

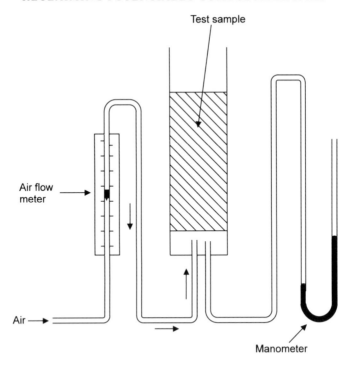

Figure 7.7 A schematic representation of apparatus for measuring air permeability of samples.

of 52 mmHg on a surface in extreme storms. A similar result is obtained by considering the maximum difference in barometric pressure between high and low pressure atmospheric zones likely to be encountered. Such pressure differences in the atmosphere would be associated with extreme wind conditions and would likely result in the de-stabilization of any embankment itself. From these analyses, a worst possible case can be accepted for the risk assessment analysis of an air pressure of 60 mmHg. The air flow that this extreme wind loading could produce can be measured, and the minimum necessary air flow needed to permit runaway combustion is known from the combustion potential test. By dividing the minimum necessary air flow to permit runaway combustion flow by that which could occur in extreme storms, a risk factor can be determined. The higher the risk factors thus obtained (Table 7.5), the lower is the risk of combustion. Typically, risk factor values in the range of 5–10 are found in stable compacts. In practice, no case has yet been found in which it has been possible to pass the critical air flow to allow combustion to take place through a fully compacted material.

The advantage of the test method is that it tests the material largely as it would exist on site and not after it has been artificially modified by sieving,

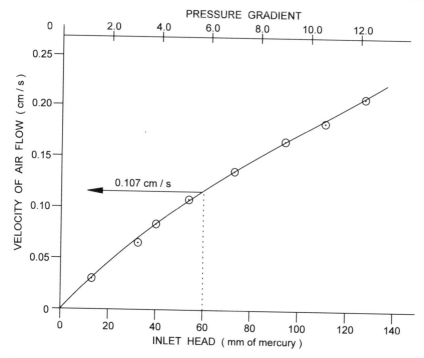

Figure 7.8 Test results for air permeability test (unreactive soil/coal admix). Critical velocity of air flow (from combustion test) = 1.04 cm/s. Maximum likely velocity of air flow (from air permeability test) = 0.107 cm/s. Risk factor = 1.04/0.107 = 9.7.

grinding or other amelioration. The data obtained are thus of practical value in assessing the risk of combustion.

7.4 Use of the combustion potential test

The test has been usefully applied in a number of cases.

7.4.1 *Case study 1*

In the case of a former coal stock yard in the North East of England still underlain by 1.2 m of waste coals, the site was investigated in order to assess its suitability for housing development. The tests showed that the site compaction necessary for the house foundations would prevent the critical air flow needed to allow combustion of the waste coal and give a high factor of safety. Provided adequate steps were taken to avoid accidental or indirect heating of the material, no removal of the waste coal proved to be necessary.

7.4.2 *Case study 2*

In the case of a former coking works stock yard, the site investigation revealed that the area was underlain by 2–4 m of coals, ash and shale, and that the entire site had a calorific value in excess of 10 000 kJ/kg. Combustion tests revealed that even the lowest CV materials would ignite at low air flows and combust at 200–260 °C. Thus the site, even if compacted to the optimum possible value, would present a significant hazard, since only very small risk factors were achieved (section 7.3.8). The owners agreed to improve the site material's compaction by a process of double washing to remove larger coal fragments. The washed discard, when returned to the site, still had sizeable calorific values, but could be compacted to higher densities and would not burn even when exposed to the air pressures anticipated in extreme storms.

7.4.3 *Case study 3*

In the case of a tip area around a former colliery at Sunderland, despite the fact that subterranean fires have been found on-site, the initial testing (section 7.3.3) showed that none of the first set of samples taken could be made to undergo runaway combustion, despite being heated to 600 °C with air flows up to 2.5 l/min. Further testing, however, showed that some later samples could be ignited at relatively low temperatures, and that, at high air flows, this could result in thermal runaway. This also revealed that the site was much less uniform than had been believed and had on it different coal materials from different mines and seams. A process of coal washing and rigorous site compaction was followed to produce conditions on the site that should prevent any subterranean heating.

7.4.4 *Case study 4*

In the case of a former colliery spoil heap in the North East of England, the spoil heap consisted of approximately 2×10^6 m^3 of colliery spoil, as a conical mound rising some 30 m above the surrounding valley site. The material was variable in density and grading, and consisted of silty sand to gravel sized fragments of shale, sandstone and coal, with soft to firm clayey material and occasional pockets of ash. The site was to be restructured and landscaped for recreational use, and might be developed for holiday flats and leisure activities. Even after restructuring the site would be relatively steep and exposed, and so particularly subject to wind loading conditions, due to the surrounding topography. Testing showed that some of the material could be ignited at low temperature and that at high air flows runaway combustion could occur (Table 7.5). Tests undertaken at site compaction levels, however, revealed that, with the exception of one sample, air flows that would lead to thermal runaway could not be established.

The exceptional sample proved to be an atypically uniform and coarse material that not only was difficult to compact but also was visibly identifiable. Tests were carried out to determine how the material compaction properties could be improved (by the addition of finer inert materials) and how the site owners could ensure that any bands of this atypical material could be identified when the site was excavated. This work showed that even in such a particularly exposed site, subterranean fire risks could be minimized and brought within acceptable limits.

7.4.5 Case study 5

The former railway sidings is a level area overlain with 1–4 m of compacted coal ash and boiler waste from the bygone era of steam locomotives. With the exception of one sample, the material showed a reluctance to burn, and thermal runaway could only be achieved weakly at high air flows (Table 7.6). The tests at difference levels of compaction were able to show that site compaction would prevent the critical air flows necessary to allow combustion of the material.

7.4.6 Case study 6

The earlier studies refer to colliery discard materials, and enough experience exists to show that such sites can be reclaimed safely. However, when materials are contaminated by fuel oils or reactive chemical wastes, more dangerous conditions can occur. In one case recently investigated, chemically reactive wastes proved to be present. In this case, the risk factors determined that the site material presented a serious hazard and it was agreed that the material would need to be removed. Thus, it is apparent that combustion risks can exist on a wider range of derelict sites than is sometimes believed, and that the decision to develop a site needs to be carefully considered.

7.5 Conclusions

The risk of subterranean heating is a real problem, which can be particularly serious to the infrastructure and fabric of a development. Calorific value determinations of the materials from such a development site have not proved adequate in assessing the potential of materials to undergo combustion, nor proved sensitive enough to identify the conditions under which spontaneous combustion might occur. The use of loss-on-ignition data can be misleading, as can tests that depend on modifying the form of the material under examination.

The combustion potential test described in section 7.3.7 does allow identification of more sensitive materials and assists in establishing which, if any, material needs to be treated or removed. A particularly useful aspect of the test is that it enables estimation of the critical air flow to allow combustion to proceed. This, in turn, provides the necessary information to ensure that the site material can be compacted to prevent this critical value being achievable, and minimizes the risk of spontaneous combustion on the site. Subterranean heating is one of the risks that can occur in the redevelopment of a disused site.

Developers and their advisers need to be able to understand the subterranean heating problems and the solutions offered so that they can judge the necessary level of work, costs and risks involved in order to give a guarantee of security for the site. What the combustion potential test cannot do is give safeguards either against poor workmanship and practice during site compaction, or against later site deterioration that permits large-scale admission of air to occur. Only good quality assurance and management can rectify such potential deficiencies (Chapter 13).

8
Gas and vapour investigations

8.1 Introduction

Subsurface occurrences of gases and vapours pose far more immediate hazards than do concentrations of solid contaminants in soils. Solid contaminants ultimately might cause chronic health risks to people or (if the substances are soluble) harm to water resources and the local environment, but they do not present acute and rapidly acting risks.

The reason this can be claimed stems entirely from the very high mobility of gaseous contaminants, many of which are lighter than air and so able to migrate up through soil profiles under quite small pressure heads. The consequences should such gases and vapours percolate into enclosed spaces, (fire, explosion, asphyxiation, toxicity and or cancer risks depending on which gaseous contaminants occur) have been emphasized by a number of very serious accidents [1,2,3] and fatalities.

The heightened awareness that such risks are feasible has made the assessment of surface gas atmospheres a crucial factor in the investigation and remediation of former industrial land.

8.2 Principal subsurface gases and their properties

While methane and carbon dioxide occurrences are usually emphasized in UK practice – simply because derelict land very often has been subjected to some degree of tipping and infilling with organic wastes – a much wider range of potentially troublesome gaseous compounds should actually be evaluated (Table 8.1). Various industrial wastes and quite commonplace underground conditions can give rise to far more than the hazards from landfill gases. Fortunately, the desk studies, undertaken to establish a conceptual model of a site (Chapter 3) will usually be enough to indicate which particular gases and vapours could be important in site-specific instances.

Other, less frequently encountered gases (such as hydrogen) could have been included in Table 8.1, as indeed could have been the risks of oxygen deficiency. Oxygen contents in uncontaminated soils are essentially similar to those in the atmosphere (20% to 22%) and are necessary for the well-being of plants and the safety of people who have to enter trenches and excavations. However, as the concentrations of gaseous contaminants increase, those of oxygen generally decline. When this leads to oxygen levels of 18% or less, health and safety concerns will be well based [4]. In very many of the contaminated sites being considered for productive re-use, far lower oxygen concentrations are apparent (see Tables 8.2 and 8.3) and call for specific evaluations and remedial action.

8.3 Gas and vapour monitoring

8.3.1 *Monitoring equipment*

Field portable gas measuring meters were developed for industrial applications (in mines, oil refineries and factories) and these earlier meters were not ideally suited for use in contaminated land studies. However, the importance of the infant land reclamation industry in the 1970s was far too slight to attract the specific attention of instrument manufacturers. Thus whatever was commercially available had to be used when subsurface gas and vapour studies were required. Only a few gas concentration portable meters then existed, and these (as well as being limited to providing concentration results) were often of dubious accuracy, particularly if soil oxygen levels were too low ($< 14\%$) to allow catalytic–thermal instrument systems to function accurately. A number of older publications [5,6,7] provide useful details on the first generation equipment and methods for gas monitoring, though these – in the light of recent improved equipment – are more of historical than practical importance. These older publications stress the need to confirm the accuracies of field meters periodically, by collecting gas samples in pressurized tubes and then analysing the contents, with the much more sensitive gas chromatographic equipment, in a laboratory.

The measurement of gas flow rates, or gas pressures, at this time, was something of a specialist activity, and so few gas surveys included information other than concentration results. An obvious consequence of this was a widespread misconception (which influenced earlier official advisory documents) that concentration readings, by themselves, were adequately diagnostic indicators of 'safe' or 'unsafe' underground gas conditions (see sections 8.3.3 and 8.7.2).

This limiting situation has abruptly improved over the last few years, as the contaminated land investigation and reclamation industry has grown in economic importance and become a profitable target for instrument providers. The increased insistence by regulatory bodies that subsurface gaseous atmospheres must be comprehensively investigated, to establish that future land users will not be exposed to risks, such as those which occurred at Loscoe [3], has also had an obvious impact.

Today, a much more complete armoury of field portable gas measuring equipment is easily available and includes:

- infra-red concentration meters, able to provide accurate measurements of the oxygen, carbon dioxide and methane contents of sampled underground atmospheres; these instruments do not suffer from the inaccuracies created by reduced oxygen concentrations and are easier to use than older meters; thus the need for laboratory based gas chromatographic analyses is much reduced;
- similar concentration measuring equipment is available for other gases (photon ionization detectors for organic vapours, and specific monitors for all the gases listed in Table 8.1); portable gas chromatographic analysers are also increasingly available;
- electronic pressure measurement methods; and
- the latest concentration meters, which can also determine higher gas outflow rates.

In consequence, if modern and properly calibrated equipment is taken on to a site today, it is possible to establish gas concentrations, gas pressures and higher gas flow rates within a few minutes of arriving at a gas observation point An obvious corollary is that investigators, who persist with the use of first generation equipment (as some do), produce less convincing and complete results, and should expect to be criticized by clients and regulators.

Useful summaries of the performances of newer monitoring equipment are available in recent CIRIA publications [8,9], though the pace of technical innovation is such that even these reports are now somewhat dated. For current gas monitoring capabilities, direct contact with instrument manufacturers is necessary.

8.3.2 Gas monitoring observation works

Several distinct types of gas monitoring holes are available. All have particular advantages but none is uniquely suitable for all conditions. Thus a careful evaluation of what gas or vapour information actually is needed is required before any choice of monitoring facility type is made.

The commonest are the traditional deeper borehole installations (Figure 8.1), initially devised to show whether landfills were causing off-site gas migrations. Typically such holes are drilled to large enough diameters (150–200 mm) to allow space for the insertion of a 50 mm internal slotted gas monitoring tube and gravel pack. This, plus the high percentage of open area (typically 6% to 12%) in the walls of the internal gas tube, allows unimpeded entry of gases from the surrounding ground. The top metre of the installation is unslotted and sealed with a bentonite grout plug to prevent atmospheric air entry (in periods of higher barometric pressures). These

Table 8.1   Principal types of gaseous contaminants found on contaminated sites (listed in order of commoner occurrence)

1. **Carbon dioxide (CO_2)**	Product of the biodegradation of organic wastes. Also produced by coal workings, limestone mines and sewers and marshes.
	Colourless and odourless. Very dense gas (specific gravity 1.53) soluble in groundwater.
	Toxic and asphyxiating to people.
	Health effects obvious >3% concentration
	Asphyxiating >12% to 25% concentrations
	Toxic >25% concentration
	Toxic to plant species, though the critical concentration varies with species (1% in most sensitive cases to 20% in least susceptible species).
2. **Methane (CH_4)**	Product of the biodegradation of organic wastes. Also produced by coal workings, sewers and marshes.
	Colourless, odourless and tasteless.
	Lighter than air (specific gravity 0.55).
	Flammable and explosive in concentrations between 5% (Lower Explosive Level – LEL) and 15% (Upper Explosive Level – UEL) in air.
	Able to expel air from soil voids and so generate asphyxiating atmospheres.
	Harms plant life if methane has expelled oxygen from plant root zones.
3. **Carbon monoxide (CO)**	Product of the incomplete combustion which typifies subsurface smouldering of heated carbon rich wastes (coal, shale, wood, papers, etc.)
	Colourless and almost odourless. Slightly soluble in groundwater.
	Lighter than air (specific gravity of 0.97).
	Highly toxic to people (>0.1% to 1%).
	High flammable (>12% to 75% in air).
	Toxic to plant life.
4. **Hydrogen sulphide (H_2S)**	Product of the degradation of sulphate rich wastes (e.g. plasterboard) and of oil refineries where sulphur rich crude oil has been refined.
	Colourless gas. Initially obvious by 'rotten egg' smell, but this becomes less noticeable as the gas in inhaled. Soluble in groundwaters. Heavier than air (specific gravity of 1.19).
	Highly toxic to people (>400 ppm). Explosive at 4.3% to 45.5% concentrations in air.
5. **Lighter 'oil' vapours (benzenes, xylenes, toluenes) (C_6H_6X)**	More volatile and diffusible components of spilled fuels. Diffusivity is high enough to permit vapour rise through soil in warmer temperatures.
	Colourless gases.

Marked 'petroleum' smells.
Lighter than air, though specific gravity varies depending on the gas mixtures.
Practically immiscible with water.

Highly flammable and explosive in air.
Suspected to be carcinogenic at low concentrations.

(Propane and butane admixtures are usually present.)

6. **Sulphur dioxide (SO₂)** — Product of the combustion of sulphur rich wastes (some coals and oils and the spent oxides on gas works sites).

Colourless gas with sharp odour.
Denser than air (specific gravity of 1.43)
Soluble in groundwater.

Toxic to humans (>12%)
Irritant to eyes and nose at lower concentrations.
Toxic to plants.

7. **Hydrogen cyanide (HCN)**

Product of the combustion of cyanide rich wastes, or the acidification of cyanide salts.

Colourless gas with faint odour (bitter almonds) soluble in groundwaters.

Flammable and explosive in air (6%)
Highly toxic to people (>100 ppm).

8. **Phosphine (PH₃)** — Product of phosphorus deposits.

Colourless gas with faint garlic odour.
Denser than air (specific gravity 1.85)

Spontaneously flammable in air.

Highly toxic to people (used as World War I poison gas).
Highly flammable and explosive.

Other hazardous gases (e.g. radon) are not caused by land contamination, though their feasible occurrence should also be considered.

boreholes are usually drilled through the full depths of gas producing materials and fills, and taken a further metre into underlying natural ground. This, for landfill monitoring purposes, is a reasonable choice, since landfill gases under high pressures need not rise or move laterally, but can migrate downwards into underlying soils, if these offer the easiest initial outflow pathways (see section 8.9.1).

Advantages of deeper borehole installations include the convenience of being able to make repeated use of the same installation over many months of a gas investigation, and the ease of taking groundwater and soil temperature observations by simply unscrewing the gas tap unit.

However, borehole costs are high (about £600 for a 4 m deep hole) and it is seldom possible to install those closer than 30 m apart. This financial limitation can pose real difficulties; not all areas of a contaminated site will usually be gas-producing and thus hazardous, yet (for the most hazardous conditions to be established) observation boreholes really have to be sited in the (unknown) areas of greatest potential concern.

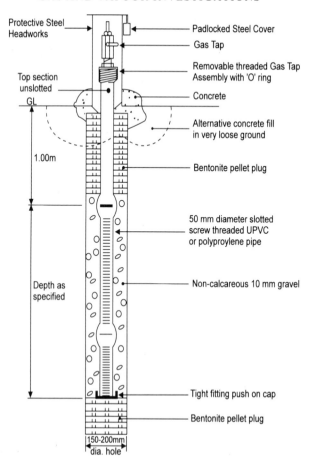

Figure 8.1 Typical borehole gas monitoring standpipe.

Because of cost constraints, use of smaller diameter boreholes (drilled at 50 mm diameter and with a 20 mm internal gas collecting tube) has become more popular in recent years. A point of caution has to be voiced if the monitoring works on a site include a mixture of boreholes of different diameters. Results in such cases will not be directly comparable, and smaller diameter holes will tend to produce far higher gas concentration measurements, than will occur in larger diameter installations.

Use is also made, on cost reduction grounds, of very small diameter drilled probe holes (usually 20 mm in diameter), and of trial pits into which gas collection slotted tubes have been installed. If trial pit installations are employed, it is necessary to centre a vertical slotted polyethylene pipe, surround it with a granular backfill to within a metre of ground surface, and then cap off the backfill with a clay, or other gas impermeable, cover. The mention of gas impermeability raises, of course, the question of whether a well-compacted clay layer will indeed be impermeable to gas migrations. If the voids in the clay are highly water saturated the required impermeability can be assumed, but if the clay (placed close to land surface and exposed to desiccation) dries out, it is likely to be very permeable to gases and will provide a poor top seal [10]. Because of this doubt over effectiveness, a uPVC sheet, set into the clay cover, is a prudent addition, to ensure collected gases do not merely migrate out to the atmosphere.

At the cheapest end of the cost spectrum are those hand-driven probes holes, to depths of 1.0 m and 1.5 m, which allow closely spaced arrays of gas measuring points. These 'spike test' holes are cheap (up to 30 can be installed for the price of a single deep borehole), and so subterranean gas conditions can be mapped out (in favourable conditions) and contouring of results can reveal specific locations where worst gassing is occurring. Sealing the tops of temporary spike test holes – to preclude atmospheric air entry – is, however, difficult, and it has to be accepted that some dilution from atmospheric air is very likely to occur.

One final choice, which often seems attractive, is to install several small diameter piezometer tubes in a single large diameter borehole, and separate each piezometer response zone with bentonite grout dividing layers (Figure 8.2). The aim of such as installation is to locate the depth at which a gassing source is located and allow reclamation work to be more precisely focused. An obvious problem is that poor construction and workmanship can give rise to less effective seals, and to gas leakages between piezometer zones.

The use of 'flux-box' installations (i.e. containers of known internal volumes set at depths of particular significance – such as the layer of soil just below planned house foundation slabs) is more commonly restricted to proof-testing (after a site has been reclaimed) to check that all gas or vapour hazards have been reduced to acceptable levels. The internal atmospheres in these flux boxes gradually come into equilibrium with gas concentrations in the soil, and if three or four samplings of flux box atmospheres are carried out, at weekly or fortnightly intervals, it can be possible to determine the inflow rates of gases [11]. This, despite advances in monitoring equipment, still remains the best method of establishing extremely low rates of gaseous migration.

None of these monitoring facilities is uniquely suitable in all circumstances, and all have their limitations. For deeper boreholes, the worst feature is that the entire internal monitoring tube atmosphere can be influenced by gas inflows from one thin layer within the investigated soil profile. The results then measured will, of course, be representative neither of the strata in the greater part of the borehole or of the atypical gassing layer. This is not important if boreholes are being used for monitoring the perimeters of landfills (where the aim is merely to decide whether gas migrations to adjoining land are occurring) but is distinctly unhelpful if the reason for monitoring is to identify which strategy is needed to reclaim a contaminated site. A recent example makes the point; a 1960s domestic waste landfill was invest-igated to establish whether the land could be used for house building. The landfilled wastes, after 30 years of degradation, had been converted to an apparently inert mixture of cinders and ashes (from domestic fires) together with broken bottles, tins and a content of stones, yet when boreholes were drilled through the fills (4–5 m) and into the underlying sandy clays, unaccep-tably high carbon dioxide and methane concentrations were found, together with very reduced oxygen contents. On these results (Table 8.2) redeveloping the site was believed to be unacceptably expensive. Fortunately some delayed trial pitting was still in progress, and this allowed spike testing as trial pit depths were advanced (from 1 m to 3 m). This second investigation revealed only entirely uncontaminated soil atmospheres in the ashy fills, and cast doubts on the validity of the deeper borehole results. To resolve the uncertain-ties, one multi-point borehole (Figure 8.2) was sunk, and revealed that (Table 8.3) landfill gases only occurred in a thin zone at the top of the sandy clay stra-tum. Later excavation proved that this had resulted from the washing down, by 30 years of rainfall, of fine organic matter. This had collected as a 10–20 mm thick black accumulation on top of the natural underlying clays and was – of course – easily removed in site reclamation. Obviously, in this case, use of traditional gas monitoring boreholes was not diagnostic.

Table 8.2　Gas concentrations revealed by standard gas monitoring boreholes drilled to natural strata (former landfill site)

Borehole number	Gas concentrations measured		
	Oxygen %	Carbon dioxide %	Methane %
1	4.1	8.3	0.15
2	1.8	9.6	1.10
3	1.3	8.0	3.96
4	trace	11.9	2.56
5	1.5	7.1	5.35

(Gas concentrations usually seen as of concern are:
Oxygen　　　　　　< 16%
Carbon dioxide　　>1.5%
Methane　　　　　>1% – section 8.7)

Table 8.3　Gas concentrations found in a multi-point gas observation borehole (former landfill site)

Depth of gas measurement, (piezometer response zone)	Gas concentrations measured on specified dates			
	14/4	21/4	2/5	16/5
1.0 m	20/0.3/0	20.4/0/0	21.1/0.6/0	20/0.1/0
1.8 m	20/0.5/0	20.5/0.3/0	21/0.4/0	20/0.3/0
2.5 m	20/0.4/0	20.1/0.2/0	20.8/0.3/0	20.1/0/0
3.5 m	18.6/0.9/0	18.4/0.7/0	18/1.3/0	17.9/1.5/0
5.0 m	0/10.4/1.3	2.1/8.3/0.6	1.3/9.4/1.0	1.0/10/2

(Gas concentrations reported in the sequence oxygen/ carbon dioxide/methane and in % terms.)

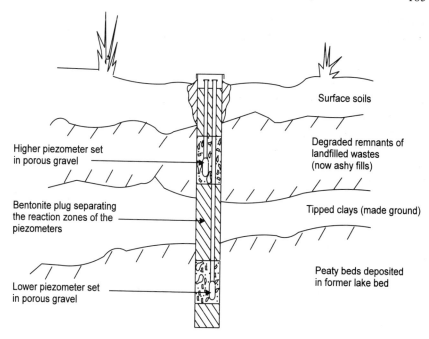

Figure 8.2 Multi-point gas monitoring borehole to establish which horizons are gassing sources.

In this example, gases in the thin organic layer were pressurized by the weight of overlying strata, and so moved rapidly into the lower pressure zones offered by the drilling of each gas observation borehole. This demonstrates that all deeper boreholes are low-pressure environments which tend to collect gases from considerable distance, and so produce anomalously high gas concentration results. Such overestimation of gas concentrations might be acceptable for lighter gases, as the higher than actual gas concentrations might indicate the riskier situations which could occur in future, more extreme, circumstances (section 8.4), but is less acceptable when denser gases are of concern. In such cases, a deeper borehole can act as no more than a 'sump' into which gases such as carbon dioxide flow and collect. Cases where very high carbon dioxide concentrations have been measured in observation boreholes, and where later excavation of sites has revealed no comparably high carbon dioxide contents, are relatively commonplace in areas where old limestone or coal mining has occurred (see section 8.9.4).

In contrast, shallow spike-tests do not produce such exaggerated gas readings, but instead tend to give lower than accurate concentrations because of dilution with atmospheric air. Additional limitations of shallower observation points are that they can fail to intersect the deeper gas producing layers (see Table 8.3) and can produce results influenced by near surface phenomena (i.e. the bacterial oxidation of methane to carbon dioxide or the solution of carbon dioxide in shallow groundwaters [12]). Despite these limitations, and particularly where sites are underlain by predominantly granular fills, spike-testing at 5 m to 10 m centres can highlight those sub-areas where gassing potential is greatest, and so allow more expensive boreholes to be sited precisely where most useful data will be obtainable.

8.3.3 *Gaseous properties of importance*

Mention has been made of the concentration, pressure and flow rate properties of gases and vapours, while these terms are well understood, it is worth considering the insights they can provide, when the intention is to identify more or less hazardous subsurface gaseous atmospheres.

Concentrations (in percentage terms or in parts per million) indicate how much of a particular gas occurs in a sampled subterranean atmosphere. As some specific concentration values (e.g. the 5% (LEL) for methane or the 400 ppm for hydrogen sulphide – Table 8.1) indicate the onset of particularly risky conditions, it is perhaps unsurprising that many regulators and some investigators react with concern when such values are measured in gas observation holes. This, however, is an illogical reaction. The cited critical concentrations are of concern if they occur in enclosed spaces, houses or structures; but can it be taken that the same risk significance is reasonably ascribed if the concentrations of gases and vapours exist deep in the ground? Obviously the answer has to be 'no'. For gaseous contaminants to present risks to people or site surface conditions, deeper gas collections somehow have to move up the soil profile and then collect above enclosed spaces at critical concentrations.

For this to occur, bulk gas transfer has to occur, and that calls for a flow of the type governed by the well-known Darcy flow formula. A pressure head (above that of the atmosphere) is needed to drive such bulk transfer flows, and such positive relative pressure heads are today quite easily measured. The flow rates, generated by the existence of higher gaseous pressures deeper in a site, are also not too difficult to quantify with modern monitoring equipment.

For risks to occur to future users of an area of land, deeper gases have to move up and out of the ground. This process inevitably exposes the original gas concentrations to dilution with atmospheric air and to the effects of ventilation in houses, and so concentrations of worrying gases will usually fall and often to levels well below those for realistic concern.

A 5% methane concentration, in soils or fills well below land surface might be a trivial matter (if no positive gas pressures and no bulk flow rates occur), or could be very serious (if very high outflow rates can be measured). The differences in these cases are that:

- the gases in the first case will not flow up the soil profile, and so cannot directly harm site users; while
- in the second case, not only will such gas outflow occur, but the flow volumes are likely to be so high that (even when dilution with the atmospheres, already present in enclosed spaces, takes place) high remnant methane concentrations will persist within homes or structures.

Reacting to concentrations of gases and vapours, found to occur at depth in a site, will thus call for a knowledge of the gas pressures and flow rates likely to exist in most hazardous conditions. Judging solely on concentration results will be a dubious business and will be likely to mean over-reacting to situations of trivial importance.

One very obvious necessity, if this evaluation approach is to be defended, is that *worst case* gassing conditions within the ground have to be measured and used in the analysis of possible risks. As noted below (section 8.4), measuring worst-case conditions is not always possible, and calls for an intelligent assessment of when to conduct subsurface gas surveys (section 8.5).

Because of earlier limitations of gas monitoring equipment, when only concentration values could be obtained with reasonable accuracies, older advisory documents (section 8.7) gave too much importance to gas concentration results as the indicators of probable future risks. This situation has changed as gas flow rates and pressure heads have become easier to establish, and it is today possible to distinguish between truly hazardous subsurface conditions and those of much lesser concern.

Barry [13] correctly makes the point that

effective ventilation (i.e. dilution) can eliminate all risks from gases, whether toxic, asphyxiant or explosive

and so has adopted the same approach to judging risks from gases and vapours, i.e. that risk only occurs when these gases move out of the ground and in such concentrations and flows that dilution will be inadequate to remove risks.

One final, and potentially confusing, type of gas or vapour movement, i.e. diffusive flow driven by differences in concentrations within a fill or soil profile, calls for comment. This is a very slow equalization process, usually taking weeks or months to become apparent, which will ultimately (in a case such as that cited in section 8.3.2) spread previously localized higher gas concentrations throughout the subsurface profile, once investigation works here pierced gas sealing layers. While this could finally give rise to quite high concentration results in near surface layers, these readings would not be accompanied by positive gas pressure heads or measurable flow rates. Thus migration of meaningfully large outflow rates, to pose risks in homes or other enclosed spaces, would not result. Diffusive flows are the usual means by which subsurface gases gradually dissipate to the above-ground atmosphere [14].

8.4 Establishing most hazardous gassing conditions

8.4.1 *Introduction*

Measuring gas concentrations, pressures and flow rates is now relatively simple, and a range of suitable gas observation facilities exists. Thus it could be believed that obtaining worst-case (and so more hazardous) gassing results would not be a problem. Unfortunately this is not the case. The analysis of measured gas and vapour results remains an art form, simply because measurements are affected to a very great extent by a wide range of factors. Table 8.4 indicates a quite commonly encountered range of variation in results, within which some could appear to be entirely safe (i.e. carbon dioxide less than 1.5% and methane less than 1% of the underground sampled atmospheres – see section 8.7), while others could be taken as signals for concern.

The factors leading to these variations and interpretation difficulties need to be separately discussed for gases created by biodegradation and those created by other mechanisms.

8.4.2 *Factors affecting landfill gas (and other degradation product gas) results*

Landfill gases dominate the UK's concerns over subsurface gaseous contamination, simply because so many old landfills exist and because tipped organic wastes occur on most derelict sites.

Table 8.4 Variability of gas measurements in a single borehole

Date	Volumetric concentrations of observations			
	Oxygen	Carbon dioxide	Methane	
05/10/91	20.4	0.3	nil	–
10/11/91	16.1	1.7	0.3	Reading after sudden barometric fall (24 mbars)
04/12/91	20.1	0.4	nil	–
03/04/92	19.7	1.4	0.1	Period of gradually
24/04/92	17.4	2.7	0.6	increasing soil temperatures
05/05/92	10.4	8.7	2.5	from 8 to 11 °C)
06/06/92	2.3	12.7	4.8	
01/07/92	0.1	2.3	5.7	Near surface soils saturated
04/08/92	0.1	1.3	4.1	by prolonged rainfall in early July
07/09/92	0.1	16.2	5.5	Near surface soils dried out
04/10/92	12.4	4.9	trace	Soil temperature falling
05/11/92	18.7	1.2	nil	Soil temperature at 8.1 °C
03/12/92	20.4	0.2	nil	Soil surface frozen

(Gas source 3 m thick refuse layer at depth of 2.3 m below land surface.)

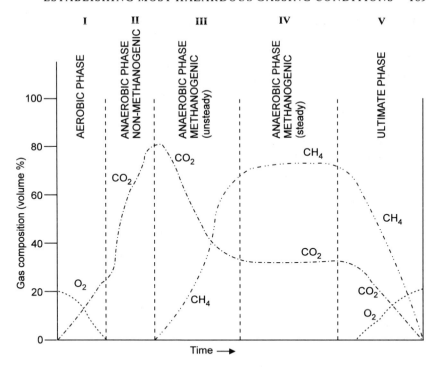

Figure 8.3   Idealized degradation phases for organic waste.

In consequence, professional associations [5,7] and government-funded studies [15,16] have developed a comprehensive understanding of biodegradation processes, and have widely disseminated their findings. Additionally a very large scientific literature has existed since the 1950s and provides extremely useful basic facts and observations.

Landfill gases are the by-products of bacterial degradation and so gas emission rates, flow quantities and concentrations depend primarily on the rate of bacterial activity and the expansion of bacterial populations. This dependent link on bacterial well-being implies that any internal or external factor which affects bacterial populations will inevitably impact on the concentrations and types of gases which can be measured in observation holes.

This stress on factors affecting biodegradation might seem surprising, especially since representations such as Figure 8.3 are widely assumed to represent the degradation process from freshly tipped wastes to the final composted remnants, and are often believed to show that biodegradation is a fixed and predictable matter. In fact, such representations are theoretical and outline what could be expected in idealized circumstances. Such idealized conditions might be close to those of very deep landfills (thicker than 6 m) and with a substantial and effective top cover, but even these –

in Western Europe – tend not to pass progressively through the various degradation phases, but, instead, usually 'stick' in the unstable phase III for decades. In the thinner and less well-engineered organic tipped deposits, which typify contaminated sites, Figure 8.3 is especially misleading, as internal factors and external climatic events impose their influences, accelerate or retard degradation, and affect the types, concentrations and emission rates of subsurface gases.

Of the internal controls, the most important ones are:

- the temperatures within the degrading wastes;
- the moisture conditions there;
- the wastes' acidity;
- the present or absence of toxic compounds; and
- the ease with which atmospheric air can intrude into the decomposing wastes [17].

While these various factors are conveniently considered separately, it should be realized that they will usually occur in various combinations, and so add a further complexity.

Temperatures of 30–40 °C particularly encourage bacterial growth and give rise to high gas emission rates. The internal heat of degradation (632 kJ/kg) is more than enough to support these temperatures, if top cover above the waste deposit provides good thermal insulation. However, the 2–3 m of soil cover thickness needed for this seldom occurs over the patches of tipped materials found in contaminated sites. If waste temperatures fall, perhaps because top cover insulation is inadequate, biodegradation slows and gas emissions decline. Even a small internal temperature drop (of 1 °C) can abruptly cause methane contents in sampled underground atmospheres to fall abruptly (from 19% to less than 4% in recorded instances). Such a chilling event is often a result of rainfall entry in winter periods. As chilling continues and waste temperatures fall below 10 °C, gas emission rates effectively halt. Thus the quite normal pattern, in thinner waste accumulations, is for methane to vanish in winter, and then reappear in hot summers (Table 8.4).

Similarly internal moisture contents dominate the gas emissions which can occur. This is because bacteria obtain their food from the moisture films on waste particles. If the wastes are very dry (<10% saturated), then food availability is essentially zero and the bacteria die off. As wastes are wetted, bacterial life and gas emissions pick up, and peak at a waste moisture content of about 60% to 80% of the totally saturated state. At this level food availability is most widespread and bacterial populations expand. However, if more water enters the wastes, this tends to wash away the organic fine materials, and so reduce food for bacteria. Total water saturation, of course, so slows biodegradation that 30-year-old wastes can appear to be nearly as fresh as the day they were tipped. In one recent land reclamation, this effect was obvious in wastes tipped in the 1950s. Because this landfill had been sited in a shallow, clay-floored depression, the basal metre or so of the wastes produced entirely readable newspapers in a matrix of totally saturated but recognizable domestic refuse.

Acidity and the presence of toxic compounds (especially metals such as aluminium) can kill off bacteria and slow degradation processes. Similarly, as is the case with the smaller volumes of poorly engineered organic wastes found on most contaminated sites, easy oxygen entry in periods of higher atmospheric pressure prevents methanogenic conditions developing, and precludes the idealized sequence of degradation phases. Together these effects so slow down biodegradation that waste accumulations persist for decades without ever reaching the high internal temperatures needed for rapid decomposition. This represents perhaps the worst of outcomes, as the wastes pose continued nuisances (organic rich polluting leachates and periodic emissions of carbon dioxide together with small concentrations of methane) for year after year.

In addition, external climatic and soil factors have their influences on measurements of landfill gases, i.e.

- abrupt atmospheric high pressures can so force oxygen deep into poorly covered wastes that no landfill gas contents can be sampled; in contrast, Figure 8.4, a deep barometric low (especially if this follows hard on the heels of high pressures) can 'pump' out enhanced landfill gas concentrations into an observation hole;
- surface climatic effect (i.e. ground freezing or thawing) can close off or open pathways via which gases migrate to observation points;
- groundwater level changes can act in a manner similar to those described for barometric changes. Sudden rises in shallow groundwaters will often force gases into observation boreholes at very high concentrations, while abrupt water level falls can have the reverse effect [18]; and
- surface effects can ensure that the gas atmospheres measured in shallow observation holes differ markedly from those which originated at a deeper gas generation location [12].

The net result of these possible influences are data sequences similar to those noted in Table 8.4. Changes at gas generation source, effects of the external atmosphere, and near-surface effects conspire to give results which are distinctly difficult to interpret. This is especially so for thinner and smaller patches of poorly covered organic wastes.

8.4.3 Factors affecting the measurement of gases not generated by biodegradation

Although the technical literature for these gases and vapours is far less comprehensive, enough exists to establish broadly why gas and vapour measurements are subject to marked variability.

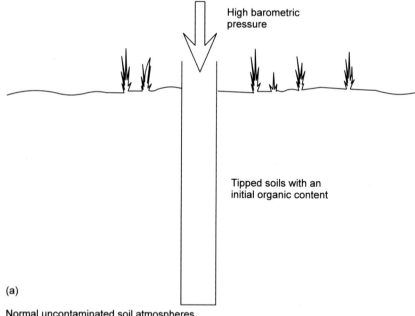

(a)

Normal uncontaminated soil atmospheres
are found as pore gases are driven out of the
borehole tube and pushed deeper into soil voids.

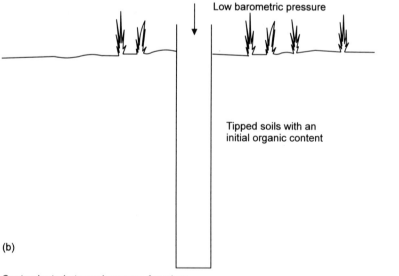

(b)

Contaminated atmospheres are found as
pore gases can move into the borehole
tube from the surrounding soils.

Figure 8.4   Atmospheric controls on pore gas concentration measurements.

The gases created by subterranean heating and smouldering (carbon monoxide, sulphur dioxide and hydrogen cyanide) can be affected by atmospheric pressure effects. Higher barometric pressures force air deeper into the ground and so accelerate smouldering and gas output rates, though this will usually take a day or so to become apparent, and will initially have the reverse effect (i.e. combustion by products will be less apparent in gas observation holes as a high pressure tracks over a site).

Similarly, variations in the proportion of combustible wastes will have obvious effects. As more abundant concentrations are exhausted, subsurface progress of a smouldering front will slow and possibly decline, though this (and the accompanying marked reduction in gas emission rates) need not mean that the problem has vanished. In one recently investigated case, where tipped sawdust and newsprint had been ignited by deep migration of a surface fire, output of combustion gases ultimately slowed and then ceased for several months. However, these suddenly reappeared (presumably because remnant soil temperatures had been enough to ignite more distant patches of combustible waste), together with high concentrations of methane. The methane, in this case, was a result of the soil warming increasing the biodegradation rate of garden wastes, which had been tipped within the combustible fills.

Soil temperatures also have very marked effects on the concentrations of volatile organic compounds (from spills of fuel soils and petroleum products). Here volatilization is the main cause of gas emissions, and meaningfully high contents of benzenes, xylenes and toluenes are most apparent in hotter summer conditions. Atmospheric pressure changes also affect the quantities of these volatile organics which are measured and in a manner similar to that noted in Figure 8.4. These volatile vapours often arise from fuel products floating on shallow groundwaters, and so can be most apparent when groundwaters are higher within the ground. Carbon dioxide (from slow biodegradation of heavier oil fractions) is often found to accompany higher volatile organic concentration results. The pattern of occurrence of these vapours and gases can usefully be employed [19] to establish the direction of flow of subsurface hydrocarbon spillages.

8.4.4 *Conclusions*

From what has been argued, it is obvious that no single gas monitoring event is likely to establish worst-case gassing conditions. Too many controlling variables affect which gas and vapour measurements can be collected in a single gas survey.

The difficulty this presents is obvious. Only the worst-case conditions are diagnostically important, since gas flow rates, pressures and concentrations can change by orders of magnitude in extremely short periods. In some manner gas surveys have to be focused to overcome this variability if readings from these worst-case gassing periods are to be obtained.

8.5 Gas and vapour survey procedures

8.5.1 *Introduction*

Two quite separate difficulties have to be addressed by well designed gas and vapour surveys, i.e.

- observation holes have to be located in those areas of a site most likely to be emitting gaseous compounds; and
- surveys then should be conducted predominantly in those periods when the variables which affect gaseous emission rates will most probably give rise to more hazardous results.

8.5.2 *Selecting observation hole locations*

If the potential for gaseous emissions were uniform within a site there would be no need for care in selecting gas observation hole locations. However, this will seldom, if ever, be the case, as most sites contain sub-areas free of gassing nuisances and others where much worse gas or vapour results occur. Thus it is important to try to locate the relatively few observation holes, which budgets permit, in areas of poorest underground atmospheres.

This situation, which typifies contaminated sites, is quite different from that on larger landfills. These, while not entirely uniform in composition, are far more consistent than are contaminated sites, where it is smaller accumulations of gas producing materials which pose the risks to future land users.

Given that small tipped areas do exist, and are usually obscured by surface fills, and that small filled quarry voids and infilled ponds are a prime source of contaminated land gassing problems, the difficulty of identifying where these worst-gassing locations exist is not inconsiderable.

Three methods can be used to position observation holes in most hazardous gassing areas, i.e.

- historical surveys will usually be able to identify where quarry voids once occurred or where small tips previously were sited;
- gas concentration measurements as investigation trial pits and boreholes are deepened can also be useful, and can show which horizons, within a site, produce greatest gas concentrations (Table 8.5); and
- spike test surveys (see section 8.9.1) together with subsurface temperature readings can identify those discrete areas where worst-gassing conditions exist.

It is usually worth the trouble to carry out such evaluations before gas monitoring holes and networks are selected, and it will seldom be prudent to site gas boreholes before preliminary site investigation results have been thoroughly analysed.

Table 8.5 In-situ gas monitoring results as a groundwater observation borehole was being drilled

Borehole	Depth (m)	Oxygen %	Carbon dioxide %	Methane %
17(W)	1.0	21.1	0.3	–
	2.0	20.3	0.2	–
	3.0	19.7	0.2	–
	4.0	18.4	0.7	<1
	5.0	18.1	1.4	1.2
	6.0	14.2	2.7	1.6
	7.0	19.7	1.3	1.0
	8.0	20.1	0.6	–
	9.0	20.7	0.4	–
	10.0	20.3	0.3	–

Horizon from 4.6 m to 6.4 m depths was an ashy layer (decomposed landfill material). Materials above and below were clayey subsoils, produced as wastes from road construction works in the area.
Subsurface conditions elsewhere in the site were found to be non-gas-producing.

8.5.3 Including more hazardous gassing conditions in gas and vapour surveys

Table 8.4 makes it clear that no single-event gas survey is likely to record worst-case gassing results in boreholes and spike test holes.

As a result, most practitioners try to include between three and five different site visits to measure subsurface gas and vapour information. These, usually at weekly or fortnightly intervals, are a balance between budget restraints (which tend to reduce gas survey number) and the need to measure in different conditions to establish trends and (hopefully) include more hazardous results. However, this practical compromise need not be technically the most effective, since gas concentrations, flow rates and pressures can change abruptly and in very short periods (of a day or less), and – ideally – continuous gas monitoring would be undertaken. However, equipment limitations and costs currently make this impractical. The general rule is clear, the more information that is collected the less difficult it is to reach conclusions which will satisfy critical audiences, and so gas surveys on more complex and troublesome sites can be continued for periods of a year or more.

Identifying those conditions most likely to yield more hazardous results is not difficult, if the effects of controlling factors are borne in mind (section 8.4).

For landfill gas surveys these more informative survey periods include:

- times when soil temperatures are higher;
- periods when atmospheric low pressure conditions occur immediately after higher barometric highs have passed by; and
- when near surface groundwater levels are rising.

Table 8.6   Adequacy of gas and vapour surveys

Reliance possible	Survey content
LOW	Gas concentrations only measured. Measurements only an a single occasion. No independent checks of measurement accuracy. Readings taken in autumn or winter.
	As above, but readings taken on at least three occasions and measurement accuracy confirmed. Atmospheric pressure and climatic conditions recorded. Readings extend into periods of warmer soil temperatures.
	As above, with gas pressure readings included and with changes in barometric conditions over prior 24 hours recorded.
	As above, with direct flow rate measurements.
	As above, with soil and gas temperatures measured.
HIGH	As above. Readings continued to include conditions especially likely to reveal highest subsurface gas emission rates (soil temperature, barometric and other more sensitive conditions adequately covered).

If combustion by-product or volatile gases are of interest, higher concentrations and flow rates are likely in times when

- low atmospheric pressure conditions occur just after a barometric high; and
- soil temperatures are at their annual peaks.

In this way, it is possible to distinguish between the adequacy of gas and vapour surveys and decide how much reliance should be ascribed to measured results (Table 8.6).

Site investigators, however, can be required to undertake gas and vapour surveys in very unsuitable conditions (say in an especially cold winter when atmospheric high pressure is dominant) and then have the problem of advising clients that while no contaminated soil atmospheres were encountered, this does not preclude the occurrence of much poorer conditions in a hot summer. Clients, naturally, are unamused by such caveats and it can be difficult to educate some individuals to understand that gas and vapour generation is a dynamic process which will peak in particular conditions. However, given the risks of gas and vapour entries into homes and structures and the concern with which insurers view this prospect, it is usually possible to persuade even the least aware of clients that risks are best identified before development decisions are taken. The costs and inconvenience of retrofitting gas protection works are extremely high.

8.5.4 Data consistency

Even a less than entirely adequate gas survey will produce a fair number of individual gas measurements. Better surveys, not uncommonly, give rise to hundreds or thousands of numerical values.

Before any use is made of this information, it is prudent to check that results are consistent, and that anomalously high values are not a consequence of instrument or human error. Such errors are not infrequent, and it would obviously be futile to invest analysis time on data which could be spurious.

One simple consistency check, when landfill gases are of interest, is to graph the measured oxygen and carbon dioxide readings. If carbon dioxide is being generated within a site then oxygen will be consumed, and a relationship between the carbon dioxide and oxygen measurements should be apparent. Departures from this relationship could be due to instrument or observer failures or could be a result of other deeper sources of carbon dioxide. Similarly the effects of gas migration can be apparent from such a graphical plot – for instance, if an abrupt rise in the measured methane concentration occurred and the accompanying soil oxygen levels were far lower than those which would correspond to the known carbon dioxide contents, it could be postulated that some oxygen had been physically dispelled by the inflow of methane under pressure. Such indicative information obviously is useful for interpretation.

Consistency checking thus not only gives an indication of data accuracy, but also offers insights into what is occurring within a site (Figure 8.5).

8.5.5 Gassing categories

A well-designed and conducted gas or vapour survey should always reveal which category of subsurface gassing conditions exists below a site.

Various categorizations are possible. ICRCL advice [20] is that a distinction can be made between low gaseous emission rates in older landfills (and most contaminated sites) and the rapid emissions of high volumes of pressurized gases more typical of modern landfills. Emberton and Parker [21] drew somewhat finer distinctions between old deep (5 m–10 m of waste thicknesses) landfills and sites more representative of much of

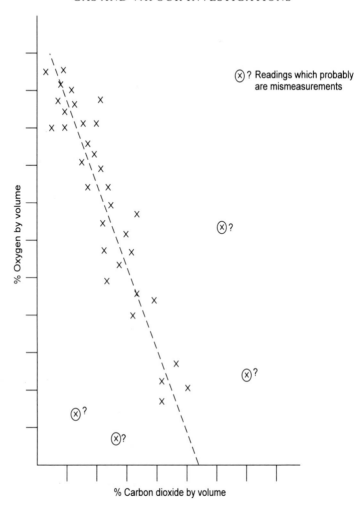

Figure 8.5   Checking the consistency of measured oxygen and carbon dioxide concentrations.

contaminated land. This last category was typified by the release of only small quantities of methane and the lack of any potential for future large-scale gas emissions. More comprehensive categorization is possible (Table 8.7) to distinguish between the more or less hazardous conditions which can occur within contaminated ground.

The most hazardous gas category (No. 4) is actually the easiest to identify and the one least open to interpretation and remediation doubts. In contrast, the least hazardous case (No. 1) is often more difficult to identify – since gaseous contaminants need not be sampled in some of the gas surveys. The larger an effect, the easier it is usually to measure and assess.

Table 8.7   Gassing categories found below contaminated land (in increasing order of hazard)

Category	Conditions encountered
1.	Gases produced intermittently, often in response to climatic/barometric changes. Flow rates/pressures usually trivial. Concentrations occasionally high.
2.	Gases produced continuously though production rate varies with climatic/barometric factors. Flow rates/pressures usually trivial (though not in highest emission periods). Gas concentrations often high.
3.	Gases produced continuously, though production rate varies with climatic/barometric factors. Flow rates/pressures/concentrations often high.
4.	Gases produced in large volumes and under significant positive head. Concentrations usually very high, and to the point where dilution with air is still likely to leave fire, explosion and toxicity risks.

8.6 Solutions to subsurface gaseous contaminants

8.6.1   Introduction

Card [16] provides the best coverage of the methods available to render gassing sites suitable for re-use; though this work is specifically directed at the risks from methane, other gas problems are solvable in the same ways.

Two basic points are stressed by Card and should be taken on board before any more detailed consideration is given to solving subsurface gassing problems, i.e.

- absolute safety can never the guaranteed for any remedial action which does not entail totally removing the source materials which are emitting the hazardous gases; all other techniques leave a risk (albeit small, if design and construction of remedial measures is well conducted) of future failure; and
- design methodologies for protective measures (other than for source removal) cannot yet be adequately quantified. Present knowledge on changes in material properties in the ground is inadequate, as is a complete appreciation of subsurface gas movement mechanisms.

Solutions include the following.

- Removal of gassing sources. This is not difficult for near surface and smaller volumes of gas emitting materials, and well conducted gas surveys ought to be able to identify such removable sources. However, for deeper gaseous nuisances (e.g. mine gases) removal is not feasible.

- Vertical barriers to stop gases migrating to areas where sensitive targets exist. Design of such barriers is somewhat uncertain, since methodologies rest on water migration flow principles; Jefferis has shown [10] that these need not be suitable in combating gaseous movements as impermeability against water flow need not mean that gas flows are similarly halted.
- Pressure relief methods to allow gases and vapours to escape into the atmosphere in areas where no risks/nuisances will be created. Essentially this is achieved by excavating vent trenches, or vent columns, which then are filled with very permeable material (single size stones or similar). Vents obviously will be more likely to prove effective when higher pressure gas conditions exist, but are unlikely to combat other gassing problems successfully.
- A combination of pressure relief and vertical barrier systems. This usually consists of a vent trench backed (on the side of the trench to be kept free from subsurface gases) with a low permeability barrier. These are widely used and appear to have been effective against both high pressure gas flows and also against the diffusive spread of gases from areas of high gaseous concentrations. Given the doubt over clay's gas impermeability, if drying out occurs, the use of a polythene or polyethylene low permeability backing layer will give a greater assurance of continued remediation effectiveness. Well-directed gas surveys should be able to identify conditions against which this solution will be satisfactory.
- Passive venting granular blankets (with or without gas collector pipes) are successful in dissipating gas concentrations from beneath houses and buildings of smaller plan area, and will combat quite high gas concentrations (methane contents of between 5 and 20% by volume) where only a low gas emission rate (less than 0.01 m/s) exists. However, their use below larger structures, or in situations where gas emission rates are higher, cannot be recommended.
- Active venting and extraction of gases and vapours is based on the proven principles of mine ventilation. As the solution depends on mechanical pumps continuing to work to specification, active venting certainly is not a fail-safe solution (as numerous coal-mine explosions have demonstrated), and it is prudent to include the installation of gas monitors and alarms in any structure so protected. Active venting, however, is the sole safe solution when larger structures have to be built over land which is generating higher outflows of gases.
- Gas proofing of house and structure foundations to preclude gas entries. As gases and vapours will, of course, still be able to affect the adjoining gardens, roads and any constructions (e.g. greenhouses) built after the gas proofing, this solution would not be appropriate in conditions where measurable escapes of gases from the ground are likely.

In terms of the widespread need to protect newly built domestic homes on reclaimed land several solutions are adoptable, and the appropriateness of each ought to be demonstrable from well designed gas surveys. Source removal, combined gas vents/barriers and gas proofing of foundations are the most reasonable solutions. Combined vents/barriers perform best when higher gas concentrations (driven by advective or diffusive mechanisms) exist, while gas proofing is more appropriate to conditions where only trivial gas flows are possible.

8.6.2 *Gas source identification*

Gas source identification is essential before any well-judged selection of protective solution is made.

In simpler cases (e.g. Table 8.5) the location, thickness and depth of a gassing source can be directly identified.

More commonly, however, no such certainty can be established and all that is known is that gases or vapours are being generated at depth, probably beyond that of gas observation holes. This (see section 8.9.2) creates difficulties; if the gaseous contaminants are methane and carbon dioxide do they arise from nearer surface biodegradation, or are they deeper coal-mine gases or emissions from deeper peaty deposits in the soil profile, or possibly from off-site sources? Additionally leakages from buried gas pipes might be another possible origin.

Distinguishing between methane and carbon dioxide atmospheres from such a range of sources is possible.

- British Gas distribution pipe records
- coal mining plans and
- geological/hydrogeological information

can suggest which sources are more or less probable. More detailed gas concentration measurements, using laboratory-based gas chromatography, can reveal the contents of ethane, propane, butane and helium trace gases and these (Table 8.8) can prove diagnostic, especially if carbon dating tests have been conducted. However, since dilution with air will often have occurred before gas concentrations have been measured in gas observation holes, it can be difficult in some cases to be certain which source is responsible for the measured gases.

If carbon dioxide is the gaseous contaminant which occurs, an equal range of sources could be the cause, i.e.

- pre-methanogenic degradation of organic wastes,
- gases rising from limestone and coal mines, and
- the slow degradation of heavier oil residues trapped in soils or contained in shallower groundwaters.

Table 8.8 Distinguishing the sources of methane and carbon dioxide measured in monitoring holes

Gas source	Properties
Landfills	Young gases (C-14 dating) Methane: carbon dioxide usually 60:40. Carbon dioxide more prevalent than methane on older landfills and on contaminated sites. Only minute traces of ethane and propane exist.
Peats	Intermediate aged gases (C-14 dating) Carbon dioxide more prevalent than methane. Ethane, propane and butane traces are significant. No odorants, of types found in mains gas, exist.
Marsh gas	As peat gases.
Sewer gas	As peat gases, albeit with younger age (C14 dating). Traces of hydrogen sulphide usually occur.
Mains gas	Old gases (C-14 dating) Significant ethane, propane and butane contents occur. Diethyl sulphide has been added to give a noticeable odour and is a diagnostic source identifier.
Mine gas	As mains gas, though without the diethyl sulphide odorant. Helium (in the range of 20 to 200 ppm) commonly occurs.

Source: after [7].

As with the methane and carbon dioxide associations, it can be difficult to be absolutely sure which of the possible types of gas emission sources is actually the cause (see section 8.9.4).

Other types of gases and vapours generally present lesser source identification difficulties (section 8.9.3).

Despite the problems often encountered in deciding which horizon or material is emitting troublesome gases, it is quite imperative to make every effort to come to a judgement. Until this is achieved it will be difficult to justify any chosen remedial strategy.

8.7 Legislative controls and official advice on redevelopment

8.7.1 Introduction

A wide range of advisory guidance and control measures has been developed for sites where landfill gas atmospheres exist below ground surfaces. So far, the same degree of guidance has not been prepared for such other gaseous hazards as carbon monoxide and volatile organic vapours, though this is likely to be produced as the scale of these problems becomes more apparent.

8.7.2 *Main restrictions*

The main restrictions on developing land where methane and carbon dioxide gases occur are contained in the following.

- The Building Regulations 1985[22] Approved Document C cites 1% methane and/or 5% carbon dioxide as requiring the provision of remedial measures. These requirements are limited only to the land below a building footprint,
- Various Department of the Environment and ICRCL guidance documents which strongly advise that housing developments should not be located on, or near, landfill sites which emit gases. To strengthen the application of this advice, a circular [23] warns planning authorities to exercise 'due caution' in granting development authorization and to insist that reliable measures must be included to manage migrating gases and minimize risks to future land users,
- Waste Management Papers [24,25], while actually devised to reduce the risks from landfill sites, have had a considerable impact on the development of contaminated land, which is affected by subsurface contaminated atmospheres. Earlier versions of these guidance documents emphasized the gas concentration values seen to be of concern, although latest versions give a more reasonable stress to gas flow rates and to the occurrence of high gas pressures as more accurate indicators of riskier conditions. Usefully, these latest editions note that residual gases are often trapped within degraded wastes, and that gas concentration measurements, *per se*, need not indicate continued gas emission or risk.

These various guidance and control documents (which are supplemented in action by the important insurers, such as the National House Building Council, attach to gaseous contamination below land used for development) can be expected to be updated and refined as further scientific information and case studies become available. Currently the consensus is that trigger levels (1% for methane and 1.5% for carbon dioxide) are the maximum allowable landfill gas concentrations in the ground. Developing sites with higher gas contents (and with meaningful gas flow rates or pressures) calls for remedial protective measures, which may not be acceptable to some more cautious planning authorities or to insurers. While the trigger levels cited above are distinctly cautious (section 8.3.3) and need not in fact indicate hazardous conditions which could affect developments at land surface, their use is justified by the fact that very rapid changes in gas flow rates, pressures and concentrations can arise, and that conditions much more hazardous than any measured in gas surveys could appear in future. However, if care has been taken to site gas observation holes and if surveys do include those conditions more likely to give rise to worst gassing effects, it can be possible to convince planners, regulators and insurers that land with gaseous concentrations can be re-used safely and without overly expensive remediation.

8.8 Gas emission predictions

While the risks from subsurface gases and vapours will usually be demonstrated by the measurement and analysis of information from gas survey facilities, some workers have attempted to calculate the ultimate volumes of gases which might be produced. This approach depends on estimating the volume of total carbon in a site and then assuming a realistic decay rate. Lord [26] initiated this approach in the UK because of the difficulty of measuring very low rates of gas flows.

While theoretically sound the method does call for a large number of assumptions (e.g. the actual amount of degradable waste in a variable tipped site, the ultimate methane emission volumes per tonne of dry waste, etc.) which are at best very difficult to make, or subject to technical uncertainties.

Given the present improvements in gas monitoring equipment, it is difficult to advise that emission predictions should be undertaken, or that they will offer additional useful insights.

8.9 Case studies

8.9.1 *Glasgow greenbelt*

A particularly simple and solvable landfill gas problem was encountered when the greenbelt around the City of Glasgow was moved further out to release land to meet a pressing housing shortage.

In one area of some 4 ha, the land was gorse and heather covered rough grazing, underlain by a thin (1 m–2 m) layer of sandy glacial clays with peaty horizons. Below this are massive coal measures sandstone rocks, which are usually highly weathered (to gravely sandy rubble) in their uppermost beds.

The presence of peat layers suggested that subsurface gaseous nuisances might occur, and a spike test survey was undertaken to confirm or deny this. Spike tests, at 15 m centres were driven to 1.5 m depths, or to rock head if this occurred at shallow depth. In addition to the concentrations of methane, carbon dioxide and oxygen in each spike hole, soil temperatures were also measured by an electronic thermocouple device.

This survey was necessarily carried out in an especially cold January, since the arrangements for the legal transfer of land ownership had been more time-consuming than had been anticipated. Thus, no significant biodegradation gaseous products were expected, and the client had been advised that the survey ought to be delayed until warmer soil temperatures developed in the spring. However, given the low probability of thin peats being able to generate gassing problems, the client's wish for an immediate gas survey was accepted.

This assumption (of no measurable biodegradation) proved to be correct *except* in two quite large areas, where carbon dioxide concentrations reached levels of up to 6%. Elsewhere only trace levels of this gas (<0.5%) were found. At the two anomalous localities, soil temperatures were found to be at, or near, 10 °C, in sharp contrast to the 1 °C to 1.5 °C subsurface conditions recorded in the rest of the site.

Shallow excavations then revealed the presence of bagged food wastes, of very recent age, obviously derived from a food processing source, and hidden below a 0.5 m thick cover of local soil. Presumably these bagged wastes represent illegal waste disposal in shallow quarries, as no official records of tipping existed.

Redevelopment of the land was delayed until the following summer, because of a downturn in the sales of private housing, and this allowed more detailed monitoring of the anomalous areas in warmer soil conditions and as atmospheric and climatic variables occurred.

This showed that, while no other gassing areas existed, landfill gases (methane) from the tipped wastes could and did migrate outwards (through the decomposed granular rubble at the top of the Coal Measures Sandstones) for distances of up to 300 m. Removal of the tipped wastes, however, entirely dispelled these gaseous contaminants.

In this case, shallow spike testing fitted the site conditions and was able to identify the buried food wastes only because active degradation was still in progress. Had the wastes been older and more fully degraded, it is probable that the initial gas survey would have revealed little useful indication. Excavation and removal of the gas source was economical because of its especially shallow location.

8.9.2 *Paper mill site, central Scotland*

This proved to be a much more complicated site, on which gas surveys were conducted at four times (April, May, July and October) to include the effects of warmer soil conditions, and also periods when barometric low pressures followed rapidly after atmospheric highs.

The site had housed a paper mill (now demolished) and included several areas where localized tipping of paper ash and wood pulp wastes had taken place. Below the surface, silty alluvial soils with thin peaty horizons were encountered over massive and open-jointed granite rock. South of the site is a deep railway cutting (15 m below site surface) and a modern local authority landfill lies some 300 m south of the railway line. Records exist to show that the landfill is within a granite quarry whose base is 12 m lower than that of the railway cutting.

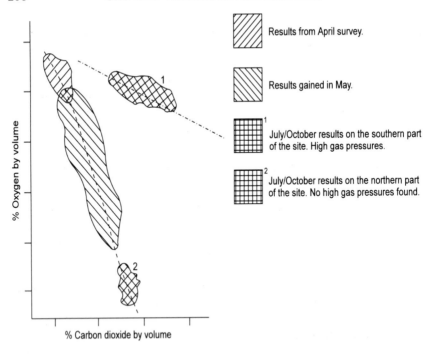

Figure 8.6 Paper mill case. Consistency checks which indicate the presence of two separate gassing sources.

Initial gas surveys (April and May) made use of spike test networks (10 m centres between spike holes) supplemented by 12 deeper boreholes in areas whose spike testing had revealed poorer subsurface atmospheres. These deeper boreholes were terminated at the upper surface of the granite rock.

The carbon dioxide and oxygen relationships revealed by these earlier surveys (Figure 8.6) followed the predictable pattern. The April survey showed little oxygen depletion and only slightly enhanced carbon dioxide concentrations. No methane or other flammable gas occurrences were recorded, and pressures in observation boreholes proved to be identical to those of the atmosphere. By May, oxygen depletion and carbon dioxide increases were more obvious, though the graphical relationship established in April still persisted. Methane contents (up to 1%) did occur in several observation holes, though gas pressures still remained minute.

This change was interpreted as a consequence of degradation of the tipped wood pulp and incompletely combusted paper ash, and appeared to indicate that only a single shallow gas source was affecting the site.

In July and more especially in October, however, matters became much more complicated. A very hot summer had occurred, and it had been assumed that this would only enhance the oxygen : carbon dioxide relationship already established. In fact (Figure 8.6) a quite different graphical pattern was obvious with very increased carbon dioxide concentrations co-existing with only slightly reduced oxygen contents. In addition, positive and high gas pressures were found (particularly in the southernmost bore-holes) and explosive concentrations of methane (up to 12%) occurred. During the October gas survey, even more anomalous results were obtained.

The pattern displayed on Figure 8.6 strongly suggests that two quite distinct gas sources affect this site, i.e.

- a shallow source of poorly degradable wastes (wood pulp and other tipped matter) and
- a deeper source of pressured gases with much higher methane and carbon dioxide contents.

Further work confirmed this view, and suggested that the deeper gases originated from the local authority's active landfill, and that these gases – when pressure conditions increased as soil temperatures rose – could force their way through joints and fissures in the granite rock to reach the paper mill site.

Resolving the problem, in principle, should not have been difficult. Shallower wood pulp and similar wastes could be identified and excavated. The migration of pressurized landfill gases through discontinuities could be intersected by a suitably located gas vent/barrier trench. Together these measures would have been adequate to permit a safe re-use for the paper mill land.

However, on practical and economic grounds, the feasible solutions could not be entertained. The vent/barrier trench would have to be excavated to depths of about 27 m, and the bulk of this would be in especially hard rock. Costs would have been out of all proportion to the final value of the land.

This case demonstrates that gassing problems can prove to be too expensive to solve in some more complex cases, and also illustrates the advantages possible from well-conducted and timed gas surveys. Had only the April gas information been available, and the site then redeveloped, substantial problems and risks would have been found at later dates.

8.9.3 *Former steel mill site, West Midlands*

Gases and vapours, other than degradation by-products, can prove to be the major concerns on sites where past industrial use has included spillages of hydrocarbon fluids.

One such case occurred on the land previously used as the by-products manufacturing area of a steelworks, where tanks and vats of lighter fuel oils, phenols and ammonium salts had existed, until the site was demolished in the 1970s.

Prior to its use for manufacturing, the area of land had been a convenient tip for excavated soils, combustion ashes, slags, scrap timber and other refuse from the adjacent steel mill. In consequence, land levels were raised by between 4 m and 5 m, and a shallow perched groundwater table had developed at the base of these fills. Natural deeper strata are mined out Coal Measures, and mining was at shallow depths (<20 m).

When the land came to be redeveloped an initial gas survey (of ten deeper boreholes located on areas of deeper and shallower fills) was undertaken. This, however, revealed a very confusing body of information, i.e.

- some boreholes showed only traces of carbon dioxide contamination without any measurable gas pressures; these instances occurred where timber and other poorly degradable materials were most obvious in the fills;
- others revealed carbon dioxide at consistently higher than atmospheric pressures, and could have originated from coal-mine workings or from subsurface geochemical effects; and
- a few, in areas where lighter oil storage tanks had existed, have rise to measurable concentrations of volatile organic compounds (mainly benzenes). These VOC contents could occur at high pressures, though when pressure readings were continued for periods of up to four hours, significant measurement variations were apparent.

This lack of consistency forced a fuller review of the available site investigation information and the siting of a large number of small diameter probe holes in areas thought to be likely to typify the various elements of the site.

In a manner similar to that described for the Dundee Paper mill site (section 8.9.2) it was possible to establish that two sources of carbon dioxide (without the presence of methane even in hotter soil conditions) did exist. The most widespread was the occurrence of low CO_2 concentrations (up to 4%) of unpressured gas, and these measured values varied in response to barometric conditions. No sizeable response to changes in soils temperatures was apparent. Wherever such conditions were encountered, easily excavatable shallow deposits of tipped timber, cloth and similar materials were obvious.

The different occurrences of pressurized carbon dioxide (at higher concentrations of up to 10% in worst cases) were found to exist only in small areas, where phenolic liquor tanks had stood. Around these surface vegetation was absent (despite the land having lain unused for more than 20 years) and both soils and shallow groundwaters proved to contain phenol contents of more than 200 ppm. These anomalous unvegetated areas seemed not to correspond to the recorded locations of coal mining. Thus it was concluded that geochemical breakdown of phenol-rich liquors was the probable cause of the measured gaseous contaminants. The depths to which phenol spillages had extended (up to 10 m), and the volumes involved if excavation were to be undertaken, were too great for an economic source removal solution. Fortunately, the land areas were not significant, and it could be possible to omit these from planned redevelopments.

The volatile organic compounds, however, proved to be much more hazardous and were found to occur whenever topographic variations brought the perched groundwater table closer (to within 3 m) to land surface. Photon ionization detector surveys revealed a wide variation in measured VOC concentrations, and these, when plotted on a site plan, outlined the probable flow paths of various zones of lighter hydrocarbon liquids. Trial pitting was able to confirm that where worst volatile organic vapour contents were found, the local perched groundwater had a visible floating 'oil' layer.

More detailed concentration and pressure measurements, conducted over several hours at a time and on subsequent days, revealed that VOC concentrations at a locality altered in amount very rapidly (0.3 to 151.6 ppm in one quite typical instance). This suggested that the VOC emissions were especially sensitive to quite small variations in ambient barometric conditions and also indicated that even less acceptable vapour concentrations would occur in warmer soil conditions.

Laboratory gas chromatographic analyses revealed that these vapours consisted of a varying mixture of benzene, toluene, xylene and naphthalene, and so might give rise to cancer risks if more sensitive human targets were exposed to the vapours.

While it was economically possible to deal with the two types of carbon dioxide contamination, full resolution of the volatile organic compound hazard would have called for the controlling, pumping and chemical treatment of the shallow groundwater and would have been a long and excessively expensive solution.

Thus the sole reasonably affordable solution was to advise that full gas-proofing measures (and the provision of vapour alarms in basements and smaller enclosed spaces) would be necessary. This, of course, precludes a safe domestic housing re-use for the land, given the difficulties in managing gas-proofing effectiveness in hundreds of individually owned properties, and forced development plans to be altered.

8.9.4 *Midlands foundry site*

Gas sources entirely unconnected with prior uses of a particular piece of land can occur and will often pose especially difficult interpretation problems.

One such case [11] was proved on an old iron foundry, where casting sands and other solid residues had been used to raise land levels by some 4 m. A few scraps of poorly biodegradable materials (mainly plywood and cloth) were visible in these fills, and this led to an initial spike test gas survey being carried out. As this revealed lowish (<4.2%) carbon dioxide concentrations and very minute traces (<0.2%) of flammable gases, more permanent monitoring boreholes were emplaced.

Initially four shallower (4 m deep) boreholes were sunk and revealed fairly widespread occurrences of carbon dioxide at concentrations of up to 3.5%. A later set of six rather deeper boreholes (8 m depths) was then drilled to check if gassing conditions in the lower sands and gravels (below the iron works fills) were acceptable. These showed a widespread and entirely unexpected oxygen depletion (down to a near zero content) associated with extremely high carbon dioxide levels (up to 19%). Given the concern these contaminant concentrations generated, it is not surprising that five deeper (15 m to 25 m deep) boreholes were then sunk into the underlying rocks. Table 8.9 shows that these produced very high (up to 27%) carbon dioxide results.

A pattern of the deeper observation holes producing far poorer subsurface atmospheres is, however, obvious from this table, and this suggests that the more worrying carbon dioxide results originated at considerable depths below the site.

More detailed monitoring (including measurements of gas pressures and flow rates) was undertaken over several months, and in various climatic and barometric conditions. Additionally flux boxes were installed at depths of 100 mm below the levels planned for domestic house foundations. This work revealed that

- in times of higher barometric pressures, sampling of shallower borehole atmospheres produced no significant carbon dioxide contents (all were less than 2.5%) and no measurable gas pressures, while,
- when very low atmospheric pressures occurred, all the boreholes gave rise to depleted oxygen and very enhanced carbon dioxide

Table 8.9   Gas survey results – Midlands foundry site

Borehole	Week 1	Week 4	Week 8
1	18/2/0.1	16/3/0.1	17/2.5/0.12
2	19/1.5/–	18/1.5/0.08	16/3/0.16
3	14/3.5/0.1	12/4.2/0.18	10/2.4/0.2
4	16/2.5/0.1	13/3.5/0.2	12/4/0.2
5	–	3.5/10/–	2.1/12/–
6	–	2.4/12/–	1.4/15/–
7	–	2.8/13/–	1.0/18/–
8	–	1.2/16/–	0.6/20/–
9	–	1.1/17/–	0.6/19/–
10	–	0.8/19/–	0.2/23/–
11	–	–	<0.1/27/–
12	–	–	<0.1/25/–
13	–	–	0.15/23/–
14	–	–	<0.1/23/–
15	–	–	0.12/26/–

Results reported in the sequence – oxygen, carbon dioxide and methane.

Boreholes 1 to 4 – 4 m deep
Boreholes 5 to 10 – 8 m deep
Boreholes 11 to 15 – from 15 to 25 m deep

concentrations, and small positive gas flows were measurable. These flow rates, however, declined as soon as the barometric pressure began to rise, and positive gas pressures then vanished, and
- flux box installations never showed any increase in carbon dioxide contents, even when abrupt barometric lows tracked over the area.

As the rocks beneath the site are the unmined shales of the Lower Carboniferous Series, any suggestion that coal workings could be the gassing source would be unsupportable. However, below these shales, and at depths in excess of 80 m, are massive limestone horizons, which formerly had been mined (1780–1830). Plans of the limestone mines still exist and showed not only that mining did take place at depths below the iron foundry site, but also that the mine entrances, in an adjacent valley, are still open. Thus a possible gas source which is in continuity with the atmosphere, and so affectable by barometric pressure changes, certainly exists.

Further investigation – including the filling up of some deeper boreholes with compacted bentonite pellets – confirmed the hypothesis, and indicated that if no stress relief holes are driven into the solid rock substratum, then only trivial carbon dioxide contents could be measured in the top 4 m of the site's profile. These carbon dioxide traces originate from poorly decomposable materials mixed in the site's fill-capping layer.

The point has already been made (section 8.3.2) that investigation boreholes alter subsurface conditions, and provide stress release zones into which gases can concentrate. With very heavy gases, this effect is especially misleading, and the use of deeper boreholes on this site increased investigation costs and durations beyond those which should have occurred.

Passive ventilation of house foundations was finally found to be the solution to gassing conditions, as toxicity risks from deeper carbon dioxide occurrences could be shown never to affect near surface soil layers.

8.9.5 *Leisure development, London Docklands*

In less easily interpretable cases, it may be necessary to carry out gas monitoring for many months before gassing conditions become understandable.

A planned leisure centre in the London Docklands proved to be a good example of this.

This site is underlain by a thin concrete and made ground cap over 8.5 m of alluvial clays with subordinate bands of peats and silts, and fronts on the River Thames.

Subsurface gas investigations were necessarily very comprehensive, given the size of the leisure centre and the commercial investment involved, and included numbers of large diameter observation boreholes supplemented by multi-point boreholes with piezometer response zones against the soil layers most likely to produce gaseous contaminants. In all some 36 boreholes were drilled over the 2 ha site.

Monitoring immediately showed that, on the eastern end of the site, up to 50% concentrations of methane could occur (though without any very enhanced carbon dioxide contents) particularly in warmer soil conditions. In contrast, boreholes on the western site of the site generally showed only uncontaminated soil atmospheres.

Since gas readings, taken as boreholes were being drilled, had indicated that measurable methane concentrations occurred to depths of up to 5 to 6 m below site surface, initial views were that perhaps peaty bands or more organic clays could be the gas sources. Because of this the multi-point bore-holes were added to the gas monitoring network. Several months of monitoring at weekly intervals took place to confirm this hypothesis, and concentration measurements were supplemented by gas flow rate estimations. Since in 1989 when this work commenced, direct measurement of smaller flow rates was difficult, use was made of the indirect nitrogen purging technique. In this, monitoring borehole internal volumes are flushed out with pressurized nitrogen, and the recovery rates of gaseous contaminants with time are measured (Figure 8.7a). From the steepest gradient of the gas (methane) recovery rate, the inflow rate of gaseous contaminant can be estimated [27] though it can in fact be difficult to entirely purge borehole volumes and then recovery curves (Figure 8.7b) can be unhelpful. Additional difficulties can arise if only small localized volumes of gases are contained in the strata around purged boreholes, and Smith [27] cites a situation where an entirely uncontaminated borehole atmosphere existed 33 minutes after the completion of nitrogen flushing, only to change to a 2% methane and 0.5% carbon dioxide atmosphere 25 minutes later, but then to revert to completely uncontaminated conditions a little later. In such circumstances, nitrogen purging flow estimates are meaningless.

After a considerable period of monitoring no unique stratum in the soil profile could with certainty be identified as the main source of the methane problem.

Attention was then directed to the geographic pattern of methane results and particularly to the fact that only in the eastern portion of the site did high methane concentrations occur almost all the time.

Historical research in the local archives revealed that this eastern area had earlier been an open stream, into which untreated sewage effluent had been disposed until the 1890s. With the stream's outflow to the River Thames controlled by the flap valve, which closed in high tide conditions, it became obvious that the former stream bed would have become silted up with coarser sewage, and that this might be the prime gassing source affecting the site. Further local enquiries showed that methane emissions from this (now infilled) stream were known to affect the inland area for distances of up to a kilometre from the River Thames junction.

With this information, the results of gas monitoring over the site's area became explicable. A north–south zone on the eastern site boundary was

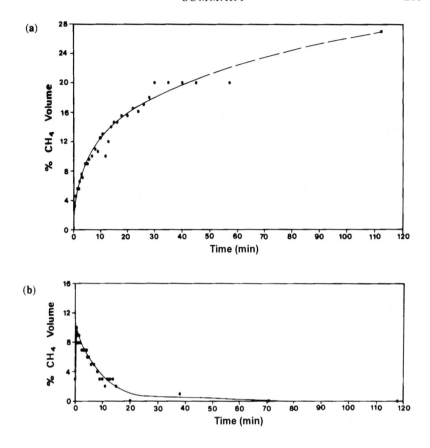

Figure 8.7 Plots of post nitrogen flush methane concentrations against time. For explanation of parts (a) and (b) see text.

the sole area where methane concentrations occurred and this corresponded with the predicted course of the infilled stream.

Exploration pits, to identify if excavatable organic wastes could be found in the infilled stream bed, were dug, and these very rapidly caused methane concentrations in adjacent gas monitoring boreholes to decline.

From this, it was concluded that gas venting works could resolve the situation and two trenches were excavated and found to remove the methane nuisance.

8.10 Summary

Not only can occurrences of subsurface gaseous contaminants give rise to especially rapidly acting risks to future users of reclaimed (and formerly contaminated) sites, but it remains extremely difficult to interpret gas survey results, and identify with certainly more hazardous situations.

This interpretation problem can never be entirely resolved, but can be reduced if properly timed and sited gas and vapour surveys are carried out, and continued for long enough periods to reveal more hazardous conditions.

Once gassing sources are known, remedial solutions can be selected, though there will be situations where remediation costs cannot be justified and where remediation effectiveness cannot be guaranteed.

Given the known catastrophic consequences which can arise if gases and vapours can collect in homes or enclosed spaces, surface gaseous contaminant occurrences have to be investigated with due care, and it always has to be borne in mind that gas surface results are indicative rather than direct measurements.

9
Establishing new landscapes

9.1 Introduction

Derelict land until recently formed a greater and greater part of our environment, as old industries ceased operation and new businesses decided that establishing their operations on a greenfield site was preferable to the possible inconvenience and additional expenditure involved in the reclamation of derelict, contaminated or despoiled sites.

However, land scarcities and the determination by local and national government to preserve greenbelt areas have changed attitudes. Today a significant proportion of redevelopment occurs on reclaimed and formerly contaminated land, and the expectations are that this source will be the dominant redevelopment opportunity in the very near future. If this is not done, then some 25% of the United Kingdom's land will be surfaced with concrete and tarmac within the foreseeable future.

Previous chapters have described the complexities of land contamination and the variety of techniques available for the reclamation of such areas. Removal of, or treatment to negate the effects of, the contaminants is, of course, only the first stage of the reclamation process. It then becomes necessary to establish a new and viable use of the land. Sites may become available for housing, commercial or industrial redevelopment, they may be required to support agriculture, forestry or recreational uses. Whatever end use is proposed, however, it is likely that some, if not all, of the site will require to be landscaped.

In order to ensure that the scheme is successful, the nature of the restoration, end use and landscaping have to be determined at an early stage of the reclamation process, and should involve the landscape architect in the reclamation team from the outset. This will enable the resources available within the site to be recognized and safeguarded for use at a later stage. Topsoil, for instance, is a valuable commodity, which can easily be damaged or destroyed by careless handling or contamination; its presence on site therefore needs to be recognized at an early stage and the resource safeguarded for use at the appropriate time. The lack of topsoil on a site need not restrict planting options, however, as will be discussed later.

No two sites are ever identical; each will require to be appraised on its own merits and a specific methodology determined, which in every case will take the following into account:

- site location: latitude, altitude, aspect;
- surroundings: urban/rural, visual quality;
- soil quality, content, alternative materials;
- vegetation type and cover, ecology, amenity;
- physical constraints: slope, drainage;
- end use;
- financial constraints;
- timescales available.

Sites can be considered as falling into two broad categories: (i) those where the contaminants have been eliminated; and (ii) others where the contaminants have either been treated in order to render them inert or encapsulated by means of a covering, or containing layer. In this latter instance, it is important to ensure that planting does not penetrate through the containment layer to the contaminated soil below; equally adverse would be the ensuing uptake of contaminants by the plants themselves, particularly if there was any possibility of these plants forming a part of the food chain.

In each case, it is necessary to consider the requirements of the types of vegetation (trees, shrubs, grasses or other herb layer plants) proposed on the site, to ensure that these needs can be fully satisfied in all respects, while still maintaining the integrity of the reclamation scheme.

9.2 Plant requirements

The basic requirements of all plant forms are well known:

- sunlight, for photosynthesis;
- the means of achieving some sort of anchorage in the ground;
- water and oxygen;
- nutrients.

9.2.1 Sunlight

Needless to say, it is unreasonable to expect plants to grow well in areas where sunlight is excluded by extensive overshadowing, from walls, under bridges or canopies. These sorts of locations are better served by treatments such as hard paving, or a shade-tolerant grass mix, although species of ivy (*Hedera*), elder (*Sambucus nigra*) and laurel (*Prunus laurocerasus*) will tolerate heavy shade.

9.2.2 *Anchorage*

Plants obtain support by anchoring their roots in the growing medium. The greater the ultimate size of the plant, the more anchorage it will require. The plant's root system is a continually growing and changing mechanism, and varies with plant species, age, season, soil structure and type. However, all root systems expand in order to maintain the supplies of essential nutrients, water and oxygen to the plant cells. To achieve this, the soil in which the plant is growing needs to have a good structure, with particle spacings of sufficient size to hold both oxygen and water, yet not so large that the soil is entirely free-draining. A very free-draining soil will fail to retain sufficient water for the plant to use, and will also give poor anchorage. It is therefore important to ensure that any soil, or soil substitute, which is loose in structure and which is proposed for use on site, has its structure and density improved by means of some moderate compaction or rolling; water retention can also be improved by the addition of substantial quantities of composts, mulches or similar products in advance of any planting taking place.

Soils are traditionally classified as either clay, sand or silt, or a combination of two or more of these categories. The texture of a soil is determined by the relative proportions of the mineral particles present (Figure 9.1); the soil structure relates to the accumulation of soil particles into larger compound units. Either subjective assessment or physical testing can determine these aspects and thereby the need for improvement to the soil prior to its use on site. Figure 9.2 indicates the range of soil types that are generally found, and their relative suitability for different end uses. From this it can be seen that the most suitable soils are those that combine two or more of the three different categories, and which are not exclusively composed of one or other of the principal soil types.

Compaction of soils, through trafficking by machinery or by the placement of an overburden, even temporarily, will damage the soil structure and restrict the diffusion of oxygen to the roots. There may also be implications for drainage, depending on the type of soil in question: sandy soils may continue to drain, as these soils have a higher percentage of macropores, whereas in clay soils the particles will coalesce on compaction and the number of larger pore spaces will be reduced. The opportunity for root growth will be restricted, therefore. Methods to reduce or remove compaction are discussed in more detail later. It is clearly most advantageous, however, if the compaction of soils that are to be used in landscaping can be avoided at the outset, by judicious site planning and careful routing of machinery.

The depth to which tree roots penetrate, and the depth therefore that is required for anchorage, is, for a wide variety of species, only the top 1 m or so of soil; very rarely will tree roots penetrate to below 2 m (Table 9.1). It is this sort of maximum depth of soil, or soil-forming material, that is required as a rule, within areas to be planted.

Table 9.1   The rooting depths of different tree species[a]

< 1.5 m	1.5–2.0 m	> 2.0 m
Ash	Cedar	Lime[b]
Beech	Fir	
Birch	Oak[b]	
Cherry	Poplar	
Crab apple		
Hawthorn		
Hazel		
Holly		
Horse chestnut		
Larch		
Maple[b]		
Pine		
Robinia		
Rowan		
Spruce		
Whitebeam		
Willow		

[a] Information taken from Gasson and Cutler, Tree Root Plate Morphology, *Arboricultural Journal* 1990 [1].
[b] Species with tap root.

Plant root penetration is, of course, a direct function of what nutrient supplies are available. With abundant and easily available food sources, root extensions are unnecessary beyond a minimal extent (as is obvious when hydroponic cultivation is practised). Thus if more fertile and physically suitable soils are provided, soil depths, thinner than those listed in Table 9.1, can

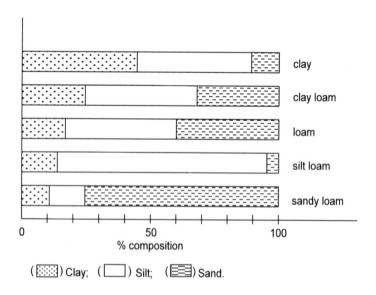

Figure 9.1   Soil texture and composition. Redrawn and adapted from Strahler [2].

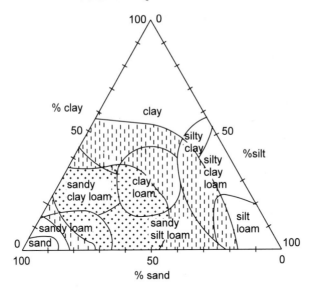

Figure 9.2   Soil texture and land use. Redrawn and adapted from Roberts and Roberts [3].

prove very suitable. The physical factor of most importance – the soil density – is considered in more detail below.

9.2.3 *Water availability*

All plants require water, which is absorbed by the root hairs, to enable photo-synthesis to occur. If the necessary water is not immediately available, the root system obviously has to expand if the plant is to survive. The water storage capacity of a soil is therefore important in ensuring the satisfactory develop-ment of the plant. This capacity is largely determined by the soil structure, and the presence of pores of a suitable size, namely macropores ($> 50 \mu$m dia-meter), which allow unimpeded penetration of roots, water and air; and meso-pores (5–50 μm diameter), which hold a reservoir of moisture available to the roots [4]. Sandy soils are, by their nature, free-draining and will store relat-ively little water (Figure 9.3), while clay soils are capable of holding consider-able quantities and can be prone to waterlogging. Too great a quantity of water in a soil can be as great a problem as too little, as it will result in the deprivation of oxygen to the root system, which will ultimately cause the plant to die. Too little water is, of course, equally detrimental to plant growth, and very free-draining soils, as previously stated, should be avoided unless large quantities of water-retaining mulches can be incorporated.

Soil water availability is related to the quantity of water held in the soil and the texture of the soil. The soil suction, or the pressure required to remove water from the soil, can be expressed in bars, and ranges from just above 0 bar, when the soil is fully saturated, to 15 bars, at which point plants are unable to remove any more water from the soil and wilting occurs. The field capacity of a soil occurs after a saturated soil has been allowed to drain (under gravity) for some 48 h and only the water held in the pore spaces remains. This will be the maximum amount of water that the soil can hold for any duration and, at this point, the soil suction is 0.5 bar. Figures 9.3 and 9.4 indicate the various characteristics of different soil types in relation to soil moisture levels, and demonstrate that while clay soils can retain more water than sandy soils, wilting can occur sooner in clay soils because much of the moisture in the soil's smallest pore spaces is under such high suction that it is unavailable to the plants.

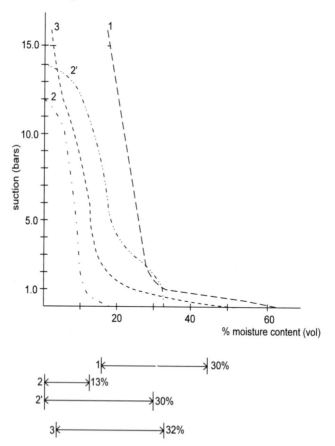

Figure 9.3 The moisture characteristics of certain soil types. Material 1: heavy London clay, low organic content, low permeability; material 2: crushed brick/rubble forming a sandy gravel, no organic content, very high permeability; material 2': material 2 with 20% organic content added, lower permeability; material 3; clean river dredging, moderate organic content, low permeability.

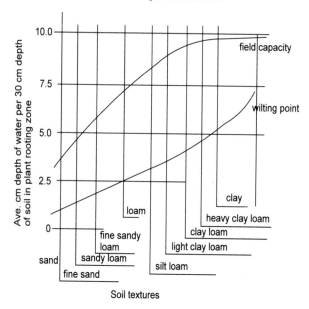

Soil textures

Figure 9.4 The relationship between soil texture and its moisture levels of field capacity and the wilting point. Redrawn and adapted from Strahler [2].

It is therefore important to ensure that any soil-forming material that is proposed for use as a planting medium is capable of both storing and releasing soil water to the plants themselves. Soils with a greater range of available water, i.e. which are not largely sand or largely clay, are therefore to be preferred.

The texture and structure of the soil, the potential for drainage provision, and the gradients of finished levels must therefore all be considered to ensure that adequate water supplies will be available. Irrigation systems and even manual watering of planting are likely to be uneconomical, therefore full consideration must be given to the movement of water across and through the site, at an early stage, to enable the scheme to be successful.

9.2.4 *Plant nutrients*

All plants need a regular and continuous supply of nutrients. Nitrogen, phosphorus, potassium and magnesium are the principal elements, together with certain positively charged ions such as calcium and sulphur and other trace elements including zinc, copper, nickel and water-soluble boron, which are only required in minute quantities.

Topsoil generally contains the majority of these nutrients by the very nature of its composition, which will include naturally degenerating plant and animal tissue. Subsoils and other generally inorganic materials, however, are much less likely to contain all of the elements necessary for plant growth. It is therefore essential that soils are chemically analysed at an early stage of the project, to determine their suitability to support plant growth and to provide an indication of the treatment necessary to rectify any deficiencies that may occur.

9.3 Soil cover

As the medium in which trees, shrubs and grasses are expected to grow, the importance of having a thorough understanding of the soil proposed for use on site, cannot be overemphasized. Evaluation at an early stage of the project will enable a variety of different options to be examined and the correct decisions to be made, taking into account factors such as end use, maintenance, financial circumstances and so on.

9.3.1 *Evaluation*

A thorough site survey, including excavations by borehole or auger, is essential at the outset of any project, to determine the nature of the substrate, which may well vary across the site. This should enable a picture to be built up of the extent and depth of the various materials, suitable and unsuitable, present on the site. Any materials that appear to have potential for re-use can then be analysed more fully, in terms of their texture, structure and chemical analysis (pH, nutrient content, salinity and so on). This should complete the picture and enable firm decisions to be taken regarding the ability of on-site materials to meet the requirements of the reclamation team.

Any material that appears to have some potential for use as a growing medium must always be considered in the light of the ameliorative work necessary to bring that material up to the standard suitable for the type of planting or other end use, for which it is required. Substances such as crushed concrete or brick rubble will be of little or no use unless they are improved by the addition of a water-retaining, nutrient-rich product. This aspect of soil amelioratives is discussed in more detail in section 9.3.3.

9.3.2 *Depth of cover*

There are three main considerations to be taken into account when determining the depth of soil cover required on a site.

Firstly, the presence (or otherwise) of contaminated materials which have been retained on site, for whatever reasons. Chapter 5 has described techniques for ensuring that there is no upward movement of these contaminants, by means of capillary break layers placed over the toxic substances. These layers are then themselves covered with a layer of subsoil, and occasionally topsoil, into which planting can be expected to take place. It will be important to ensure that there is no possibility of the plant roots penetrating through the break layer and absorbing the contaminants through their root systems. Not only would this be harmful to the plant but there could also be implications for the food chain, in addition to the loss of integrity of the break layer itself.

Such unwanted root penetration is, however, avoidable if underlying layers are compacted to an adequately high density (Chapter 5, section 5.5.3). When this is done, it should be realized that, with all a plant's roots confined within a single soil layer, this has to be adequately thick to ensure wind-firmness of the planting and suitably fertile to support the plant cover. Secondly, the type and nature of the planting proposed, which will itself relate to the proposed end use of the site, and the landscape into which the development has to fit. Areas that are merely to be restored as grassland will require a shallower depth of soil cover than will areas of tree planting, although agricultural grassland will require a better quality of soil than, for instance, rough grass for informal open space. Soil depths can also be reduced if only shallow-rooting species of trees and shrubs are proposed for the site, although it is then necessary to ensure that future management of the planting does not allow deeper-rooted species to encroach or invade the planted areas.

Nevertheless, an optimum overall depth of approximately 1 m of soil cover is generally accepted as being adequate for most planting; this can be reduced to 200–300 mm where grassed areas only are proposed. Of these overall depths of soil, at least 30% should comprise a proper topsoil that has the necessary humus, or moisture-retentive matter, providing a supply of nutrients to the plant roots. Where good grass cover is needed, on playing fields, agricultural land, golf courses and so on, a minimum topsoil depth of 150 mm is essential, and more will be required if arable crops are expected to be produced.

If the site cannot supply a full 1 m depth of soil cover (topsoil and subsoil, or subsoil only) then the species and size of planting proposed requires consideration. If the ground conditions below the soil layer are not expected to be harmful to plant growth, small whips and transplants can be planted (in pits if necessary). If the reverse is the case, then shallow-rooting species are to be preferred. Table 9.1 indicates the different rooting depths of some of the principal species of trees likely to be used in any planting scheme.

The third consideration is financial; where budgets are limited, then there is likely to be a need for minimal depths of soil cover. This should not, however, be at the expense of the overall integrity of the scheme. It would be preferable to revise the end requirements, if finances are constrained, rather than to produce a second-rate environment that is likely to require further restoration at some later stage.

9.3.3 *Soil ameliorants*

Because of the expense and general scarcity of good quality topsoil, all landscape schemes are having to examine ways in which a suitable growing medium can be obtained for use on site. It is clearly most cost-effective to make use of materials that are available, either on site or close at hand, or which can be imported at little or no cost. As previously stated, the ability of a substitute material to reflect good soil characteristics is all important: soil texture, structure, drainage capacity, pH and nutrient content all need to be considered.

Most soils found on reclamation sites will contain many of the general requirements stated above. The areas in which deficiencies generally occur are in the soil pH, its nutrient status, and occasionally in its water-retentive capacity.

The acidity or alkalinity of a soil can drastically affect plant growth as well as the microbial activity within the soil. Soils with a low pH are acid soils, which will support gorse and heathland vegetation; extremities of acidity can often be rectified by the addition of lime, although it is essential to ensure (through the process of soil evaluation) that there are no on-going chemical processes (such as the oxidation of iron pyrites in a colliery spoil material) that would give rise to continuing acidity.

More frequently, however, soils found on contaminated sites are alkaline in nature. There is no easy solution to excessive alkalinity. Where the pH is around 8, alkaline- or lime-tolerant plants can be grown, and can often form rich communities containing otherwise rare species such as orchids. Where soils exhibit such extremes of pH that they cannot be readily treated, the simplest solution is often to bury them.

The improvement of a soil's nutrient content is generally the most critical aspect in ensuring good plant growth. It is most important that the soil contains materials that will retain water, and which contain the necessary nutrients for good plant growth. Several alternative options are available, which may well be worth considering. The quantities of ameliorants to be incorporated will vary with the precise requirements of the particular soil: on average, a quantity of 25–30% of the total soil mass should be incorporated (equivalent to a 50 mm thick layer on top of a 200 mm thickness of soil); where soils are particularly poor then this figure might be increased to 40–50% of the total mass.

Sewage sludge. The addition of composted sewage sludge to generally infertile soil can be of benefit, as it is a good source of organic matter, nitrogen and phosphorus. The nutrients are generally released slowly and will provide an improved soil structure, with good water retention. The use of sewage sludge has been developed over recent years, with much improved techniques now available, making the handling of the material more acceptable.

To be of benefit as a mulch or water-retentive additive, the sludge needs to be incorporated in a dry form, to prevent the development of anaerobic conditions, and needs to be well mixed with either the infertile soil, or a binder such as straw, which will assist in improving the handling qualities of the material. The mixed sludge is stored under cover to dry, for a period of 6 months, which improves its spreading ability.

Composted sewage sludge is now available as a commercial product [5] and is a good means of recycling an otherwise unacceptable waste product. Where a site has an infertile soil medium, the use of such a product should always be considered after discussion with the relevant control bodies. However, in urban areas the sewage may well be heavily contaminated with heavy metals and will therefore not be suitable for use. It is essential that a full chemical analysis is carried out to confirm the acceptability of the sewage product and its compatibility with the soil with which it is to be incorporated. There is strict legislation in place nowadays that controls the use of sewage sludge.

Sludges are more usually available as a wet product than a dry one, simply because they are costly to dry out. However, when dried, they are significantly reduced in volume, and are accordingly cheaper to transport. Consequently, the two costs may balance each other out. Wet sludges may be difficult to incorporate into soils and will not add the water-retentive bulk of dried or composted sewage. This must be borne in mind when specifying this type of material.

Spent mushroom compost. Another waste product that may be available locally is spent mushroom compost. This material is also a good source of nutrients, particularly nitrogen, and can either be used as a surface mulch or can be mixed into the soil to improve the nutrient content. It is a very lightweight material, however, and is not recommended for spreading as a surface mulch in windy conditions!

Shredded bark. This is another recycled product that is widely available. As a soil ameliorant, it is of limited nutrient value, but will improve the structure of clay soils, is a useful mulch for suppressing weed growth and gives valuable protection against moisture losses, if laid as a 50 mm thick surface dressing.

Farmyard manure. This is a nitrogen-rich material which, like sewage sludge, is a good source of nutrients for plant growth. However, its use may be limited in that it is unlikely to be available in sufficiently large quantities to be a viable option.

River dredgings. These can be a useful soil substitute if they are available locally at a minimal cost, and if they are free from toxic metals. It will, however, be necessary to analyse the dredgings at the outset as they could have a

high saline and/or metals content, if the river is tidal or has been contaminated from another source; in this case they may be unsuitable or
uneconomical (or both). As with sewage sludges, the material will be highly
saturated and will need to be allowed to drain over a permeable base for a
few months before it is able to be readily handled.

Whichever type of soil ameliorant is used, it is essential to ensure that its
qualities are suited to the type of soil that is to be improved. Sandy soils will
require a different formulation to clay soils, for example, and this may
require careful consideration to ensure that the correct balance of soil properties is achieved.

9.4 Soil fertility

It is important to consider soil fertility in both the long term and the short
term. Soil ameliorants will provide the necessary nutrients over a specific period of time (18 months or so) but if conditions are largely hostile to the
encouragement of microbial activity, then the nutrient supply could eventually be seriously depleted. Depending on the proposed end use of the
site, it will be necessary to assess future plant needs in order to ensure that
these can be adequately met.

The use of land for agriculture or forestry will be the most demanding in
terms of plant nutrient requirement. Grassland established for pasture or
silage will require regular and frequent applications of fertilizer, particularly
nitrogen. The number and size of applications will depend to a large extent
on the quality of the land in question, and will be determined by analysis.
Forestry plantations will also benefit from annual applications of nitrogen,
unless nitrogen-fixing plants such as alder, gorse, broom, or lupins are
planted or sown to assist in raising the nutrient levels of the soil. It is important to note that the presence of specific organisms is essential for lupins
to carry out nitrogen-fixation; these organisms may not always be present in
poorer soils.

Conversely, the presence of fertile soils is disadvantageous when establishing 'wild flower' meadows or grasslands. These species-rich herb layers are
best established on poorer soils, where the more competitive weeds
and grasses will be less inclined to overpower and smother the slower-grow-
ing varieties. Applications of fertilizer will not, therefore, be necessary, unless
these are applied solely to any trees or shrubs planted within the wild flower
areas. With this in mind it is important to remember that grass cuttings from
areas sown with a wild flower mix should be raked off and removed from
site, to maintain low fertility levels within the soil. Such wild flower
vegetations have become popular on many reclaimed sites in recent years,
simply because fertile soils are locally scarce and all that is available is
sandy or gravely subsoils. The sole disadvantage of these often extremely

attractive landscapes is the very high costs of wild grass and flower seed mixtures, which tend to be twice as expensive as those for ordinary plantings.

Soil fertility is not only related to its nutrient content, however. Equally important is the establishment of good microbial activity within the soil, to restore a more natural balance to the lifecycles therein. The earthworm is particularly important in this respect, as it will mix the soil, aerate it and provide root and drainage channels to the benefit of any plant life, as well as being of importance in the wildlife chain generally.

British Coal's Opencast Executive have carried out research which has demonstrated that soils with a good earthworm population are vastly superior to soils where there are few, or no earthworms present. Typically, such soils are found on newly reclaimed land [6]. This is often because the earthworms normally present in the topsoil are largely killed off when this soil is stockpiled in large mounds; the only worms to survive will be those present in the topmost layers of these mounds. When this soil is eventually spread back over the area, the remaining earthworms are likely to all be concentrated in one place, leaving the rest of the site barren. It can take years for the worms to migrate across a large area. In the absence of a good earthworm population such areas will probably drain badly, and may be anaerobic.

British Coal's solution to this problem is to take thin slices of the top layers of stockpiled topsoil, and lay them in narrow strips, up to 50 m apart, across the restored land. This allows the earthworm population to multiply naturally, to the benefit of a wider area; numbers reach a near normal level more rapidly than would otherwise be the case.

The research described has also demonstrated that land management practices have an influence on the development of earthworm populations. Drainage by subsoiling is important for deep-burrowing species; food supply is also a controlling factor and, where the grass sward is grazed rather than cut (with organic manure applied rather than mineral fertilizers), and cultivation is avoided, then conditions are generally more favourable to the establishment of an earthworm population.

9.5 Site preparation

The design of any reclamation proposal may well include the creation of new levels and contours across the site area. This is often the case where contaminated materials have to be buried within the site. It is important to create natural-looking mounds wherever possible, and the gradients and heights of such features should always reflect the topography of the wider area.

The reclamation of contaminated sites will inevitably involve the movement of machinery across much, if not all of the area concerned. This can be particularly damaging to the soil structure as the weight of the machinery will cause severe compaction of the surface layers of the soil, reducing the size of pore spaces and leaving the land more susceptible to waterlogging. The use of machinery across wet soils is likely to worsen the problem and this should therefore be avoided as far as possible.

In order to relieve compaction, and also to improve the soil's drainage capacity, the technique of subsoiling, or ripping, is used. This involves the use of single, double or winged tines, pulled through the subsoil layer (in advance of any topsoil being spread) in a down-slope direction at a depth of between 450 and 750 mm, and at centres of between 1.5 and 3 m, depending on the type of soil in question. Surface water drainage is also improved when the gradient of the surface is at about 6° or 1 in 10. The length of slope that can be adequately drained in this way varies with the soil type but 50 m has been found to be acceptable, with a ripped channel, a gravel or rubble toe drain, or a ditch installed to carry the water off the site.

Erosion and slope stability are problems that can often arise on slopes where the soils have recently been replaced. The establishment of a vegetation cover is useful in preventing surface erosion; ensuring that slopes do not exceed gradients of 20° or 1 in 3 is also helpful in this respect. Problems associated with slope stability are generally water-related, so it is advisable to ensure that water drains away at the surface rather than below ground. It is particularly important to maintain the soil cover at the desired gradient where this is overlaying and protecting contaminated materials beneath.

A number of commercially available mattings and geotextiles can be useful in assisting vegetation establishment while ensuring slope stability. These can consist of natural materials such as coconut fibre, coir or straw combined with a polymer mesh, which will degrade slowly over a period of years, or of more permanent materials such as polypropylene, which will add durability to surfaces while still allowing vegetation to grow through. The cost implications of the use of these types of materials need to be set against the potential cost of any reconstruction of slopes and vegetation, should erosion prove to be a problem. This is particularly important where the incorporation of steep slopes is unavoidable.

9.6 Establishing grass cover

Having decided on the ultimate end use of the site, it is likely that some, if not most of the site will require to be grassed. The type of grassland to be established will vary, from agricultural pasture to golf course fairway or playing field, to rough grassland and open space or wild flower meadow. Each type of sward will require different establishment and management techniques, if it is to be successful.

Cultivation of the surface layer is essential in all instances to create a tilth in which the grass seed can germinate. Stone picking may also be necessary, to a greater or lesser degree, depending again on the end use. Areas where the grass will be closely mown should have all stones larger than 30 mm removed, whereas with areas of rough grass this figure can be increased to 50 mm. The quality of the cultivation can also be varied according to the requisite end-product. A rough tilth is adequate for wild flower grasses while a fine tilth is necessary for playing fields and other quality grassed surfaces.

A pre-seeding fertilizer should be incorporated where good and rapid establishment and growth is required. The principal exception to this use of fertilizers, both pre- and post-seeding, is with wild flower mixtures and their associated grasses, where any sort of fertilizer will encourage the incursion of the coarser, more competitive grasses, which will ultimately eradicate the original species and reduce the diversity of the sward.

The selection of an appropriate grass seed mix will depend on the desired end use, the site location and the soil condition. Agricultural swards must contain productive and nutritious cultivars, while playing fields or fairways must be durable and have a good colour. Mixes are available for the entire spectrum of soil and site conditions, from acid to alkaline and from damp to shady; low-growing, low-maintenance cultivars are increasingly popular where maintenance costs need to be kept to a minimum. Where soil fertility levels are low, or where the soil pH is low, then species such as flattened meadowgrass, browntop bent, sheep's fescue and hard fescue can be grown.

There are also cultivars of the various grass species commercially available that are tolerant of a variety of contaminated soils; the use of these may be appropriate in some instances. Cultivars of perennial rye grass, such as Wendy, are tolerant of the leachates produced by controlled waste; cultivars of slender creeping red fescue, such as Merlin, will grow on soils with a wide range of pH values and particularly on alkaline soils contaminated with heavy metals such as lead, zinc or copper. Browntop bent cultivars such as Parys Mountain, on the other hand, grow well on contaminated neutral to acidic soils. As research and development continues, these cultivars will be continually refined; it is important therefore to seek qualified advice to obtain the most suitable grass seed mix for any one site, particularly where the soils are atypical.

Sowing rates will also vary according to the location and the requisite end-product. Agricultural land should be sown at around 50 kg/ha; playing fields and closely mown swards at around 300 kg/ha; and wild flower mixes at around 50 kg/ha. Rates may need to be increased by up to 50% where ground conditions are poorer than normal, and increased levels of clover can also be included to improve the soil's nitrogen content.

The timing of operations can be critical to the good establishment of grassed areas. In the majority of instances, sowing should take place in spring (April–June) or autumn (September). These periods are preferred because soil conditions will be warm, without being too wet or too dry.

Sowing in late September can delay the first growth until the following spring, by which time excessive damp, snow and so on may well have reduced the chances for good establishment. The exception to this is again for the wild flower seed mixes, which are likely to contain species that require a cold period for vernalization prior to germination the following spring.

The correct management of grasslands is essential to the success of any scheme. It will inevitably take some years for substitute soils to reach normal levels of fertility and organic content, and returns from agricultural land will be correspondingly low for the first few years. The continuing application of organic manures and fertilizers is therefore essential, as is the on-going analysis of the soil to monitor its improvement and enable the management techniques to be adjusted accordingly.

Where a close sward is required, for playing fields and so on, regular cutting is essential to encourage tillering. Grass cuttings can be left on the surface for the benefit of the soil where the cuttings are not so long that they smother the grass underneath, and where improving the soil fertility is important. As grass cutting can be an expensive maintenance item, the use of low-growing and low-maintenance species should always be considered. Areas of wild flower grasses, after the initial establishment period when up to ten cuts may be required in the first year, are generally only cut twice annually, in mid-July and late September, and the cuttings removed to maintain low fertility levels.

9.7 Establishing trees and shrubs

Areas of tree and shrub planting within a reclamation scheme will add height, colour and pattern to the environment, with seasonal changes contributing to this variety and interest. The choice of species will vary according to the location, soil condition, and the function of the planting – for screening, protection, amenity and nature conservation value. Planting will ideally reflect any existing vegetation so that the reclaimed area will ultimately blend in with its surroundings.

The use of native, indigenous species of trees and shrubs is increasingly popular as a result of a greater awareness of the need to protect our environment. Such species are generally of greater value in providing food and shelter for birds and insects, as well as providing autumn colour and fruits. Nevertheless, ornamental and amenity-type planting can be important in certain locations to add more varied colour, flowers and so on, around buildings and in an urban context.

If woodland planting is proposed then it will be necessary to consider the natural succession of the vegetation, and to plant species that reflect all the layers of plants found therein, from shrub understorey up to the forest canopy. Careful planning at the design stage can reduce the amount of management ultimately needed by allowing nature to take her course. It may also be preferable to phase planting over a number of years, to give a mixed age structure to the woodland.

Successful planting begins with the use of good quality plant material, although careful handling and planting is also essential. The size of plants to be used should also be given due consideration, as smaller material will ultimately catch up in growth with larger plants, and can often adapt better to any inhospitable site conditions. Small plants will also be cheaper and can therefore be planted in greater numbers, so that any early losses will be less noticeable. For this reason, whips and transplants are the most commonly planted sizes of trees, although where a more immediate impact is required then light or selected standard trees can be specified.

Whips and transplants are generally planted either into pits, or by 'notching' into the prepared ground. In every instance it is important that the root system is spread out fully and that the plant is set into the ground at the same level that it was growing in the nursery. If a pit is excavated for the planting hole then it is essential that the sides and base of the hole are loosened, to enable the root system to expand outwards. The pit should be backfilled with a mixture of soil, compost and slow-release fertilizer to help establishment; often a water-storing polymer, in granular form, is also incorporated to provide water to the root system during dry periods, avoiding the need for watering during the maintenance period, which can otherwise be costly.

Depending on the size of plant material to be used and the finances available, whips and transplants can be planted at between 1 m and 2 m centres; shrubs are planted at between 1 and 2 per m^2, depending on their ultimate height and spread. Planting at 3 per m^2 will ensure a rapid covering of an area by the shrubs, although thinning may be required ultimately, depending on the nature and form of the material planted.

Stakes are usually only required for plants taller than 1200 mm, and on exposed, windy sites. It has been generally established that the use of stakes for longer than the first year can be detrimental to the planting, as the plant stem fails to thicken at the base. Short stakes are preferred, as they permit the crown of the tree to flex in the wind.

The timing of planting operations is also important. Although the planting season is usually from November to late March, planting in November or early December is often more successful as the soil is still fairly warm and the plant roots can establish somewhat prior to the post-December rain and snow. Soil conditions often deteriorate into January and beyond, and can become totally saturated by the end of the planting season, which again will limit working on the land due to the greater risk of compaction. Planting should not be carried out in frosty conditions, or when ground temperatures are below 3 °C.

Any plants grown in containers can be planted at virtually any time of the year, although they will need to be well watered through any dry periods if

they are planted in the summer, unless a water-storing polymer is incorporated into the soil around the base of the plant. Evergreen species are usually
best planted between April and October, and again will require regular
watering to ensure good establishment, if water-storing polymers are not
used. It is good practice to water in areas of new planting at the time of planting, particularly as the soil should be dry for the planting to be able to take
place. Water-storing polymers also require a thorough soaking at the outset
if they are to work efficiently.

In many areas, it will be essential to protect new planting from possible
damage caused by rodents or small mammals, or by vandalism if the area
is prone to trespass or adjacent to public rights of way. Large areas of planting are best protected by enclosing them with fencing, to which can be
attached rabbit-proof mesh if required; deer will need to be controlled by
higher fencing. Individual plants can also be protected by the use of guards
or tubes, where this is more economical than fencing.

9.8 Maintenance

The correct maintenance, or on-going management of any planting scheme is
essential to the success, or otherwise, of the project. Without this, all monies
expended on the planting itself are likely to have been wasted. The first 3–5
years of any new planting are critical to the future development of the landscape; it is during this period that the majority of plants will fail if they are
not adequately maintained. This is particularly important when the planting
comprises smaller material such as whips and transplants.

The young plants, as stated earlier, require a good supply of water and
nutrients to promote growth. Any other vegetation in the vicinity of the
plant will therefore be competing for the soil's water and nutrients, and reducing the quantities available to the plant itself. Regular weeding, to remove
such competing vegetation from around the base of the plant, is therefore
essential; a 300 mm diameter of clear soil is the usual recommended area.
The application of a layer of mulch or compost can be effective in keeping
weed growth to a minimum. Where plants are spaced at 1 m centres (or
less) there could be a conflict between the need to clear the soil around the
base of the plant, and the need to maintain surface vegetation cover, particularly on sloping sites, to reduce the risk of erosion. With plants spaced at 1.5
or 2.0 m centres, however, this becomes less of a problem.

If conditions during the summer are dry, then ample quantities of water
will need to be applied across the planting area. Twenty-five litres per square
metre or per plant is recommended where a water-storing polymer has not
been incorporated into the soil during planting. Additionally, plants should
be inspected to make certain that they have not lifted out of the ground, and
firmed back in if necessary. Where soils are likely to be generally infertile,

then the regular application of compost, fertilizer or manure, once or twice yearly, is advisable.

Depending on the density of the planted material and the ultimate form of the planted area, any significant numbers of dead trees and shrubs should be replaced during the subsequent planting seasons. Should widespread losses be encountered in any particular area, then it may be necessary to investigate the reasons for this, as it may indicate problems within the soil related to the original contamination.

As time goes by and the planting becomes better established, provided that the soil fertility levels have improved, then regular maintenance should become less important. Areas may only require the occasional visit, to remove litter or other debris, strim unwanted weed growth and so on. If the scheme has been designed to provide a natural succession of plant growth, using native species, then the planting will develop naturally without a significant need for human input. Areas planted for forestry may require thinning to enable the ultimate tree crop to flourish. More ornamental planting may, on the other hand, require annual pruning to promote stem colour or flowering. Whatever the particular needs of any one area, a long-term management scheme can be drawn up to highlight the necessary work and evaluate the financial implications.

9.9 Species selection

When designing the landscape of a site, the type and form of planting will be determined by such factors as end use, the nature conservation value of the site, its relationship with its surrounding environment, amenity and visual qualities, and the degree of management to which the site will be subjected.

The actual species selected for use on the site will also depend on a variety of different factors: the location of the site (latitude and altitude) and the degree of exposure to wind and sun; the soil type, pH and drainage; visual interest such as autumn colour, flowering and berrying; value for nature conservation as food sources or habitat; low or high maintenance requirements and so on.

Exposed coastal sites in the north of Scotland, for example, will be greatly restricted as to the species of trees and shrubs that may be successfully specified. These could be restricted to the forestry-types of conifers such as pine and spruce, native broadleaves such as ash, whitebeam and poplar, and to gorse, broom and the Scottish roses (spinosissima). A site in the Midlands or South of England, on the other hand, which is sheltered from the prevailing winds, will have a much wider range of species to choose from, including the major native hardwoods such as oak and beech. The planting of elm is still not recommended, as although this tree is starting to regenerate naturally in the southern half of the country, it is still too early to be certain that the trees will be resistant to the Dutch elm disease.

9.10 Natural regeneration

The methods for establishing vegetation on contaminated or reclaimed sites, outlined in the previous sections, assume that there is a positive requirement for a particular end result on the site in question, within a short timescale. However, it is relevant to mention that nature will generally take a hand in the revegetating of land, if the site is left undisturbed. Depending on the particular location of the site, the soil type and nutrient content, regrowth can be expected to commence very quickly, with a low, herb layer forming the initial vegetation cover. All sites below the tree line will ultimately revert to woodland, the original vegetation cover across the United Kingdom, provided there is no interference from outside agencies.

The natural succession of vegetation within an area can be of great significance to nature conservation, supporting a wide variety of plant and animal life. Contaminated sites in particular, where the soils can often be very different to the norm, frequently give rise to rare and unusual plant habitats [7].

Alkaline soils, such as result from sodium waste, power station ash or other lime-rich waste products (e.g. those produced by iron smelting and steel-making), give rise to a species-rich flora that can often contain a large number of rare and unusual species, particularly orchids. These plants, in turn, are host to a wide range of insects, particularly butterflies and moths, which feed and breed on the flora. The time span involved in the creation of these sorts of habitats is usually in the order of 10–20 years, however.

Acidic soils are generally the result of colliery works and mining spoil heaps, which may frequently contain high levels of such metals as lead, copper and zinc. The natural regeneration of these types of soil is often more difficult as the wastes are generally toxic to plant growth until the metals have been broken down by the gradual process of soil weathering and leaching. Any plant growth will generally resemble heath or moorland, with plants such as heathers and gorse establishing. Some species of grasses and other plants are tolerant of particular metals and may also establish across the area.

Regeneration is less of a problem on soils with more normal pH levels, and soils with a higher nutrient status will rapidly establish a sward of coarser, more competitive rough grasses, such as are regularly found on roadside verges or beneath hedgerows, and so on.

Where drainage problems create permanently wet or damp areas, a different habitat will establish, with reeds, rushes and other marsh or aquatic vegetation developing; this can also attract a variety of interesting fauna. Wetland areas that lie above the water table and which do not contain a natural lining of impervious clay soil will eventually dry up, however, as the growth of rushes develops across the area and the quantity of dead vegetation builds up within the pond or marsh, unless management techniques are exercised to remove this material at regular intervals.

9.11 Conclusion

The establishment of a successful landscape across a previously contaminated, reclaimed site is one of the principal keys to ensuring a successful future for the site in question, both environmentally and economically.

To ensure that any proposals are successful, site planning must be undertaken from the outset in order to pay full regard to the materials and other qualities available within the site, the surrounding environment, the nature of the proposed end use, and any financial constraints that may exist. This will allow the preparation of solutions that fit both site and budget.

The flexible nature of landscape techniques will allow for a number of alternative solutions to be prepared, to fit the range of budgetary or other constraints, using a variety of materials, plant sizes and species and planting methods as necessary. In all instances, however, the ultimate impact on the local environment must be the most important factor to be considered. If the new environment fails to blend with its surroundings or to suit the type of use to which it is to be put, it will not be successful.

Landscaping is a process that takes all of the environmental features and rearranges them into a shape and form appropriate to the location. To enable this to be achieved, the landscape architect should be involved in the reclamation process from an early stage; otherwise the end treatment is likely to be merely cosmetic: a mask that may, in time, slip and then require additional, and possibly costly, reparative treatment.

10
UK legal framework

10.1 Introduction

Since the late nineteenth century when Victorian society began to take an interest in the health of the populace, the body of law relating to public health and in particular to contaminated land has steadily increased.

Concern over the well-being of the environment and a wider appreciation of the world ecosystem in our present century, has given added impetus to legal controls with the result that the contaminated landowner and remediation practitioner are now faced with a complex legal framework. Not only is that framework complex, it is continually changing as the law evolves. As a result it cannot readily be presented in simple check-list form for use as an accessible guide by landowners and their consultants.

Such a basic guide, however, is needed and it is the objective of this chapter to provide a guide to the major aspects of the relevant law relating to England and Wales. Readers will appreciate the need to obtain their own legal advice when dealing with specific circumstances, as this chapter is not intended to be more than a general and necessarily much simplified guide to the subject as a whole. The law is stated as at mid 1996.

10.2 Sources of law relating to contaminated land

In the United Kingdom, there are two primary sources of law, Statute and Common Law, and two secondary sources, delegated (or subordinate legislation) and case law. Non-statutory Guidance Notes and Codes of Practice also exist; while not law in themselves, these often have a bearing on the outcome of court action when the question arises as to whether or not reasonable steps had been taken by the defendant.

10.2.1 *Legislation: primary and subordinate*

Primary legislation. The primary source of law is legislation by Parliament in the form of Acts of Parliament. Law arising from this source is also referred to as Statute Law and is said to be 'on the Statute Book'. Acts of Parliament generally arise at the instigation of the Government of the day, which will lay a Bill before Parliament for its consideration. That Bill is essentially a draft of the proposed law and, following debate and perhaps modification during its passage through Parliament, it will eventually emerge as an Act of Parliament. Once 'on the Statute Book' an Act of Parliament may become law immediately or may be implemented in sections over a period of time. Bringing it into effect may occur over several years and parts of the Act may never be brought into force. This process makes the law difficult to discern as it is no simple matter to determine at any one time which parts of an Act of Parliament are in force and which are not.

Subordinate or delegated legislation. Acts of Parliament commonly include provision for Ministers to provide the detailed requirements of parts of the Act or indeed to bring into effect parts of the Act itself at some time in the future. These powers are exercised through subordinate legislation and are used to provide the many detailed rules and requirements that are necessary to allow the Act to function as intended. Subordinate legislation takes the form of orders or Statutory Instruments.

10.2.2 *Case law*

Case law is the description of that body of law that arises as a result of the decisions handed down by judges in individual court trials. (These decisions interpret both Statute and the Common Law and 'fill in the gaps'.) For example, the rules for negligence were set down by Lord Atkin in the famous negligence case of *Donoghue* v *Stevenson* and similarly the rules that apply to 'things kept on land' arise from the decision in the nuisance case of *Rylands* v *Fletcher*.

Decided cases (i.e. case law) are then used by the Courts as precedents for determining the outcome of subsequent cases of a similar factual background. The process provides some certainty in the pretrial prediction of the decision likely to be delivered by the courts in individual cases.

10.2.3 *Common law*

Common law is the name given to that body of law that sets out those rights and remedies that are founded in long-established principles and are not embodied in Acts of Parliament. This area of the law may be considered to be static in that new rights and remedies are not constantly being created or removed. It is not, however, stagnant since, in applying the common law, the Courts have a limited ability to adapt existing rights and remedies to accommodate new circumstances. Actions at common law relating to contaminated land generally arise in nuisance (see section 10.3.3).

10.2.4 *Guidance Notes and Codes of Practice*

Guidance Notes, Practice Notes and Codes of Practice are issued by Central Government in relation to virtually all aspects of contaminated land. Well-known examples are the Waste Management Paper series issued by the Department of the Environment and Her Majesty's Inspectorate of Pollution and the Guidance Notes issued by the Interdepartmental Committee on the Redevelopment of Contaminated Land (ICRCL).

These documents have no legal standing except that they may be adduced as evidence to establish whether or not a party exercised due care. It should be noted, however, that compliance with these documents does not necessarily ensure discharge of a legal obligation. This principle was clearly stated in the introduction for the Draft Code of Practice on Duty of Care issued in September 1991 by the Department of the Environment, The Scottish Office and the Welsh Office:

> The purpose of this code is to set out practical guidance for waste holders subject to the duty of care. It recommends a series of steps which would normally be enough to meet the duty. The code cannot, however, cover every contingency; the legal obligation is to comply with the duty of care itself rather than the code. Breach of the duty of care is an offence.

10.2.5 *European Union regulations, decisions and directives*

Membership of the European Union (EU) by the United Kingdom carries with it the obligation to conform with law made by the Union. The Union utilizes three types of legislation:

(a) *Regulations*. These are directly applicable law in Member States and are therefore enforced in the National Courts of each State. Regulations are principally used to give effect to uncontroversial or precise measures that require no further legislation to give them practical effect.

(b) *Decisions*. These again are directly binding but, unlike regulations, are generally addressed to specific bodies and not the Member States at large. Decisions are often used in environmental policy with regard to international conventions and procedural measures.

(c) *Directives*. These are the main vehicle for the implementation of Union policy and are binding on Member States. They specify the results to be achieved and timescale for compliance while leaving the precise form and method to the Member State. In the United Kingdom, most EU Directives are given effect by Ministers using Statutory Instruments under delegated powers derived from Statute. A recent example is the EU Industrial Plant Emissions Directive 84/360 brought into the UK legal framework by Statutory Instrument SI 1991 No. 836, the Environmental Protection (Prescribed Processes and Substances)

Regulations 1991, made under powers conferred on the Secretary of State by Section 2 of the Environmental Protection Act 1990.

Figure 10.1 gives a simplified illustration of the process by which European Policy becomes law in the UK.

It is apparent from the above that while an EU Directive is not law at the time of issue, it provides a direct pointer to what the law will be within the ensuing 18 months to 3 years. All those involved with contaminated land thus need to be able to keep abreast of Union policy to enable them to make decisions in the present and thus avoid expense and/or liability in the future.

10.3 The law relating to contaminated land

10.3.1 *General outline*

In sections 10.3.2 and 10.3.3 the principal areas of law that affect contaminated land are identified and briefly reviewed. Specific aspects are examined in greater detail in section 10.4 placing emphasis on practical guidance rather than legal analysis. It will be appreciated that in a book of this nature, it is not possible to deal with every detailed legal requirement and facet, and when particular proposals or transactions are contemplated it will be necessary to seek specific legal advice. Cognizance of the areas covered, however, should assist in identifying that need.

10.3.2 *Principal legislation*

The principal legislation relevant to contaminated land may be identified as follows:

(a) Health and Safety at Work Act 1974
(b) Control of Pollution Act 1974
(c) Special Waste Regulations 1996
(d) Occupiers Liability Acts 1957 and 1984
(e) Building Regulations 1985
(f) Collection and Disposal of Waste Regulations 1988
(g) Control of Pollution (Amendment) Act 1989
(h) Water Act 1989
(i) Town and Country Planning Act 1990
(j) Environmental Protection Act 1990
(k) Controlled Waste (Registration of Carriers and Seizure of Vehicles) Regulations 1991
(l) Water Resources Act 1991
(m) Environment Act 1995
(n) Landfill Tax Contaminated Land Order 1996.

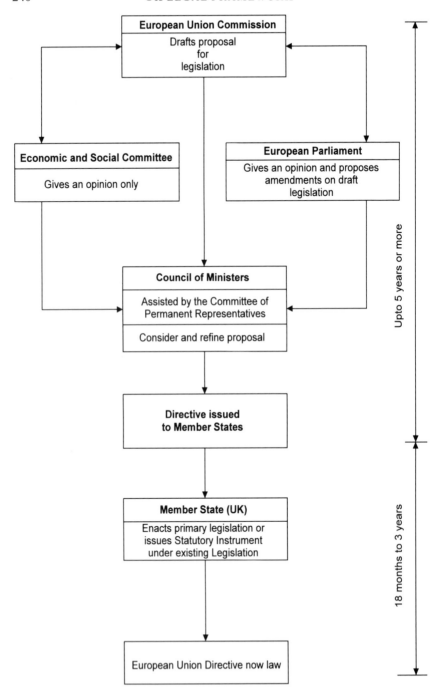

Figure 10.1 Implementing European Union policy.

(a) *Health and Safety at Work Act 1974.* This Act requires employers to ensure that their workplace is as safe as is reasonably practical for both employees and visitors. In the context of contaminated sites this may not prove to be of direct relevance to landowners but it is obviously important for developers who will need to carry out reclamation works on the land.

(b) *Control of Pollution Act 1974 (COPA 1974).* This was the major enabling Act for the control of waste disposal in the United Kingdom and it provided the administrative framework for the collection and disposal of waste materials. It is now largely superseded by the Environmental Protection Act.

(c) *The Special Waste Regulations 1996 SI 1996/972 as amended by SI 1996/2019.* These regulations, made by the Secretary of State under powers conferred by the European Community Act 1972, the Control of Pollution Act 1974 and Environmental Protection Act 1990, define certain wastes considered difficult to deal with as 'special waste'. The regulations define the term 'special waste' shall apply to any controlled waste which is either:

(a) one for which a six-digit code in the hazardous waste list set out in Part 1 of Schedule 2 and which displays any property (toxic, carcinogenic, etc.) specified in Part II of that Schedule; of particular interest is a new property 'ecotoxic' substances and preparations which present or may present immediate or delayed risk for one or more sectors of the environment;

(b) any other controlled waste, other than household waste, that displays properties, H3A 'Highly flammable', or H4 'Irritant', H5 'Harmful', H6 'Toxic', H7 'Carcinogenic' or H8 'Corrosive' specified in Part II of Schedule 2 if it satisfies the threshold in Part III of that schedule; or

(c) is a medicinal product, as defined, in Section 130 of the Medicines Act 1968 (c) (meaning of a 'medicinal product', etc.) of a description or falling within a class specified in an order under section 58 of that Act (d) (medicinal products on prescription only).

Any testing for properties given in Part II of Schedule 2 is to be carried out in accordance with Annex V to Council Directive 67/548/EEC, as amended (c).

The regulations require the use of a consignment note system to provide an audit trail for the disposal of the waste and the consignor, carrier and consignee each to maintain, respectively, a register of each consignment of special waste removed, transported and received. The depositor of any special waste on land is to keep site records of the location of each such deposit.

(d) *Occupiers Liability Acts 1957 and 1984.* A general duty of care is placed upon occupiers of land by these Acts to ensure that visitors are kept as reasonably safe from injury as practicable. It is to be noted that the duty applies to 'occupiers' and not simply landowners and that the duty, which is one of reasonable care, extends in a qualified form to trespassers.

(e) *The Building Regulations 1985.* As subordinate or delegated legislation under the Building Act 1984, the Building Regulations describe in detail standards of construction to be adhered to in the construction of buildings. They are principally aimed at securing reasonable standards of health and safety for persons in or about a building and others who may be affected by any failure to comply. It is therefore at the time of application for Building Regulation approval that the limited requirements would affect contaminated land. One would normally anticipate that the amelioration of contamination had been achieved by means of planning conditions, but in the event of permitted development, for example, the Building Regulations provide a further opportunity for control.

(f) *Collection and Disposal of Waste Regulations 1988 SI No. 819.* These Regulations were made on 3 May 1988 under powers conferred by the Control of Pollution Act 1974, and came into effect fully on 3 October 1988. They provide guidance as to the classification of waste, and the circumstances when a disposal licence is not required but are not applicable in Scotland. Of particular interest to owners and practitioners in land restoration is Schedule 6 to the regulations, which itemizes instances in which a waste disposal licence is not required for the deposit of specified controlled waste.

Power to prescribe cases when controlled waste may be deposited, treated, kept or disposed of without a licence is conferred upon the Secretary of State by S33(3) Environmental Protection Act, and detailed is the Waste Management Licensing Regulations 1994.

(g) *Control of Pollution (Amendment) Act 1989.* This Act provides in Section 1 for it to be an offence 'for any person who is not a registered carrier of controlled waste, in the course of any business of his or otherwise with a view to profit, to transport any controlled waste to or from any place in Great Britain'. Sections 2 to 4 of the Act make provision for the registration of carriers and Section 6 makes provision for the seizure of vehicles used for the unlicensed disposal of waste. The Act was given effect on the 14 October 1991 by Statutory Instrument 1991 No. 1624. The Controlled Waste (Regulation of Carriers and Seizure of Vehicles) Regulations 1991 require all carriers to be registered on 1 April 1992.

(h) *Water Act 1989 (WA 1989).* This Act created for the first time a single authority (The National Rivers Authority) responsible for water quality in England and Wales. Section 107 of the Water Act 1989 (re-emphasized in S85 of Water Resources Act 1991) makes it an offence to cause or knowingly permit the pollution of ground or surface water and the National Rivers Authority is charged with enforcing this and other such provisions. The NRA (now the Environment Agency) is given an important power under S115 of the 1989 Act (S161 of the 1991 Act) to require the clean-up of the sites that are causing water pollution.

(i) *Town and Country Planning Act 1990*.   With certain exceptions, 'development' (as defined by TCPA) requires the prior approval of the local planning authority. Development is often construed as including remedial work on contaminated land, whether or not subsequent built development occurs. When called upon to determine a planning application, the authority is required by S70(2) TCPA 1990 to have regard to the development plan, so far as material to the application, and to any other 'material considerations'. The Planning and Compensation Act 1991 has inserted a new S54A into the TCPA, which requires the Authority to make the determination, in such cases, in accordance with the Development Plan unless material considerations indicate otherwise. It is under the head of 'material considerations' that the presence and effect of contaminated land must be addressed. Circular 21/87 and Planning Policy Guidance Note 1, January 1988 refers. Four decisions are possible:

- refusal;
- unconditional permission;
- permission subject to conditions;
- permission (with or without conditions) subject to the applicant first entering into an agreement with the authority under S106 TCPA 1990.

The Grant of Permission for development of a contaminated site usually involves the imposition of a condition that no construction work commences until the local Planning Authority is satisfied with the remedial treatment. Circular 21/87 gives guidance to authorities on this point.

In the event that a refusal is issued on the grounds that the site is contaminated the owner may serve a purchase notice on the authority under Sections 137–8 of the TCPA 1990 on the grounds that the site has become incapable of reasonably beneficial use in its present state. Acceptance of the notice or its confirmation by the Secretary of State would result in the authority being obliged to purchase the land.

(j) *Environmental Protection Act 1990 (EPA 1990)*.   Enacted to make provision for the improved control of pollution arising from certain industrial and other processes, the Environmental Protection Act 1990 contains three sets of provisions of relevance to contaminated land. Part II introduces a new regime for waste management to replace Part I of the Control of Pollution Act 1974. The most important new provision imposes a duty of care on all those who handle waste at each stage of the journey from production to disposal.

(k) *Controlled Waste (Registration of Carriers and Seizure of Vehicles) Regulations 1991 SI 1991 No. 1624.* Made by the Secretary of State on 17 July 1991 under powers conferred by the Control of Pollution (Amendment) Act 1989, these regulations came into force on 14 October 1991. The Regulations bring into effect the 1989 Act and set out the requirements and mechanics of registration of carriers and seizure of vehicles. All carriers of controlled waste are required to be registered on 1 April 1992.

(l) *Water Resources Act 1991.* This consolidation Act, which draws together existing legislation, replaced the Water Act 1989 on 1 December 1991. The corresponding sections of both Acts of particular relevance to contaminated land are set out Figure 10.2

(m) *Environment Act 1995.* Section 57 of this Act creates a new regime for the control of the environmental problems from contaminated land, and – for the first time in UK law – provides a legal definition for 'contaminated land'. This requires land remediation only where contamination presents unacceptable risks to human health, property or the wider environment (S. 78A(2)). Local authorities are required to inspect and identify contaminated land within their areas and base any such identification on a defensible risk assessment which takes note of reasonably predictable pathways via which contaminants might affect specified receptors or targets. The test that 'significant harm' (S.78A(5)) must be probable before any land may be identified as 'contaminated' is especially severe, and will inevitably limit the numbers of sites so designated. Section 10.4 examines in more detail the probable consequences of this latest strengthening of the UK's environmental legislation.

(n) The Landfill Tax Contaminated Land Order 1996. Landfill tax was introduced on 1 October 1996 under The Landfill Tax Regulations 1996 which implemented the provisions of the Finance Act 1996 for the taxation of landfilled wastes. The collection of the tax is being undertaken by the Customs and Excise who have issued a number of guidance notes.

The landfill tax is currently set at £2.00 + VAT per tonne for inert wastes and £7.00 + VAT per tonne for active wastes. As this tax is considered a 'green tax' and will be easy to collect it is to be anticipated that these taxation rates will rise in the future both to raise government tax revenue and encourage recycling of wastes.

In respect of contaminated land the government has made an important concession by exempting from tax landfill wastes resulting from the redevelopment of historic contaminated sites where polluting activities have ceased. Formal application with supporting documentation has to be submitted to Customs and Excise at least 30 days in advance of the works commencing. The area to be treated and timescale for the works have to be clearly defined

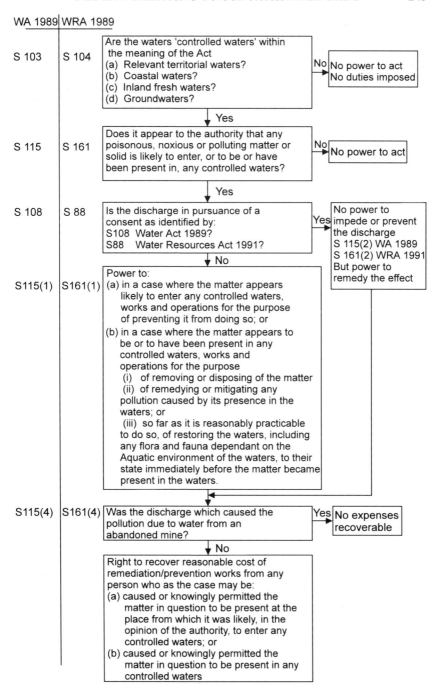

Figure 10.2 Powers over water quality preservation.

and names of the landfill sites to be used provided. Within the defined area even uncontaminated wastes can be disposed of tax free. However any subsequent development infrastructure construction wastes disposed to landfill, even if contaminated, will be taxed.

The government made this concession on contaminated land to avoid blighting brownfield sites by increasing development costs. The remediation industry has however complained that this exemption has lost an opportunity to encourage research and development of in situ contamination treatment technologies by penalizing landfilling. If the government decide to encourage recycling in the future this exemption may be revoked.

10.3.3 *Common law*

The most important common law duty arising from the ownership of a contaminated site is based on the tort of nuisance. There is private nuisance (which involves the unlawful interference with another person's use or enjoyment of his or her property) and public nuisance (which means an unlawful act that endangers the lives, safety, health or comfort of the public). There are some fairly subtle distinctions, all of which need not be pursued here. Of more immediate relevance is the species of nuisance represented by the rule in *Rylands* v *Fletcher*. This involves strict (i.e. no fault) liability and therefore warrants special attention. The rule was stated thus:

> The person who for his own purposes brings on his land and collects and keeps there anything likely to do mischief if it escapes must keep it at his peril and if he does not do so is prima facie answerable for all the damage which is the natural consequence of its escape.

Therefore in order for this principle to apply there must be:

1. non-natural use of the land;
2. an escape;
3. damage to the plaintiff's property.

This has obvious applications to land contamination, although it is arguable that conventional industrial use lies outside the ambit of the case. It has been asserted that for *Rylands* v *Fletcher* to apply: 'it must be some special use bringing with it increased danger to others and not merely to be the ordinary use of the land or such use as is proper for the general benefit of the community'. However, a relatively recent Canadian case held that the escape of landfill gas (which caused an explosion in an adjoining housing development) arose from the 'non-natural' use of the landfill site.

This suggests that *Rylands* v *Fletcher* may have wider application to environmental hazards than at first appears. If this is true a plaintiff will only need to establish that loss or damage was caused by the non-natural use. Strict liability means that it will avail the landowner nothing to show that he took all prudent steps to prevent the loss or damage occurring. One could, on the other hand, consider *Cambridge Water Company* v *Eastern Counties Leather plc and* v *Hutchings & Hardings Ltd* (Queens Bench Division, 31 July 1991) in which it was held that the storage of solvents was not an unnatural use of the land in question and therefore the rule in *Rylands* v *Fletcher* had no application. It would appear that this area of law is in a state of flux.

The other important common law duty arises from the tort of negligence. Since the 'neighbours' to whom a duty of care is owed have been widely construed, and since the presence of dangerous contaminants demands a high standard of care, negligence might perhaps be thought to offer plaintiffs more scope than nuisance. However, it is essential to establish fault (i.e. a failure to act with reasonable prudence) and although the principle of 'res ipsa loquitur' (the thing speaks for itself) can be invoked in an environmental context (i.e. the escape of toxic material would not have occurred but for a failure to take appropriate precautions), the adherence to technical and professional advice should afford the landowner an effective defence. Even if negligence could be established, there has to be some physical damage for the plaintiff to succeed. Economic loss alone is not sufficient.

10.4 The Environment Act 1995

The importance of this legislation, which specifies how land qualities are to be assessed and the processes by which the parties responsible for necessary remediation costs (or some part thereof) will be identified is obvious and justifies as detailed a consideration of the 1995 Act's intentions and probable applications as is possible at present.

Questions likely to be raised by interested parties are posed and answered below to highlight the more salient provisions.

Is this new Act really important? Yes. It will be far easier for regulators to use than were the Environmental Protection (1990) and Water Resources (1991) Acts, and owners or occupiers of suspect land are far more likely to encounter regulatory interest than was previously the case.

Does the Act intend to compel the clean-up of every industrially affected site. No. The aim is to achieve a reasonable balance between necessary environmental protection and the proper re-use of land for new productive purposes. Only land likely to generate significant risks will have to be remediated.

What type of land is to be remediated? Only land which can be deemed to be 'contaminated'. The Act (S. 78A) defines this as land which appears to the local authority in whose area it is situated to be in such a condition, by reason of substances in, on or under the land, that

(a) significant harm is being created or there is a significant possibility of such harm being caused; or
(b) pollution of controlled waters is being, or is likely to be caused.

Land which earlier housed industrial activities will usually contain enhanced chemical concentrations in soils and waters (because of poor past practices over waste disposal and process linkages) but this alone will not be enough for the 'contaminated land' judgement to be made. Unless the land is likely to generate 'significant harm' or create water pollution, regulators will be unable to make use of the 1995 Act's provisions.

If 'significant harm' is the crucial test, precisely what does this mean? The word 'harm' is itself very loosely defined (S. 78(4)) as: 'harm to the health of living organisms or other interference with the ecological systems of which they form part and, in the case of man, includes harm to his property'.

If this were the sole guidance, landowners and occupiers would have reason for concern, since human occupancy inevitably interferes with natural ecosystems. Fortunately, the legislation goes on to specify that any actual 'harm' (or the potential for its future occurrence) has to be 'significant', and insists that regulators must follow the Secretary of State for the Environment's specific guidance in determining what might be 'significant'.

While this guidance has yet to be published in its final form, draft guidance (issued in September 1996) does exist and provides a revealing insight into governmental intentions.

Instead of the vast range of adverse occurrences which could feasibly arise, the draft guidance allows for *only* four cases in which regulators can claim that 'significant harm' already exists, as follows.

- Where humans have experienced chronic or acute toxic effects, serious injury or even death. Quite obviously if any land creates those conditions its use is entirely unsafe and remediation is long overdue.
- Where irreversible or other adverse changes to the functioning of natural ecological systems have occurred. However, since this *only* is to be applied to locations which already possess environmental protection in law (i.e. under the Wildlife and Countryside Act 1981, or the European Union's wild bird (EC Directive 79/409/EEC) and habitat (EC Directive 92/43 EEC) conservation regulations, i.e. the Conservation (Natural Habitats, etc.) Regulations of 1994) no actual extension of legislative protection is intended.
- Where buildings, plant and equipment suffer such substantial damage that they become unfit for their intended usages. As with the human health risk test, this surely is reasonable. If land is of such poor quality

that it destroys building foundations or allows large-scale settlement to occur and this leads to structural collapse, then that land's deficiencies call for immediate correction.

- Where the chemical quality of land causes such substantial damage to crops or livestock established there that these are no longer fit for their intended purpose. Again this must be uncontentious; land so toxic that large-scale animal death or crop failure result requires remediation.

The draft guidance specifically denies regulators the right to consider harm to other possible targets, or to take note of other (lesser) damage to the receptors noted above.

If the final guidance document reflects these initial views it is obvious that the Secretary of State intends that regulators only should focus on land of such poor quality that its re-use, without remediation, is quite unacceptable.

While it seems simple to identify 'significant harm' where this already has occurred, how is the potential for future 'significant harm' to be judged?
Obviously this has to be less clear cut. Contaminant assemblages in soils are extremely variable, as are contaminant concentrations. Thus the Secretary of State's draft guidance only indicates that the probability of 'significant harm' arising is more likely in the following circumstances.

- When contaminant concentrations are very high compared to guidance values, and likely to adversely affect a large human population. The situation where a new housing estate is planned for land which contains very enhanced levels of metals, cyanides, polyaromatic hydrocarbons or sulphates in near surface soil layers is probably a case in point.
- When contamination is of a type known to be able to generate risks within very short times. Gas entries into homes can reach explosive or asphyxiating levels in a matter of a very few hours or days, thus land emitting meaningfully high gas or vapour flows would be likely to be seen as having a high probability to cause future 'significant harm'.
- When the contaminants, while not at very enhanced concentrations, would be likely to generate a serious risk even if only short duration exposures occurred. An obvious example of this would be the presence of abundant asbestos fibres in a friable surface soil, since more sensitive human subjects would then be prone to carcinogenic ailments.

If, as is obviously intended by the Secretary of State, regulators accept the allowable 'significant harm' situations specified earlier (for cases where harm has already arisen) and use similar criteria in identifying where future risk is probable, many 'brownfield' developments should be exempt from regulatory interest, after an initial inspection.

Since the former National Rivers Authority developed a reputation for initiating legal actions, will the frequencies of these become more numerous, under the 1995 Act? Probably not. Pollution of controlled waters will be assessed exactly as was the case under the Water Resources Act 1991. Since the Secretary of State has clearly indicated how 'significant harm' is to be judged, for non-aqueous contamination, it can be assumed that water quality regulators will adopt a very similar set of judgmental tests, which should not result in more frequently mounted legal actions.

What does all this mean in practice? Government intends to compel the remediation of more hazardous areas of land, where contaminants could enter the food chain or where very rapidly acting or especially mobile contaminants exist. Land which is more likely to be targeted should include:

- sites emitting measurable outflows of flammable, asphyxiating and toxic gases and vapours;
- land where subterranean smouldering and burning is probable;
- areas which contain especially harmful health-affecting substances in near surface layers (i.e. asbestos fibres and cancer inducing polyaromatic hydrocarbons);
- sites where mobile or soluble substances could affect adjacent water courses or deeper aquifers; and
- land which would be more likely to contaminate garden produce or commercial crops, and harm people or animals who eat these crops.

What, presumably, will be exempt will be these very commonplace sites where only very slightly enhanced near surface contamination occurs.

Who will actually be making these judgements on land quality? The local authorities (advised by the Environment Agency when necessary) will judge whether or not land is 'contaminated'.

How will the judgements be made? Local authorities are charged with the duty to periodically inspect land in their areas, but are advised both to deal with 'pressing and serious' problems soonest, and not to duplicate the work of other regulators. Logically this suggests that local authorities will look for gas emission problems adjacent to landfills and areas where uncontrolled tipping has taken place (both of which an authority should already have identified under earlier legislation) and for toxicological risks on land which was used for metal working, asbestos tipping or other similar heavy industrial use.

Local authorities are required by the Act to note the pollution linkages, i.e.

contaminant source → pathway → contact with receptor or target

and so must defend their judgements that land is actually 'contaminated' on a rigorous risk assessment basis (Chapter 2).

Does this mean that local authority personnel are likely to enter and inspect private land?   Yes, but precisely the same powers of entry featured (for England and Wales) in the Environmental Protection Act 1990 and the Water Resources Act 1991. Regulators were also granted the right to collect samples for analysis in these earlier laws and will continue to have this power. Landowners and occupiers in Scotland who – previously – were not exposed to these entry powers will be faced with a more intrusive control regime.

Can it be expected that local authorities will aggressively seek out areas of land which might be contaminated?   No. authorities' resources are finite and subject to other demands. The Secretary of State obviously intends that authorities should inspect the worst (i.e. more hazardous) sites first, and pay less attention to other potential targets. Since the 'pollution linkage' test has to be applied, large areas of land will be exempt simply because their present usage does not expose one of the targets, allowed by the draft guidance, to 'significant harm'. For example, an area of land, well away from a river course and not above an aquifer, can be capped with soils which could generate chronic health risks to humans and still be outside of the reach of the 1995 Act, because no people currently live there.

How will local authorities decide if land is 'contaminated'?   Only four categories of 'contaminated land' are legally allowable, i.e.

- where actual significant harm has been identified;
- where the probability of significant harm arising is high;
- where controlled waters are being polluted at present; and
- where the potential for water pollution is probable.

As, in each case, the local authority's judgement has to be supported by a risk assessment which demonstrates that the

$$source \rightarrow pathway \rightarrow target$$

linkage is unbroken, landowners or occupiers will be provided with documentary evidence to substantiate any 'contaminated land' decision, and will have the right to appeal against this within 21 days of receiving the decision.

Quite obviously, this process places substantial demands on a local authority's resources, and will be difficult for many to undertake since few have staff adequately experienced in land contaminated matters. It thus is most unlikely that authorities will casually expose themselves to these difficulties.

Will regulators take note of existing pollution abatement works?   Yes. Local authorities, specifically, are denied the right to claim that a significant possibility for harm exists to controlled waters if adequate risk management arrangements (e.g. bundings around oil storage tanks or leachate cut-off works on a landfill's perimeter) already exist and are in good working condition.

Can landowners expect that local authorities will carry out intrusive site investigations on private land? No. Local authorities could only expend scarce funds on site investigation works if it is plausible that land is contaminated *and* if no other source of detailed information is available. Landowners can, and should, seek to obtain information on their land's subsurface conditions and contaminated state, and provide this to local authorities. This has the advantage of the landowner exercising a measure of control over a process which ultimately could lead to a large land reclamation bill.

If, from desk studies and available site investigation results, a local authority judges that land is 'contaminated', what then will happen? Local authorities, at this point, have two tasks to undertake, i.e.

- to establish who is the appropriate person(s) responsible for the contamination; and
- commence discussion with that person(s) to establish what necessary land remediation should be carried out.

Remediation Notices should not be served until these processes have been completed.

Is the imposition of a Remediation Notice inevitable once an authority identifies that land is 'contaminated'? No. Most landowners or other 'appropriate' persons will be sensible enough to discuss the position with the regulators and negotiate the least costly necessary land quality improvement and so avoid the imposition of a Remediation Notice.

Once a Remediation Notice has been served, it will be a breach of statute law for a landowner or other 'appropriate' person not to obey its requirements within the time specified in that Notice. If landowners or other responsible parties ignore a Remediation Notice, the local authority itself can carry out the remediation and reclaim its costs. Since these will usually be for higher than those the private sector would charge, 'appropriate' persons would be very ill-advised to follow such a course.

Can a local authority itself carry out remediations in any other situations? Yes. If emergency action is deemed imperative (say, where an encapsulation cover is visibly failing and could suddenly collapse), or if no appropriate person can be found to take liability for the contamination, or if – because of financial hardship – the responsible party/parties could not bear the necessary costs.

Who might be an 'appropriate' person to accept the costs of land remediation? The basic intention is that the 'Polluter Pays' principle should apply. Section 78F states that an 'appropriate person' is the individual 'who caused or knowingly permitted' the substances, or any of the substances, by reason of which the land is now contaminated. Such a person, or persons, is specified as a 'Class A liability group'.

If no such original polluter can be found, liability then can fall onto the current landowners or occupiers ('Class B liability group').

Is the new Act clear on who will be liable for land remediation costs? No. In many ways this is the most worrying aspect of the new legislation. The simplicity of the 'Polluter Pays' is more apparent than real in many cases, and it can be predicted that the legislation is likely to generate legal arguments rather than to resolve contaminated land problems in some situations. For example, an individual could admit to tipping tarry wastes on land, but at a time when that area was uninhabited. Although apparently a Class A appropriate person, the tipper would appear not to be exposed to any liability (since the pollution linkage, which local authorities have to recognize as a critical test did not exist when the tipping took place). If a later landowner or occupier or developer then re-engineered the site, this could permit the contaminants to come into contact with adjacent scheduled receptors. Thus an original polluter can be exempted while later individuals who were merely attempting to improve the land's state could be deemed the parties responsible for necessary remediation costs. This seems not to represent obvious justice.

Landowners and occupiers, who had no part in causing the original contamination, could be liable for remediation costs if they had permitted the polluter's liability to be transferred with the land ownership. Likewise developers and their professional advisers would be at risk if construction activities opened up new pathways, and so permitted existing contaminants (such as ponded seepages of oils) to interact with one or other of the scheduled receptors.

Why will case law be necessary? Given the complexities inherent in land ownership transfers and whether or not past liabilities have been transferred with the land, it is apparent that courts will have to decide who is liable for land remediation costs in many cases.

The fact that the legislation employs words – i.e. 'knowingly permit' and 'knowledge' over the existence of soil contamination, and 'found' and 'reasonable inquiry' when local authorities seek to identify who is a responsible person – which will be open to various interpretations makes it more certain that legal judgements will be necessary.

Government accepts that it will ultimately be for the courts to decide the meanings of these words in particular instances and whether the tests and procedures carried out by local authorities have been fair and legal.

Given this uncertainty landowners or occupiers who may be faced with land remediation costs should seek legal advice, since a local authority's judgements may not have compelling force.

What limits are there on the remediation costs which a local authority can recover? As above, this is an area of considerable uncertainty at the time of writing.

The basic principles, while clear and reasonable, i.e.

- that polluters should pay;
- that the public purse should not cover the costs of a guilty private individual; and
- that financial hardship should be recognized as a reasonable basis for reduced or zero costs,

are complicated by the very understandable reluctance on government's part to allow businesses to be bankrupted by land restoration costs (as this itself would add substantial costs to the national exchequer) and by the inevitable legal arguments on liability and liability transfers.

Thus a complicated system of recovering remediation costs, where no 'appropriate person' comes forward to carry out the work at his own cost, is predictable. Some more obviously innocent individuals (such as owners of homes built over a poorly reclaimed area of contaminated land) are very likely to be exempted from all or most of whatever costs arise, particularly if their financial positions are such that there is little or no hope of their being able to afford the demanded money. Other obviously responsible parties, such as the determined polluters who cheerfully continue to wash out process vats and then tip the residues on adjacent land, will be expected to pay for all the remediation costs, unless they can demonstrate that this is financially impossible.

However, between these extremes is a grey and confused area which calls for case law to adjudicate.

Since land remediation costs invariably are extremely expensive, in terms of normal family budgets, landowners and occupiers of suspect sites certainly have reason for proper concern, if they have liability for land which is identified as 'contaminated'.

Does this new Act extend the law substantially? No. The Environmental Protection Act 1990 (S79 on statutory nuisances and S.81(4) on cost recovery if local authorities had to remediate land) and the Water Resources Act 1991 (S161 and S161(3) on the National Rivers Authority's right to carry out clean-up works) gave essentially the same powers over land contamination.

So why should the new Act be of concern? Two good reasons for concern exist.

The first is that the 1995 Act will be far easier for regulators to employ, as they will no longer have to risk, in advance, their own funds on remediations in the uncertain hope that courts will find their actions necessary and proper, and then order that recompense be made by the polluter or landowner/occupier. In sharp contrast, the person who caused or knowingly permitted contaminations (under the 1995 Act) will be under a liability to comply with the terms of a Remediation Notice, and will be responsible for whatever costs arise.

Earlier legislation (the 1990 and 1991 Acts) seems never to have been employed to remove statutory nuisances or to force the clean-up of land which caused water pollution. In contrast the 1995 Act can be expected to be the catalyst for a substantial number of clean-up actions.

A second, and equally worrying, cause for concern is that it is the Secretary of State for the Environment's guidance which will control the severity with which the 1995 Act is applied. Politicians change their views and governments change, thus it is not too cynical to anticipate that the present government's draft guidance (which restricts regulatory action to preserve a reasonable balance between protecting human health and the environment and the necessary recycling of land in a small and overcrowded island) could become far more onerous at some future time.

What should a well-advised landowner/occupier do if his land appears to be of suspect quality? The most prudent action is to find out what is on, in or under the land and have the risks assessed. While this will incur site investigation costs of a few thousand pounds, it certainly will save on defensive legal action and will give the landowner/occupier advance warning of possible regulatory interest. Site investigations and risk assessment costs will always be higher if the work has to be carried out as an emergency response after local authority inspectors have shown unwelcome interest.

Who should be worried about the new Act? Although the legislation is written with specific reference to the original polluter and the present landowner/occupier, a wider audience could be at risk. Development which increases contaminant migration pathways is not an improbable prospect and could expose developers, contractors and consultants to possible action. One test, currently suggested by the Secretary of State's guidance to regulators, is on the 'introduction of pathways'. Many building and civil engineering works could, in some circumstances, cause localized and restricted contamination to become more widespread and harm sensitive targets (e.g. spreading spilled oils into deeper aquifers by installing piled foundations or expelling landfill gases from a source site into adjoining land by compacting the source soils).

The only parties, among all of these involved when old industrial land is being developed, who will escape regulatory interest appear to be funders, insurers and others who can prove that they hold no beneficial interest in, or control over, the ownership of the land. However, local authorities are advised to adopt the following prioritization when seeking to compell land clean-ups, i.e.

- target first those Class A responsible persons who caused the pollution or who knew it was being created;
- target Class B persons (landowners and occupiers) *only* when these should reasonably have known that the pollution existed on, in or under the land.

Current owners or occupiers who made reasonable efforts to establish the state of their land pre-purchase should not be at risk.

10.5 Other relevant legislation

10.5.1 *Introduction*

Despite the stress given to the Environment Act 1995, it should not be forgotten that other legislation exists, and may be more suitable for some regulatory uses. In particular, planning law and waste licensing regulations have obvious importance for contaminated land and its remediation.

10.5.2 *Government overall policy*

Government's overall policy remains to bring derelict and polluted land back into beneficial use, i.e.

> The aim is to protect and enhance an environment regarded as being high quality, and to improve a poor environment, for example, by reclaiming contaminated and derelict land [1]

and its advice to local authorities is that 'Development plans provide an opportunity to set out policies for the reclamation and possible use of "contaminated land" ' [2].

'Contamination' is a material consideration for planning purposes [3] and this permits a planning authority either the right to restrict the use to which land is put, or to require that reclamation during development work is adequate to ensure the future safety of land users and occupiers. Quite obviously this power complements the Environment Act 1995; land where anthropogenic chemicals exist very often will not be badly enough contaminated to create, or threaten, 'significant harm' (to the prescribed receptors), but still will be less clean than actually is desirable. Thus it is commonplace for planning permission to be granted subject to specific conditions. These – in land contamination terms – can include the investigation detail necessary to establish if future risks to land users are probable and the remediation required to minimize these. For most mildly contaminated areas of land conditional planning approval will remain the frontline environmental safeguard.

As a local planning authority consults other statutory bodies (including the Environment Agency) before specifying planning conditions, consistency with the philosophy of the Environment Act 1995 is predictable. Enforcement of planning conditions compliance has been distinctly less rigorous than are the 1995 Act's powers, since it is not an actual legal offence to breach planning conditions. The offence, should one be identified, is for a landowner, occupier or developer to fail to obey a later enforcement notice. As in all other areas of UK law, however, legislation has been strengthened by secondary Acts (the Planning and Compensation Act 1991) to make compliance more the norm.

10.5.3 *Building regulations*

Building Regulations (1991 Regulations made under the Building Act 1984) provide a further check on the suitability of land, though only on that directly covered by a building's footprint. Land quality in adjacent garden and public open space areas is not subject to these regulations.

10.5.4 *Waste management powers*

Waste management powers under the Environmental Protection Act 1990 are onerous and carry severe penalties if breached.

Essentially these powers (S.34(1) EPA 1990) aim to impose a 'duty of care' on all producers, carriers and disposers of waste, to ensure that unlicensed waste disposal is precluded. An essential element in the 'duty of care' is that full documentation must be retained to prove that wastes have been removed and carried only by properly licensed contractors and disposed of only at facilities authorized to receive such wastes.

This duty compels a landowner, should he or she choose to retain wastes on site, to take appropriate measures to contain the wastes, prevent the escape of contaminated dusts, noxious liquids or gases and to prevent any entry of the wastes into controller waters. Figure 10.3 summarizes the procedures necessary to discharge this duty legally.

In terms of contaminated land, the more important of the duty of care provisions are:

- unwanted excavated contaminated soils are legally 'wastes', unless the material (up to a maximum of 20 000 cubic metres/hectare) can be employed beneficially *or* the screened waste soils can be employed in the manufacture of bricks, building blocks and other construction materials, *or* the excavated wastes can be employed as fuels;
- that any activities/uses which might be seen as exempt from the waste licensing regulations are registered with the Environment Agency's Waste Management Officer.

Steps	*Section in Waste Management Licensing Regulations 1994*
(a) Is your material a 'waste'?	Schedule 4, part II
Is it a 'controlled waste'?	Schedule 4, part II
If so, protect and store it properly while it remains in your care.	
Do you have a proper written description of the waste, including the waste name, its production process, and a complete chemical analysis?	
If not, have this prepared without delay.	
(b) If your waste is to be removed, ensure that only a licensed carrier is employed and that a licensed facility has agreed to receive the waste.	Schedule 4, part III
Check the credentials of the selected carrier.	
Ensure that vehicles are properly sheeted, to avoid waste spillages, and that waste description/chemical analysis accompanies the load.	
Surrender the wastes to the carrier after he has signed a transfer note.	
Obtain from the carrier the transit docket, stamped by the licensed receiving facility. Keep this for two years in case regulators inspect.	
Be alert for suspicion that illegal waste transfer/disposal may be occurring. If in doubt advise the Environment Agency's Waste Management Officer.	
(c) If you receive wastes ensure that the person who delivers these is licensed for their carriage.	
(d) Are any activities exempt from the waste licensing regulations?	Schedule 3
43 specific exemptions are allowable, mainly when waste recovery/recycling is intended. These include:	Schedule 4, Part IV

- waste glass
- scrap metals at furnace plants
- use of waste as fuels
- cleaning of containers
- burning waste in small appliances
- burning waste oils
- spreading wastes to improve soils

- sludge spreading on land
- land reclamation work
- sewage works
- bulk reduction of waste
- composting
- manufacture of construction materials
- manufacture of finished goods
- beneficial use without disposal
- diseased animal carcasses
- waste storage in a secure place
- wastes in secure containers
- wastes for construction use on the same site
- recovery of textiles
- preparatory waste treatment
- silver recovery
- animal by-products
- crushing of concrete and bricks
- waterways dredgings
- recovery as part of a production process
- baling, compacting or pulverizing
- storing returned goods
- incineration
- burning wastes in the open
- discharge from trains
- campsite waste
- peat working
- railway ballast
- prospecting waste
- garbage from ships
- pet burial
- waste samples for testing
- medicines
- storage at place of production
- storage elsewhere
- transitional exemptions allowing for the change from the 1974 Act.

Figure 10.3 The duty of care: a summary check-list (after [4]).

Since off-site disposal of contaminated soil remains a preferred option in many land reclamations, particular care has to be exercised by the waste producers to ensure that a legally acceptable disposal takes place.

10.5.5 *Summary*

Overall the current legislative powers are likely to be employed in the manner described below:

- if a site is especially hazardous, the powers of the Environment Act 1995 would be exercised by a local authority;
- however, in less serious cases, the Planning Authorities will usually insist that full subsurface information is obtained and that a necessary level of land remediation is undertaken as part of the development process;

- during reclamation, the Waste Licensing Regulations will be in force and will compel all the parties involved to ensure that illegal waste tipping is not allowed; the sole departure from this regime will be if exempted activities can be claimed and accepted by the regional Waste Management regulators;
- landowners and occupiers, while conscious of their obligation to always act in a legal manner, will seek the least expensive compromises, while regulators will usually give lesser importance to the costs which could arise. Thus prolonged discussions are increasingly a commonplace consequence of the strengthened legislative system.

10.6 Conclusions

Increasing regulations at all levels and the intention to enforce the 'Polluter Pays' principle seem likely to give rise to two distinctly conflicting outcomes.

- On the positive side, integrated pollution control, the establishment of a unified Environment Agency and the powers of legislation since 1990 seem certain to restrict the creation of yet more areas of contaminated land. Proven occurrences of pollution incidents predictably will result in immediate damage limitation followed by necessary remediation of the contaminated land source.
- On the negative side, however, increasing concern over liabilities and legal consequences must force the costs of land remediation to rise. Purchasers already tend to insist on receiving 'clean' land and require warranties that the land is suitable for a particular re-use. Since professional indemnity insurance is both costly and increasingly difficult to obtain, this obviously impacts on reclamation costs. In some worse cases, landowners are likely to be unable to dispose of their property and may have inadequate financial resources to cover necessary quality improvements. Predictably areas of land too contaminated for re-use will be left derelict, since neither public nor private funds for remediation are likely to be adequately available. In an overcrowded island, where 20% of the land area will soon be covered with buildings, concrete and tarmac, it seems most unfortunate that permanently unusable areas of land are likely to be one result of what otherwise is desirable environmental protection legislation.

11
Introduction to waste management practices in the United States of America

11.1 Introduction

The remediation of contaminated soil has had a short, but tumultuous history in the United States of America. The remediation business is almost entirely regulation driven; the important sets of regulations include the Resource Conservation and Recovery Act (RCRA) and the Comprehensive Environmental Response, Compensation, and Liability Act (CERCLA), which governs the Superfund programme. A myriad of other legislative acts and sets of regulations impact remediation activities. Most of these regulations were written and are enforced by the United States Environmental Protection Agency (EPA). The Superfund programme involves the largest and most important remediation projects in the US and is heavily publicized. As this chapter was being written, the CERCLA legislation had expired. While remediations in progress were continued, this brought to a standstill much of the planning for future projects. In allowing the CERCLA to expire, the legislative branch of the federal government expressed a sentiment shared by many that the Superfund programme is too expensive and is not sufficiently effective. The legislative and executive branches of the federal government agree that reform is needed, but to date have not reached agreement on the nature of that reform. Other remediation programmes, including RCRA corrective action and underground storage tank clean-ups, are in place and are being continued.

11.2 Regulatory impetus for remediation

As shown in Table 11.1, several sets of regulations cause companies and federal agencies to initiate remediation activities. Under the RCRA, when a permitted treatment, storage, or disposal facility (TSDF) is shut down, it becomes subject to corrective action. This often requires remediation activities. Releases of contaminants to the environment can also trigger remediation. In a 1993 report [1], the EPA estimated that approximately 5100 TSDFs, containing 80 000 solid waste management units, required corrective action programmes.

Table 11.1   Regulations impacting remediation activities in the US

Name of regulation/programme	Nature or impact
RCRA	Requires corrective action upon shutdown of permitted treatment, storage or disposal facility.
Superfund (through the CERCLA)	Requires investigation and remediation of sites placed on the NPL.
USTs (regulated under RCRA and by individual states)	Requires remediation of soil and groundwater contaminated by hazardous chemicals or petroleum products leaking from tanks.
ECRA (employed by New Jersey; similar programmes are used in 17 other states)	Requires clean-up of property upon change of ownership.
State Hazardous Waste Programs (patterned after the CERCLA)	Requires investigation and remediation of sites not on the NPL.

As is discussed later, many clean-ups are initiated by private companies, not because of current regulations but in anticipation of stricter regulations or to limit future liability.

Obviously, US industry must pay a substantial cost for this clean-up effort. This negatively impacts the ability of these companies to compete on the international business scene. At least one large company (The Johns–Manville Corporation) was driven into bankruptcy because of its environmental problems. However, the worst appears to be past. By and large, US companies have identified their major environmental liabilities and have set aside funds to deal with them.

11.2.1 *Background on regulations*

Remedial actions carried out during the past two decades in the US have either been carried out under the provisions of the CERCLA, enacted in 1980 and its reauthorization in 1986, known as SARA, or have been heavily influenced by these federal acts. Treatment standards for remediation were influenced strongly by the RCRA (1974) and the subsequent amendments in 1984. The latter are known as the Hazardous and Solid Waste Amendments (HSWA). The RCRA includes provisions for developing criteria to determine which wastes are hazardous; establishing standards for siting, designing and operating treatment and disposal facilities; and encouraging the individual states which comprise the US to develop their own regulatory programmes. The HSWA reauthorized RCRA and established the land disposal restrictions (LDRs), also known as landbans. The

CERCLA and SARA established a massive programme for the remediation of previously contaminated (often closed or abandoned) sites. The programme is funded by a tax on industry, but the EPA attempts to recover the expenditures from the people and firms which contaminated the site. These people and firms are known as the Potentially Responsible Parties (PRPs). SARA contained an interesting provision 'that requires that remedial actions meet all applicable, relevant, and appropriate public health and environmental standards. Therefore, the Superfund programme must be consistent with the BDAT approach when disposing of contaminated soils and debris from Superfund sites.' BDAT refers to Best Demonstrated and Available Technology and will be discussed later in this chapter. Another key element of SARA was that it made it clear that federal facilities, including those operated by the Department of Energy and the Department of Defense, were covered by the authority of the CERCLA, requiring them to comply with the law in the same way that the private sector must comply.

Remediation under the Superfund programme must consider the hazardous substances which are or may be present and must consider the concentrations of the hazardous substances jointly and severally to applicable requirements. These applicable requirements are commonly referred to as Applicable, Relevant and Appropriate Requirements (ARARs) and will also be explained later in this chapter.

From the earliest days of its environmental awareness era, the US has tended to base clean-up standards on concentration levels for specific chemical compounds, as opposed to toxicity measurements or risk assessment. Debate on this issue began in the 1970s with the passage of the Water Pollution Control Act Amendments of 1972 [3]. The EPA was created to implement the Water Pollution Control Act Amendments. Under the settlement terms of a private party lawsuit (*Natural Resources Defense Council* v. *Train*, 1976), the EPA was required to set chemical-specific clean-up standards. A list of 127 chemicals known as the 'Priority Pollutants' was assembled, and became the basis for the specific chemicals subsequently controlled by RCRA. These chemicals appeared on the Hazardous Substance List and then on the Target Compound List.

The Target Compound List included both organic and inorganic (heavy metal) compounds, with the inorganic compounds referred to as the Target Analyte List, or TAL, metals. In 1980, the Superfund legislation, CERCLA, combined the treatment standards from these programmes, plus the Toxic Substance Control Act (TSCA) and the Solid Waste Disposal Act, the Clean Air Act and the Clean Air Act Amendments to create a comprehensive list of contaminants referred to as 'hazardous substances'. The CERCLA defines 'hazardous substances' as:

(A) any substance designated pursuant to section 311(b)(2)(A) of the Federal Water Pollution Control Act, (B) any element, compound, mixture, solution, or

substance designated pursuant to section 102 of this Act, (C) any hazardous waste having the characteristics identified under or listed pursuant to section 3001 of the Solid Waste Disposal Act (but not including any waste the regulation of which under the Solid Waste Disposal Act has been suspended by Act of Congress), (D) any toxic pollutant listed under section 307 (a) of the Federal Water Pollution Control Act, (E) any hazardous air pollutant listed under section 112 of the Clean Air Act, and (F) any imminently hazardous chemical substance or mixture with respect to which the Administrator has taken action pursuant to section 7 of the Toxic Substances Control Act. The term does not include petroleum, including crude oil or any fraction thereof which is not otherwise specifically listed or designated as a hazardous substance under subparagraphs (A) through (F) of this paragraph, and the term does not include natural gas, natural gas liquids, liquefied natural gas, or synthetic gas usable for fuel (or mixtures of natural gas and such synthetic gas).

Land disposal, i.e. placement in a constructed landfill, is not the method of choice of the EPA. Rather, it is the least preferred of the various options available. In an attempt to minimize the quantity (and the toxicity) of hazardous wastes which are landfilled, Congress passed the landban regulations.

11.2.2 *The landbans*

As part of the 1984 HSWA, the EPA was required to implement certain regulations that would ultimately ban the land disposal (placing in a landfill) of hazardous waste, including liquids, sludges, solids and debris, and especially untreated hazardous waste. The consequences of the landban legislation included:

8 May 1985	Bulk hazardous waste liquids were banned from landfill disposal, whether or not the waste had been treated with absorbents. However, treatment by stabilization, involving a 'chemical reaction', followed by landfilling was acceptable.
8 November 1985	Nonhazardous bulk liquids, including water, were banned from land disposal unless it could be demonstrated that landfilling was the only safe alternative.
8 November 1986	Dioxins and solvents were banned from land disposal, with some exceptions. The EPA issued a schedule for implementation of regulations which would ban the land disposal of all remaining hazardous wastes.

The EPA published a proposed schedule for the statutory landbans, including the designation of three groups of hazardous compounds, with each group to become regulated at a specific time. These groups became known as the 'first-third', 'second-third' and 'third-third', with starting regulation dates of 8 August 1988, 8 June 1989 and 8 May 1990, respectively. The 'first-third' group of wastes contained the wastes which are the most toxic and were generated in the largest volume at the time.

8 July 1987   A group of wastes called the 'California list' were banned from land disposal. These included sludges containing cyanide, arsenic, cadmium, lead, mercury, nickel, selenium, PCBs, thallium and halogenated organic compounds.

The 'first-third', 'second-third' and 'third-third' regulations were subsequently implemented, albeit with delays for certain aspects. More recently, the EPA added LDRs for debris contaminated with hazardous waste.

The intent of this legislation was to reduce the amount of waste, whether it be liquid, sludge or solid, which is being landfilled. This is in fact what has happened. Prior to 8 May 1985, liquids were co-disposed with solid wastes in landfills. Subsequent to that date, most of these liquids, if aqueous, were diverted to commercial hazardous aqueous waste treatment facilities or to deep wells. Organic streams were diverted to solvent recovery facilities, cement kilns or incinerators.

The quantities of solids and sludge being landfilled have also declined, as was the intent. The requirement for treatment prior to landfilling raised the disposal costs substantially, providing incentive for generators of wastes to implement waste minimization, or pollution prevention, programmes which have been very successful.

11.2.3 *BDATs (Best Demonstrated and Available Technologies)*

As previously stated, SARA required that the Superfund programme 'be consistent with the BDAT approach when disposing of contaminated soils and debris from Superfund sites'. BDATs are set by the EPA. The methodology to be employed in the selection of the technology and the achievement level is as follows [4].

EPA promulgated a technology-based approach to establishing treatment standards under section 3004(m). Section 3004(m) also specifies that treatment standards must 'minimize' long- and short-term threats to human health and the environment arising from land disposal of hazardous wastes. [...] the intent is 'to require utilization of available technology' and not 'process which contemplates technology-forcing standards.' [...] EPA has interpreted this legislative history as suggesting that Congress considered the requirement [...] to be met by application of the best demonstrated and achievable [*i.e. available*] technology. [...] Accordingly, EPA's treatment standards are generally based on the performance of the [...] (BDAT) identified for treatment of hazardous constituents. [...] The treatment standards, according to the statute, can represent levels or methods of treatment, if any, that substantially diminish the toxicity of the waste or substantially reduce the likelihood of migration of hazardous constituents. Wherever possible, the Agency prefers to establish BDAT standards as 'levels' of treatment (i.e. performance standards) rather than adopting an approach that would require the use of specific treatment 'methods'.

The EPA has set general methods of treatment, such as incineration or stabilization, as BDAT, but not specific processes under these general categories. The technology actually used on a remediation project has only to achieve a treatment level, such as a concentration in a leaching test, to meet BDAT requirements. The EPA has set the treatment requirements by collecting and analysing data from actual treatment operations. The statistical and other data analysis techniques are given in reference [4].

11.2.4 *Characteristic wastes versus listed wastes*

The EPA identifies hazardous wastes as either listed or characteristic. Listed wastes are those wastes specifically identified according to the source. Each is given a specific code (F006, K081, etc.). Characteristic wastes are those found to be hazardous because they possess one or more of four characteristics: corrosivity, reactivity, ignitability or toxicity. Toxicity is determined by measuring the leachability of specified metal, pesticide or organic compounds.

The EPA recognized that its procedures for listing hazardous wastes might not be applicable in all cases. To provide for these cases, it created a 'delisting' process, under which the EPA can be petitioned to exclude a waste from management as a hazardous waste. The petitioner must prove to the EPA that the waste is not hazardous because of facility-specific variations in raw materials, processes or other factor [5]. Delisting is done on a case-by-case basis and is rarely granted. If a waste is delisted for one generation facility or one treatment process, it is not automatically delisted for other facilities or other processes. As applied to remediation, waste removed from a remediation site can be treated with a process for which a delisting petition has been submitted and approved. The treated waste is then no longer considered a hazardous waste. As an example of delisting, the EPA has granted delisted status to the products of treatment of spent aluminum potliner, RCRA code K088. The delisting applies to a specific treatment facility and applies only to potliner from specific aluminum production plants. Treatment of spent aluminum potliner will be required prior to landfilling beginning in January 1997.

In 1990, the EPA implemented the toxicity characteristic (TC) rule, which established treatment standards for 26 additional organic waste constituents [6], [7]. The standards then in place for characteristic wastes are shown in Table 11.2. In this table, the 'regulatory level', often referred to as the TC level, was the concentration limit, according to a leaching test known as the toxicity characteristic leaching procedure, for defining a material as a hazardous waste. As applied to sludges and solids, the test requires size reduction to less than 3/8th inch (9.525 mm) and extraction of the waste sample with a dilute solution of acetic acid. The extraction is carried out by tumbling in a canister for 18 hours at 23 plus or minus 2 °C. The acetic acid extraction is intended to simulate conditions in a mismanaged landfill in which treated hazardous waste is co-disposed with municipal solid waste. Organic acids

Table 11.2 Toxicity characteristic regulatory levels

USEPA hazardous waste number	Constituent	Regulatory level (mg/L)
D004	Arsenic	5.0
D005	Barium	100.0
D018	Benzene	0.5
D006	Cadium	1.0
D019	Carbon tetrachloride	0.5
D020	Chlordane	0.03
D021	Chlorobenzene	100.0
D022	Chloroform	6.0
D007	Chromium	5.0
D026	Cresol (total or individual)	200.0
D016	2,4-D	10.0
D027	1,4-Dichlorobenzene	7.5
D028	1,2-Dichloroethane	0.5
D029	1,1-Dichloroethylene	0.7
D030	2,4-Dinitrotoluene	0.13
D012	Endrin	0.02
D031	Heptachlor and its hydroxide	0.008
D032	Hexachlorobenzene	0.13
D033	Hexachloro-1,3-butadiene	0.5
D034	Hexachloroethane	3.0
D008	Lead	5.0
D013	Lindane	0.4
D009	Mercury	0.2
D014	Methoxychlor	10.0
D035	Methyl ethyl ketone	200.0
D036	Nitrobenzene	2.0
D037	Pentachlorophenol	100.0
D038	Pyridine	5.0
D010	Selenium	1.0
D011	Silver	5.0
D039	Tetrachloroethylene	0.7
D015	Toxaphene	0.5
D017	2,4,5-TP (Silvex)	1.0
D040	Trichloroethylene	0.5
D041	2,4,5-Trichlorophenol	400.0
D042	2,4,6-Trichlorophenol	2.0
D043	Vinyl chloride	0.2

would be produced by biological action under such conditions, and the acetic acid is intended to represent these acids. This test appears to be more severe than the leachability tests used in other countries, including that used in the United Kingdom, which uses carbonated water.

Characteristic waste remains hazardous until treated to reduce the leachability to below the TC levels. Characteristic wastes containing the heavy metal constituents in excess of the TC limits in Table 11.2 can be placed in a hazardous waste landfill after treatment to reduce the leachability of the metals to below the TC limit. This would be allowable under the Land Disposal Restrictions. The TC limits for metals and pesticides have been employed as clean-up standards for Superfund projects, as well as for private and state-managed remediations.

For listed wastes, the situation is more complex. As shown in Table 11.3, treatment standards can vary depending on which code the waste carries. For a waste with multiple codes, multiple standards could apply. In that case, the most stringent standard of all the possibilities would become the treatment standard.

Obviously, leaching tests are performed frequently when dealing with hazardous wastes, and a considerable cost can be incurred for these tests. Costs per test range from $100 US for only the eight RCRA heavy metals to about $100 US for heavy metals, volatile organics, semi-volatile organics, pesticides and herbicides. These costs represent a financial burden well in excess of that which applies elsewhere.

11.2.5 *Phase II disposal rule – Universal Treatment Standards*

On 19 September 1994, the EPA published the final rule, 'Land Disposal Restrictions Phase II – Universal Treatment Standards for Organic Toxicity Characteristic Wastes and Newly Listed Wastes' [8]. This final rule promulgated federal treatment standards for several newly identified wastes. It also simplified the federal LDR programme by implementing universal treatment standards (UTSs). The effective date was 19 December 1994.

UTSs were promulgated for organic, metal and cyanide constituents in wastes, with these standards replacing previous federal treatment standards. The UTSs provide a single treatment standard for each wastewater and non-wastewater constituent, regardless of the source of the hazardous waste containing the constituent. Therefore, whereas a constituent may have been listed under two different waste codes and previously had two different treatment levels, that constituent now has a single treatment level regardless of the waste code with which it is associated.

The UTSs do not replace treatment standards specified as a method of treatment. Wastes with treatment standards of specified technologies (BDATs) must still be treated using those required technologies. However, a small number of wastes can now be treated to meet the UTSs as an alternative instead of utilizing the specified technology.

The UTSs are shown in Table 11.4 and give the requirements for both wastewater (WW) and non-wastewater (NWW) wastes.

11.3 Regulation escalation and backlash

These are by no means the only regulations affecting remediation. Regulation after regulation has been promulgated, creating a tangled web that no one can comprehend. Some indication of the complexity of the situation is shown in Figure 11.1, which shows the number of environmental laws and amendments enacted versus year for the period of 1900 through 1993 [9].

Table 11.3 First-third landban BDAT treatability levels

Waste code	Description[a]	Concentration of constituent (mg/l) in waste or leachate								
		Arsenic	Cadmium	Chromium	Lead	Mercury	Nickel	Selenium	Silver	Cyanide
F006	Electroplating sludge		0.066	5.200	0.510		0.320		0.072	b
K001	Wood-preserving WWT sludge				0.510					
K022	Distillation bottom tars			5.200			0.320			
K046	Lead-based initiating compound WWT sludge				0.180					
K048	Petroleum refining DAF float	0.004		1.700			0.048	0.025		
K049	Petroleum refining slop oil emulsion solids	0.004		1.700			0.048	0.025		
K050	Petroleum refining heat exchanger cleaning sludge	0.004		1.700			0.048	0.025		
K051	Petroleum refining API separator sludge	0.004		1.700			0.048	0.025		
K052	Petroleum refining leaded tank bottoms	0.004		1.700			0.048	0.025		
K061	EAF emission control dust		0.140	5.200	0.240		0.320			
K062	Steel finishing spent pickle liquor			0.094	0.370					
K071	Mercury cell brine purification mud					0.025				
K086	Ink manufacturing washes and sludges			0.094	0.370					
K087	Coking decanter tank tar sludge				0.510					
K101	Arsenic-containing pharmaceutical distillation tar residue		0.066	5.200	0.510		0.320			
K102	Arsenic-containing pharmaceutical activated carbon residue		0.066	5.200	0.510		0.320			

a All non-wastewaters unless otherwise specified.
b Reserved for future action.

Table 11.4   Universal Treatment Standards

Regulated constituent/common name	Wastewater Standard Concentration in mg	Non-wastewater Standard Concentration in mg/kg *unless noted as* mg/l *TCLP*
A2213	0.003	1.4
Acenaphthylene	0.059	3.4
Acenaphthene	0.059	3.4
Acetone	0.28	160
Acetonitrile	5.6	38
Acetophenone	0.010	9.7
2-Acetylaminofluorene	0.059	140
Acrolein	0.29	NA
Acrylamide	19	23
Acrylonitrile	0.24	84
Aldicarb sulfone	0.056	0.28
Aldrin	0.021	0.066
4-Aminobiphenyl	0.13	NA
Aniline	0.81	14
Anthracene	0.059	3.4
Aramite	0.36	NA
alpha-BHC	0.00014	0.066
beta-BHC	0.00014	0.066
delta-BHC	0.023	0.066
gamma-BHC	0.0017	0.066
Barban	0.056	1.4
Bendiocarb	0.056	1.4
Bendiocarb phenol	0.056	1.4
Benomyl	0.056	1.4
Benzene	0.14	10
Benz(a)anthracene	0.059	3.4
Benzal chloride	0.055	6.0
Benzo(b)fluoranthene (difficult to distinguish from benzo(k)fluoranthene)	0.11	6.8
Benzo(k)fluoranthene (difficult to distinguish from benzo(b)fluoranthene)	0.11	6.8
Benzo(g,h,i)perylene	0.0055	1.8
Benzo(a)pyrene	0.061	3.4
Bromodichloromethane	0.35	15
Bromomethane/Methyl bromide	0.11	15
4-Bromophenyl phenyl ether	0.055	15
n-Butyl alcohol	5.6	2.6
Butylate	0.003	1.4
Butyl benzyl phthalate	0.017	28
2-sec-Butyl-4,6-dinitrophenol/ Dinoseb	0.066	2.5
Carbaryl	0.006	0.14
Carbenzadim	0.056	1.4
Carbofuran	0.006	0.14
Carbofuran phenol	0.056	1.4
Carbon disulfide	3.8	4.8 mg/l TCLP
Carbon tetrachloride	0.057	6.0
Carbosulfan	0.028	1.4

Chlordane (alpha and gamma isomers)	0.0033	0.26
p-Chloroaniline	0.46	16
Chlorobenzene	0.057	6.0
Chlorobenzilate	0.10	NA
2-Chloro-1,3-butadiene	0.057	0.28
Chlorodibromomethane	0.057	15
Chloroethane	0.27	6.0
bis(2-Chloroethoxy)methane	0.036	7.2
bis(2-Chloroethyl)ether	0.033	6.0
Chloroform	0.046	6.0
bis(2-Chloroisopropyl)ether	0.055	7.2
p-Chloro-m-cresol	0.018	14
2-Chloroethyl vinyl ether	0.062	NA
Chloromethane/Methyl chloride	0.19	30
2-Chloronaphthalene	0.055	5.6
2-Chlorophenol	0.044	5.7
3-Chloropropylene	0.036	30
Chrysene	0.059	3.4
o-Cresol	0.11	5.6
m-Cresol (difficult to distinguish from p-cresol)	0.77	5.6
p-Cresol (difficult to distinguish from m-cresol)	0.77	5.6
m-Cumenyl methylcarbamate	0.056	1.4
Cycloate	0.003	1.4
Cyclohexanone	0.36	0.75 mg/l TCLP
o,p'-DDD	0.023	0.087
p,p'-DDD	0.023	0.087
o,p'-DDE	0.031	0.087
p,p'-DDE	0.031	0.087
o,p'-DDT	0.0039	0.087
p,p'-DDT	0.0039	0.087
Dibenz(a,h)anthracene	0.055	8.2
Dibenz(a,e)pyrene	0.061	NA
1,2-Dibromo-3-chloropropane	0.11	15
1,2-Dibromoethane/Ethylene dibromide	0.028	15
Dibromomethane	0.11	15
m-Dichlorobenzene	0.036	6.0
o-Dichlorobenzene	0.088	6.0
p-Dichlorobenzene	0.090	6.0
Dichlorodifluoromethane	0.23	7.2
1,1-Dichloroethane	0.059	6.0
1,2-Dichloroethane	0.21	6.0
1,1-Dichloroethylene	0.025	6.0
trans-1,2-Dichloroethylene	0.054	30
2,4-Dichlorophenol	0.044	14
2,6-Dichlorophenol	0.044	14
2,4-Dichlorophenoxyacetic acid/2,4-D	0.72	10
1,2-Dichloropropane	0.85	18
cis-1,3-Dichloropropylene	0.036	18
trans-1,3-Dichloropropylene	0.036	18
Dieldrin	0.017	0.13
Diethylene glycol, dicarbamate	0.056	1.4
Diethyl phthalate	0.20	28
p-Dimethylaminoazobenzene	0.13	NA

Table 11.4   (*Cont.*)

Regulated constituent/common name	Wastewater Standard Concentration *in* mg	Non-wastewater Standard Concentration in mg/kg *unless noted as* mg/l *TCLP*
2-4-Dimethyl phenol	0.036	14
Dimethyl phthalate	0.047	28
Dimetilan	0.056	1.4
Di-n-butyl phthalate	0.057	28
1,4-Dinitrobenzene	0.32	2.3
4,6-Dinitro-o-cresol	0.28	160
2,4-Dinitrophenol	0.12	160
2,4-Dinitrotoluene	0.32	140
2,6-Dinitrotoluene	0.55	28
Di-n-octyl phthalate	0.017	28
Di-n-propylnitrosamine	0.40	14
1,4-Dioxane	12.0	170
Diphenylamine (difficult to distinguish from diphenylnitrosamine)	0.92	13
Diphenylnitrosamine (difficult to distinguish from diphenylamine)	0.92	13
1,2-Diphenylhydrazine	0.087	NA
Disulfoton	0.017	6.2
Dithiocarbamates (total)	0.028	28
Endosulfan I	0.023	0.066
Endosulfan II	0.029	0.13
Endosulfan sulfate	0.029	0.13
Endrin	0.0028	0.13
Endrin aldehyde	0.025	0.13
EPTC	0.003	1.4
Ethyl acetate	0.34	33
Ethyl benzene	0.057	10
Ethyl cyanide/Propanenitrile	0.24	360
Ethyl ether	0.12	160
bis (2-Ethylhexyl) phthalate	0.28	28
Ethyl methacrylate	0.14	160
Ethylene oxide	0.12	NA
Famphur	0.017	15
Fluoranthene	0.068	3.4
Fluorene	0.059	3.4
Formetanate hydrochloride	0.056	1.4
Formparanate	0.056	1.4
Heptachlor	0.0012	0.066
Heptachlor epoxide	0.016	0.066
Hexachlorobenzene	0.055	10
Hexachlorobutadiene	0.055	5.6
Hexachlorocyclopentadiene	0.057	2.4
HxCDDs (All Hexachlorodibenzo-p-dioxins)	0.000063	0.001
HxCDFs (All Hexachlorodibenzofurans)	0.000063	0.001
Hexachloroethane	0.055	30
Hexachloropropylene	0.035	30
Indeno (1,2,3-c,d) pyrene	0.0055	3.4

Iodomethane	0.19	65
3-Iodo-2-propynyl n-butylcarbamate	0.056	1.4
Isobutyl alcohol	5.6	170
Isodrin	0.021	0.066
Isolan	0.056	1.4
Isosafrole	0.081	2.6
Kepone	0.0011	0.13
Methacrylonitrile	0.24	84
Methanol	5.6	0.75 mg/l TCLP
Methapyrilene	0.081	1.5
Methiocarb	0.056	1.4
Methomyl	0.028	0.14
Methoxychlor	0.25	0.18
3-Methylcholanthrene	0.0055	15
4,4-Methylene bis(2-chloroaniline)	0.50	30
Methylene chloride	0.089	30
Methyl ethyl ketone	0.28	36
Methyl isobutyl ketone	0.14	33
Methyl methacrylate	0.14	160
Methyl methansulfonate	0.018	NA
Methyl parathion	0.014	4.6
Metolcarb	0.056	1.4
Mexacarbate	0.056	1.4
Molinate	0.003	1.4
Naphthalene	0.059	5.6
2-Naphthylamine	0.52	NA
o-Nitroaniline	0.27	14
p-Nitroaniline	0.028	28
Nitrobenzene	0.068	14
5-Nitro-o-toluidine	0.32	28
o-Nitrophenol	0.028	13
p-Nitrophenol	0.12	29
N-Nitrosodiethylamine	0.40	28
N-Nitrosodimethylamine	0.40	2.3
N-Nitroso-di-n-butylamine	0.40	17
N-Nitrosomethylethylamine	0.40	2.3
N-Nitrosomorpholine	0.40	2.3
N-Nitrosopiperidine	0.013	35
N-Nitrosopyrrolidine	0.013	35
Oxamyl	0.056	0.28
Parathion	0.014	4.6
Total PCBs (sum of all PCB isomers, or all Aroclors)	0.10	10
Pebulate	0.003	1.4
Pentachlorobenzene	0.055	10
PeCDDs (All Pentachloro- dibenzo-p-dioxins)	0.000063	0.001
PeCDFs (All Pentachlorodibenzofurans)	0.000035	0.001
Pentachloroethane	0.055	6.0
Pentachloronitrobenzene	0.055	4.8
Pentachlorophenol	0.089	7.4
Phenacetin	0.081	16
Phenanthrene	0.059	5.6
Phenol	0.039	6.2
o-Phenylenediamine	0.056	5.6
Phorate	0.021	4.6

Table 11.4   (*Cont.*)

Regulated constituent/common name	Wastewater Standard Concentration in mg	Non-wastewater Standard Concentration in mg/kg unless noted as mg/l TCLP
Phthalic acid	0.055	28
Phthalic anhydride	0.055	28
Physostigmine	0.056	1.4
Physostigmine salicylate	0.056	1.4
Promecarb	0.056	1.4
Pronamide	0.093	1.5
Propham	0.056	1.4
Propoxur	0.056	1.4
Prosulfocarb	0.003	1.4
Pyrene	0.067	8.2
Pyridine	0.014	16
Safrole	0.081	22
Silvex/2,4,5-TP	0.72	7.9
1,2,4,5-Tetrachlorobenzene	0.055	14
TCDDs (All Tetrachlorodibenzo-p-dioxins)	0.000063	0.001
TCDFs (All Tetrachlorodibenzofurans)	0.000063	0.001
1,1,1,2-Tetrachloroethane	0.057	6.0
1,1,2,2-Tetrachloroethane	0.057	6.0
Tetrachloroethylene	0.056	6.0
2,3,4,6-Tetrachlorophenol	0.030	7.4
Thiodicarb	0.019	1.4
Thiophanate-methyl	0.056	1.4
Tirpate	0.056	0.28
Toluene	0.080	10
Toxaphene	0.0095	2.6
Triallate	0.003	1.4
Tribromomethane/Bromoform	0.63	15
1,2,4-Trichlorobenzene	0.055	19
1,1,1-Trichloroethane	0.054	6.0
1,1,2-Trichloroethane	0.054	6.0
Trichloroethylene	0.054	6.0
Trichloromonofluoromethane	0.020	30
2,4,5-Trichlorophenol	0.18	7.4
2,4,6-Trichlorophenol	0.035	7.4
2,4,5-Trichlorophenoxyacetic acid/2,4,5-T	0.72	7.9
1,2,3-Trichloropropane	0.85	30
1,1,2-Trichloro-1,2,2-trifluoroethane	0.057	30
Triethylamine	0.081	1.5
tris-(2,3-Dibromopropyl) phosphate	0.11	0.10
Vernolate	0.003	1.4
Vinyl chloride	0.27	6.0
Xylenes–mixed isomers (sum of o-, m-, and p-xylene concentrations)	0.32	30
Antimony	1.9	2.1 mg/l TCLP
Arsenic	1.4	5.0 mg/l TCLP

Barium	1.2	7.6 mg/l TCLP
Beryllium	0.82	0.014 mg/l TCLP
Cadmium	0.69	0.19 mg/l TCLP
Chromium (Total)	2.77	0.86 mg/l TCLP
Cyanides (Total)[4]	1.2	590
Cyanides (Amenable)[4]	0.86	30
Fluoride	35	NA
Lead	0.69	0.37 mg/l TCLP
Mercury – Non-wastewater from Retort	NA	0.20 mg/l TCLP
Mercury – All Others	0.15	0.025 mg/l TCLP
Nickel	3.98	5.0 mg/l TCLP
Selenium	0.82	0.16 mg/l TCLP
Silver	0.43	0.30 mg/l TCLP
Sulfide	14	NA
Thallium	1.4	0.078 mg/l TCLP
Vanadium	4.3	0.23 mg/l TCLP
Zinc	2.61	5.3 mg/l TCLP

Note: NA means not applicable.

Further tightening of standards is meeting strong opposition. The people of the US still want a clean and safe environment, and it is highly unlikely that they will ask for or accept any lessening of environmental standards. However, a large fraction is now asking for more realism and more sensible environmental management. Many of the regulations shown in Figure 11.1 were implemented to solve a specific environmental problem or to solve problems in a specific medium without taking a comprehensive view of the environment and the ecosystem.

Figure 11.1 shows that growth of the number of environmental regulations in the US has been exponential. As with nearly every situation in which exponential growth occurs, forces are emerging which will slow this growth. For remediation activities, these include cost and liability brought about by the command and control approach taken to date by the federal government. The public in the US has begun to question why additional regulations are needed. The most obvious environmental problems have now been managed. Meeting additional regulations will require the allocation of resources – time, money and energy – which could be spent in other areas of high visibility and awareness, including unemployment, AIDS and health care. This will very likely result in fewer new regulations and in the use of risk assessment to select, as carefully as possible, any additional environmental problems for management.

11.4 The remediation process under CERCLA

As might be anticipated, remediation of a Superfund site is a highly complicated, lengthy and expensive process. In the early years of the programme, which started in 1980, progress on actually completing remediation was

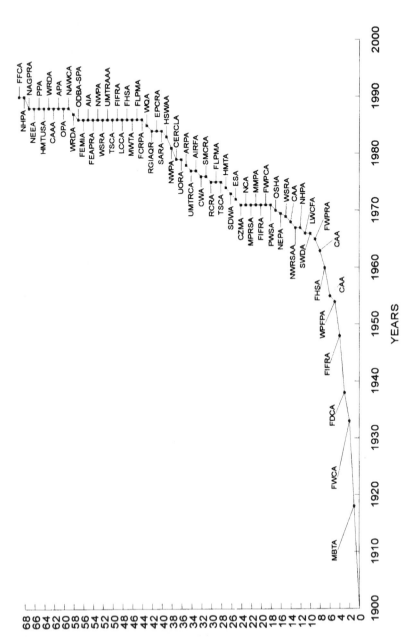

Figure 11.1 Federal environmental legislation. Graph Courtesy of Army Environmental Center, Aberdeen Proving Ground, MD, USA

slow. The pace accelerated during the early 1990s, but very recently has slowed again with the expiration of the regulations.

The basic steps in the process are:

- a preliminary assessment/site investigation (PA/SI) to determine the extent and severity of contamination;
- a determination of whether or not the site is to be placed on the National Priority List (NPL);
- a remedial investigation/feasibility study to more accurately determine the extent and severity of contamination and generate information on applicable treatment;
- the remedial process selection leading to the record of decision (ROD);
- the remedial design/remedial action under which the actual remediation is carried out;
- post-closure activities.

11.4.1 *Preliminary assessment/site investigation*

A facility subject to CERCLA is any site where a hazardous substance has been stored, placed, disposed or deposited at any time in the past. CERCLA jurisdiction is triggered when a 'reportable quantity' of a hazardous substance has been released. The EPA is usually the lead agency for managing site activities. If the site is owned by a federal agency, that agency may also have some jurisdiction. Also, the EPA may allow the Potentially Responsible Parties (PRPs) to lead site activities under certain circumstances. A preliminary determination of the extent and severity of contamination is carried out by a contractor to the EPA under the preliminary assessment/site investigation.

11.4.2 *Assignment to National Priority List*

To determine the need for remediation under the Superfund programme at a specific site, the data from the preliminary assessment/site investigation are employed to determine a hazard ranking system score. If the score is sufficiently high, the site is placed on the NPL. If not, the site might still be subject to a state-required remediation, RCRA-based or otherwise.

11.4.3 *Remedial investigation/feasibility study*

The remedial investigation is an in-depth study of the nature and extent of the contamination, usually carried out by an engineering consulting firm.

Early in this process, areas of contamination are delineated. For areas of contamination containing material which fits the definition of a hazardous waste, the RCRA regulations serve as ARARs. By and large, RCRA rules serve as ARARs for CERCLA operations.

Remedy selection is an important part of the remedial investigation/feasibility study process. Remedy selection, the choice of a clean-up technology or technologies plus the clean-up standards, must take into account (1) the ARARs, (2) the extent and severity of contamination, (3) risk to human health and the environment and (4) site specific variables [10].

The National Contingency Plan requires clean-up levels at Superfund sites to be based on applicable or relevant and appropriate requirements (ARARs). ARARs include federal and state clean-up standards such as maximum contaminant level goals, water quality criteria and national ambient air quality standards. The National Contingency Plan also states that, for chemicals exhibiting systemic, non-carcinogenic health effects, clean-up goals should protect for a lifetime of exposure, protect sensitive subgroups and incorporate an adequate margin of safety. Chemicals that are potential or known human carcinogens should have clean-up goals that are associated with maximum excess lifetime cancer risk of 10^{-6} to 10^{-4}. These are risks of contracting cancer, not the risk of death from cancer. To put these numbers in perspective, 25% of the US population eventually contracts cancer. The overall risk to an individual exposed to a 10^{-6} excess risk increases from 25% to 25.0001%.

Involved human health risk assessment procedures are therefore employed in setting treatment standards at Superfund sites. In most cases, a conservative risk assessment procedure is used to ensure adequate protection. This approach involves using reasonable worst-case exposure scenarios to protect the most sensitive individual likely to be exposed.

The steps used to determine risk-based clean-up criteria are [11]:

- identify chemicals of concern;
- identify exposure pathways;
- calculate the acceptable human exposure level for each chemical of concern;
- determine the acceptable contaminant concentration at the point of exposure for each exposure pathway;
- calculate the clean-up level at the contaminant source.

In the first step, all substances or chemicals present in appreciable concentrations are identified. The concentrations, physical and chemical properties, fate and persistence, toxicity and environmental mobility of each is then evaluated to determine which pose the most significant risk to human health and the environment. Those presenting the highest risk, called indicator chemicals, are used in the risk assessment calculations.

Next, all existing and potential sources of contaminant release(s) that will exist after remediation are identified. For each source, the corresponding environmental transport medium (e.g. air, surface water, ground water, soil) and human exposure routes (e.g. inhalation, ingestion, physical contact) are determined.

From this information, human exposure levels, contaminant concentration at points of exposure and, finally, clean-up levels for given risk levels are calculated. Detailed calculational procedures are available [11], [12].

Upon completion of this exercise, the clean-up concentration limits implied by the ARARs can be compared to those determined by the risk assessment. Logically, the more stringent requirement will be selected for the actual remediation. It should be kept in mind, however, that remedy selection is a negotiation process between the potentially responsible parties for the site and the EPA. It is therefore not an absolute requirement that the most stringent standards become part of the remedy selection.

11.4.4 *The remedy selection process*

Remedy selection, the choice of a technology or technologies, plus the clean-up standards, must take into account (1) the ARARs, (2) risk assessment, (3) the extent and severity of contamination and (4) site-specific variables [10]. The latter include such factors as depth to groundwater, rate of groundwater movement and soil characteristics such as permeability. The presence or absence of non-aqueous phase liquids can be important. This information can then be used to carry out the iterative process that leads to selection of a remedy.

Determination of the extent and severity of contamination can be a difficult and expensive process. It is often carried out in stages, with initial measurements and mapping carried out during the remedial investigation and additional studies carried out in the remedial design phase. It is not economically feasible to analyse at all the locations and depths at which one might wish. Therefore, the history of the site is considered in choosing initial sampling sites, and screening studies are carried out. Techniques such as magnetic measurements, soil resistivity measurements and soil vapour measurements are employed. Field analytical methods such as those employing portable gas chromatographs and X-ray fluorescence spectroscopy are used. Mobile laboratories and devices for rapidly obtaining below grade soil samples are used ever more frequently [13]. This is a more comprehensive and expensive process than the UK's site investigation procedures (Chapter 3).

The ultimate objective of the extent and severity investigation is to prepare accurate three-dimensional maps of the location and concentration of each contaminant. Geostatistical methods are employed to ensure that sufficient data have been collected [3]. After it has been determined that sufficient data have been collected and that geostatistical methods may be applied reliably, then geostatistical tools are used to prepare a conservative map or maps of the extent and severity of contamination.

As previously stated, site-specific variables must also be identified or measured. These fall into two categories: (1) those necessary for assessment of potential risk to human health and the environment and (2) those needed to support the implementation of potential remedies. Certain information on geometrical, hydrogeological and chemical variables is needed for both categories.

11.4.5 *Remedial action selection*

The CERCLA and its reauthorization, the SARA, have resulted in guidance from the EPA which indicates that remedial action selection includes, (1) development of a list of alternatives, (2) screening of the alternatives and (3) conducting a detailed analysis of the remaining alternatives. Treatability studies are employed to confirm the applicability of a treatment technology to a given contamination situation, as well as to obtain data for the full scale remediation.

In a first step, an attempt is made to list the entire array of treatment technologies that may be applicable to the site. The list is arrived at by considering the remedial objectives and knowledge of the extent and severity of the contamination. No single source exists for obtaining candidate technologies. Rather, the experience of consulting firms involved, literature searches, books, journals, EPA publications and various databases all come into play. Information, the more the better, is gathered for each technology, with emphasis on applicability to the media and type of contamination involved in the current study and on the history of the application of the technology to similar situations. Among the available databases is one maintained by the EPA, the Vendor Information System for Innovative Treatment Technologies (VISITT). The ReOpt™ database is sponsored by the Department of Energy and is managed by Battelle Memorial Institute, which runs the DOE's Pacific Northwest Laboratory in Richland, Washington. The ReOpt database focuses on available technologies, whereas the VISITT database emphasizes innovative or emerging technologies.

A second step in the remedial action selection process is to determine which technologies are applicable to the problem at hand. Starting with a list of essentially all remediation technologies, it is decided which are applicable to the media to be remediated (sand, clay, fractured rock, etc.) and the type of contaminant (PCBs, pesticides, heavy metals, etc.). Those technologies which are not applicable are eliminated. To this point no judgement has been employed in the remedial action selection process.

The next step is to rank the applicable technologies. A long list of factors, including cost, acceptance by the regulators and the public, *in situ* versus *ex situ*, and the nature of the treatment residues, is considered. A procedure is employed to arrive at a numerical score for each technology, with the highest scores assigned to the highest ranked technologies. Certain technologies or groups of technologies, those with the lowest scores, can be eliminated from further consideration at this step in the process.

A second round of screening is then generally employed. One applicable technique is to conduct a detailed analysis of the application of each technology, leading to an estimate of remediation time and cost, or ranges for these items. Additional treatability studies might be carried out.

11.4.6 *Remedial process selection*

All information gathered to this point is summarized in a feasibility study report. This report is a key factor in the subsequent negotiation process leading to a record of decision which specifies the treatment levels to be attained and the technology to be employed. Historically, clean-up of soil and sludges has been to the levels given the ARARs, which, in turn, are often concentration-based RCRA requirements. Ground and surface water treatment levels have generally been based on maximum contaminant levels established under the Safe Drinking Water Act. However, many standards have been set on the basis of calculated potential risk to human health. Ecological risk assessments and background concentration levels have also served as bases for setting clean-up goals [12]. Overall, establishment of the clean-up standards is an extremely involved and, at times, very subjective procedure.

11.4.7 *Remedial design/remedial action*

Upon issuance of the record of decision, a contractor prepares a scope of work and carries out other work to prepare for the actual remediation. This may include preparation of contracting documents. The gathering of additional data on the nature and extent of the contamination might be required. The actual clean-up of the site is then carried out under the remedial action phase.

11.4.8 *Post closure*

Upon completion of a remediation and 'closure' of a site, the CERCLA calls for a review of the environmental situation at that site every five years.

11.4.9 *Public participation*

As the record of decision is developed, the public is invited to participate in each process. The CERCLA contains a very specific public participation programme that includes interviews with local officials, community educational activities, opportunities for public input and activities to promote public involvement. Public participation is encouraged from the remedial investigation/feasibility study phase through the remedial action phase. An information repository near the site is also required.

11.4.10 *Cost*

As might be gathered from the above discussion, all-in costs for investigation and clean-up of a Superfund site are extremely high. As an example, the total cost projection for investigation and remediation of the 12 acre Fink/Artel site in West Virginia is $109 million US. Fifty-six firms will share the cost.

11.4.11 *Reforms needed*

It is widely understood and accepted that major reforms are needed in connection with remediation activities in the US, most especially for the Superfund programme. This programme is perceived by most to be a failure, with progress far short of expectations. Arguments and legal battles over who pays for clean-up have been lengthy and costly. The initial policy of 'joint and several liability', under which any party which placed waste at the site, no matter how little, can be responsible for any or all of the clean-up cost, must be replaced with a more rational concept. Relief on this point is expected if and when the programme is reauthorized. Streamlining of the myriad of regulations impacting the selection of a remedy is clearly also needed. It is not clear that this will happen any time soon.

11.5 Trends in remediation technology

Although clean-up under the Superfund programme is not the biggest effort underway, very careful records are kept of the technologies mandated for the clean-up of the various sites and of the actual status of the remediation efforts. The EPA's Technology Innovation Office issues semi-annual status reports on these subjects [2].

The EPA considers incineration and solidification/stabilization to be established technologies; approximately 55% of the technologies listed in the Records of Decision in 1994 and earlier listed these two treatment methods. Congress and the EPA encourage the use of innovative technologies for the clean-up of Superfund sites, and close attention is paid to the reported use of the various innovative technologies. As shown in Table 11.5, soil vapour extraction, bioremediation, and thermal desorption are the innovative technologies most frequently listed in Records Of Decision. Those listed less often include soil washing, in situ flushing, dechlorination and solvent extraction. Table 11.5 also shows that relatively few Superfund projects have actually been completed using innovative technologies.

Another factor that must be considered is the fact that much technological development is driven by the perceptions of the public and the media. Landfilling and incineration are generally considered as the technologies to be replaced. Those technologies that are readily accepted and encouraged by the public, the media, and the regulatory community include soil washing,

Table 11.5 Superfund remedial actions: Status of use of innovative treatment technologies as of August 1995

Technology	Predesign/in design	Design complete/being installed/operational	Project completed	Total
Source control technologies				
Soil vapour extraction	52	70	13	135
Thermal desorption	21	13	17	51
Bioremediation (ex situ)	17	19	4	40
Bioremediation (in situ)	11	12	2	25
Soil washing	10	1	1	12
In situ flushing	11	7	1	19
Dechlorination	2	0	2	4
Solvent extraction	4	2	0	6
In situ vitrification	0	1	0	1
Cyanide oxidation	1	0	0	1
Phyto-treatment	1	0	0	1
CROW	0	1	0	1
Plasma high	1	0	0	1
Total	**131 (44%)**	**126 (42%)**	**40 (13%)**	**297**
Groundwater technologies				
Air sparging	10	6	0	16
Bioremediation (in situ)	6	6	0	12
Passive treatment wall	3	0	0	3
Dual phase extraction	2	0	0	2
Surfactant flushing	1	0	0	1
In situ oxidation	1	0	0	1
Total	**23 (66%)**	**12 (34%)**	**0**	**35**

Note: Data are derived from Records of Decision (RODs) for fiscal years 1982–94 and anticipated design and construction activities as of August 1995. CROW = Contained Recovery of Oily Wastes.

bioremediation, thermal desorption and phyto-treatment. The latter term refers to the cultivation of specialized plants which are capable of taking up specific soil contaminants into their roots or foliage. Those that occupy a middle-of-the-road position include soil vapour extraction, stabilization, solvent extraction, and non-treatment techniques such as capping and natural attenuation.

11.6 Remediation under RCRA

Remediation under the RCRA is similar to remediation under the CERCLA. The RCRA applies to sites that are still in operation, whereas the CERCLA applies primarily to closed sites. In some situations, both sets of regulations apply [14]. Most RCRA regulations affecting remediation were implemented under the HSWA in 1984, which established the RCRA corrective action programme. Corrective action is required in response to releases of hazardous waste or constituents from any solid waste management unit at a TSDF seeking or holding a RCRA permit. This requirement applies regardless of when the release occurred. A solid waste management unit is any discernible unit into which solid wastes have been placed at any time, regardless of whether the unit was intended for such use. The EPA is authorized to require corrective action beyond the boundaries of the facility as necessary to protect human health and the environment.

The steps to be taken are essentially the same as those under the CERCLA, although the terminology and acronyms involved are different. These steps are (1) site investigation, (2) site prioritization and delineation, (3) site investigation and remediation planning, (4) the actual remediation and (5) post-closure activities. Some significant variations from the CERCLA process are:

- prioritization under RCRA corrective action is based primarily on exceeding 'action levels' of hazardous constituents, versus the use of a hazard ranking system score for CERCLA;
- standards governing remediation are incorporated into the facility's permit for RCRA corrective actions, whereas these are established by the ARARs (primarily) under CERCLA;
- post-closure review procedures are incorporated into the site's permit for RCRA corrective action projects, whereas a site remediated under the CERCLA is to be reviewed every five years.

11.7 Radioactive wastes

Radioactive wastes are regulated by the Nuclear Regulatory Commission (NRC), the individual states, the Department of Energy (DOE) or the Department of Defense (DOD). The NRC regulates the management and disposal of radioactive waste from commercial nuclear power plants, hospitals and all non-government sources. In many instances this authority has been transferred to individual states.

The DOE has several nuclear weapons production sites with massive amounts of radioactive soil, sludges and other media which will need to be remediated. The DOE waste management system differs from that of the NRC. DOE waste management facilities conduct performance assessments to determine waste acceptance criteria, which are based on the ability of the facility to manage and dispose of the waste safely. Most of the DOE sites have areas designated for the disposal of radioactive waste and materials.

The DOE's efforts to date have focused on identification and characterization of contaminated sites and wastes. Some low level wastes, such as uranium mine tailings, were at one time scattered throughout communities, and they have now been retrieved and consolidated. The DOE has also funded a major research and development programme on new technologies to solve the problems unique to managing radioactive wastes. Many of these employ techniques for separating the radioactive components from the rest of the waste stream, resulting in two residual streams. One of these streams is non-radioactive and can be handled by conventional means. The other is radioactive only, and management emphasis is placed on converting it to a form which can be stored safely while its radioactivity declines. By and large, however, only a very small portion of the DOE's remediation programme has been completed. Clean-up will be a massive and very complex task. Several decades and expenditures in the tens to hundreds of billions of dollars US will be needed to complete this task.

The DOD is assessing its sites through the Installation Restoration Program. To date only a few sites on the DOD's 7000 facilities are known to have radioactive contamination. Most DOD radioactive waste management activities are regulated by the Nuclear Regulatory Commission, the EPA or both.

11.8 Remediation markets

Some rough estimates of the size of the remediation market in the US are shown in Table 11.6. The figures are based on 1991 estimates by the EPA [1].

Again the Superfund programme is the best defined. The EPA has 1232 sites on the National Priorities List. The total cost is not known, but the EPA estimates their cost for the clean-up of these sites to be $16.5 billion.

Under the RCRA corrective action programme, the EPA estimates that between 1500 and 3500 TSDF sites will eventually require correction. A broad range is given for the cost of undergoing this remediation ($7.4 billion to $41.8 billion). The UST programme, with its 295 000 sites and 43 million cubic metres of waste to be treated, will cost approximately $30 billion.

For the approximately 7000 contaminated military facilities under the management of the Department of Defense, the clean-up cost is estimated to be $25 billion. The Department of Defense is aggressively pursuing its remediation programme, and good progress is being made.

The DOE estimates that about 4000 sites will require clean-up. Many of these sites are large and the extent of contamination is not fully known. Also, many of the sites are contaminated with radioactive components.

Table 11.6 1991 Estimates of remediation market size [1]

Programme agency	No. of sites	Clean-up cost
Superfund	1235	$16.5 billion (EPA's cost only)
RCRA corrective	1500 to 3500	$7.4 billion to $41.8 billion
USTs	295 000	$30 billion
DOD	7 000	$25 billion
DOE	4 000	1994–98: $ 12.3 billion
Civilian federal	350	Total: 'Hundreds of billions'
State programmes	19 000	1991–95: $1 billion for further
Private	Unknown	1991: $1 billion

This causes much uncertainty in the numbers, but cost estimates are always large. The EPA simply estimates the total cost for remediation to be 'hundreds of billions of dollars'. Of this, $12.3 billion is expected to be spent during 1994–98.

The federal government also has contaminated sites under the control of various civilian federal agencies, including the Departments of Agriculture and the Interior. In all, 16 agencies have a minimum of 350 sites to be remediated. These agencies requested $1 billion for further investigation and clean-up during 1991–95.

In general, the various states have programmes that are patterned after the Superfund programme. More than 19 000 individual sites are now listed under state programmes. No dollar estimate is available for the clean-up of these sites, but it will undoubtedly be large.

Another important part of the clean-up effort is handled directly between private parties and remediation companies. The number of sites to be treated under this scenario is unknown, but the EPA reported an estimate for 1991 of $1 billion.

11.9 Remediation business outlook

Remediation activities under the Superfund programme, which have never moved swiftly, have slowed further as potentially responsible parties (PRPs) await reauthorization of the programme by Congress. The PRPs are, of course, hoping that clean-ups will be less involved and less expensive after the reauthorization. Other programmes tend to follow the Superfund programme. With the exception of clean-ups at Department of Defense sites, which appear to be moving forward steadily on clean-ups, many owners or managers of sites are taking a 'wait and see' position while awaiting the reauthorization of Superfund.

11.10 Conclusions

1. Remediation of contaminated soil in the US is impacted directly and indirectly by a complex set of regulations. The Resource Conservation and Recovery Act (RCRA) and the Comprehensive Environmental Response, Compensation, and Liability Act (CERCLA) are of most importance.
2. Remediation under the CERCLA, the Superfund programme, has moved slowly, and the programme is widely perceived to be a failure. The Act has now expired, and reauthorization is being considered by Congress. Major reforms are expected, especially on liability issues.
3. The set of regulations impacting remedial activities and especially those affecting selection of clean-up standards has grown to be incredibly com-

plex. A cessation in the promulgation of new regulations and simplification of current regulations is needed.

4. The Department of Energy has the responsibility for remediation of vast quantities of radioactive soil and debris. Most effort to date has focused on identifying and characterizing the waste. Decades of effort costing tens to hundreds of billions of dollars US will be required.

5. The Department of Defense is aggressively addressing its remediation needs. Actual clean-ups are underway, and good progress is being made.

12
Netherlands' reclamation practices

12.1 Introduction

The well-known Netherlands emphasis on multifunctionality (Chapter 1, section 1.2.3) has tended to obscure more fundamental differences from the contaminated land practices adopted both in the UK and the USA.

Probably the most striking of these distinctions is the overwhelming social acceptance that removing the legacies of past industrial malpractices from soils is an essential national aim, with which industry and the owners of contaminated sites co-operate to a quite remarkable extent. Defensive legal actions, to avoid or reduce clean-up costs (which so typify USA practices, and which seem certain to grow in the UK as the 1995 Environment Act begins to bite), have until now been extremely infrequent in The Netherlands.

12.2 Legal and administrative controls

Like the UK and The USA, the Netherlands has evolved an increasingly strict body of legislation intended both to preclude the formation of more contaminated land by enforcing good waste disposal practices, and also to ensure that existing sites which are hazardous to human health or the wider environment are remediated as rapidly as possible. The differences which do exist (e.g. the willingness to employ public funds where no polluter can be identified to bear necessary remediation costs) essentially reflect a national determination to improve the Dutch environment.

Soil protection was first recommended in 1962 when groundwater protection policies were being devised for the Ministry of Social Affairs and Public Health, and a draft Act (1971) did advocate that soil qualities around groundwater abstraction zones should be controlled. This draft Act, however, was in advance of its time and was not brought into force. The national mood changed in 1980 when the well-publicized Lekkerkerk incident forced a governmental reaction. The Soil Cleanup Interim Act (Interimwet Bodemsanering. IBS) was enacted to address such serious historic soil contamination instances and allow government to use public monies for necessary remediation and then recover costs from the guilty polluter at a later time.

While the Interim Act was successful, it was obvious that its scope (limited to already contaminated land) was too narrow. Thus the Soil Protection Act (Wet Bodembescherming, WBB) was brought into force on 1 January 1987 to introduce a 'duty of care' (cf. Chapter 10, section 10.5.4) to prevent future soil contamination. This placed the legal responsibilities on both the landowners and occupiers, who had to undertake any necessary remedial actions. Unlike similar UK legislation, the safety net of government action, in situations where landowners and occupiers cannot bear the costs, is more easily available.

For older (i.e. pre-1 January 1987) cases where land has been contaminated, the more serious and urgent cases are tackled in one or other of the following ways:

- private parties (owners or occupiers) voluntarily undertake necessary remediations;
- private parties are compelled by Cleanup Orders, Investigation Orders, or Temporary Safety Orders to do what is required; or
- the government itself carries out the work and then seeks to recover its costs from polluters.

The 1987 Soil Protection Act is an evolving framework legislation to which more detailed regulations are added by General Administrative Orders (GAOs) and Ministerial Decrees. Consequently an interim policy stage has been reached with the publication of a document entitled 'Circulaire inwerkingtreding sanerings regeling Wet Bodembescherming' (Circular on the coming into force of the regulations on the soil cleanup provisions of the Soil Protection Act)[1]. This became law on 1 January 1995 and requires:

- the assessment of contaminated levels in soils;
- determination of the severity of this contamination;
- prioritization of identified contaminated sites;
- remediation targets which have to be achieved;
- assessment of the practicality of cleaning soils to these target levels;
- isolation, containment and monitoring requirements for those cases where clean-up is not technically achievable; and
- various financial and legal matters.

Since soil protection quality levels ('Intervention Values') [1] above which soil multifunctionality is or will be severely diminished, and 'target values' which reflect negligible human toxicological or ecotoxicological risks – (Appendix I) have been produced, the Netherlands' regulations are in advance of those which exist today in the UK.

The responsibility of enforcing the regulations rests with the provincial authorities, and with the larger City Councils of Amsterdam, Rotterdam, The Hague and Utrecht. Specialist advice is also available from the Technical

Soil Protection Committee, established under the Soil Protection Act, and set up to issue scientific and technical recommendations to the Minister of Environment.

In addition to the Soil Protection Act and its regulations, other legislation is relevant. In particular the Environmental Management Act 1993 and the Housing Act are important.

These mirror the UK's Environmental Protection Act 1990 and the Town and Country Planning Acts, and exist to ensure that wider social concerns are properly addressed.

12.3 Private-sector initiatives

The point has already been made that private industry has co-operated, to an extent, almost unthinkable in the UK or the USA, with national soil quality ambitions.

This is most apparent in the 1993 (BSB) private-sector investigation and clean-up programme, which arose after a joint committee of industry, local, regional and central government had concluded, in 1991, that the exact nature and extent of soil contamination on active industrial sites was unknown and should be established without delay [2,3]. The private-sector BSB (Bodemsanering Bedrij Festeireiner) programme aims to prioritize sites on the basis of the risks these could pose to the environment and public health, and so far has identified some 25 000 sites on which serious soil contamination can be expected [1].

Table 12.1 lists the priorities and planning approach which has been adopted.

That such a scheme should be administered and staffed by private industry (in one or other of 12 provincial foundations), with government's role restricted only to observer status, will seem strange to UK and USA readers and is a striking confirmation of how strong is the public will in The Netherlands to rectify the adverse effects of industrial activities. Only in Denmark has a similar programme has been supported by industrialists.

Table 12.1 The priorities adopted in the BSB initiative

Urgency class	Initial soil investigation required by	Future (detailed) soil investigation required by	Remediation to be undertaken by
1 to 3	1996 or earlier	1993 to 1998	1994 to 2004
4 to 6	1999 or earlier	2004 or earlier	2009 or earlier
7 to 8	2004 or earlier	2014 or earlier	2024 or earlier
9 to 10	2009 or earlier	2024 or earlier	2034 or earlier

Classes 1 to 3 urgent
Classes 9 to 10 not urgent
Other classes intermediate in importance

12.4 Decision-making on contaminated sites

12.4.1 *Introduction*

The essential principle in The Netherlands is that soil is a valuable resource whose quality should be maintained. Thus the preservation or restoration of soil multifunctionality is the dominant emphasis.

Target values (Appendix I) have been devised on background levels or available ecotoxicological grounds and are expressed so that clay and/or organic contents (which influence the mobilities of many metallic contaminants) are noted. Similarly intervention values (Appendix I) have been identified to indicate the points at which remediation is needed to remove possible risks to human beings and ecosystems. It should be noted that these target and intervention values are *the average contaminant concentrations over specific areas and to specified depths* (i.e. over a $50\,m^2$ area and to 1.5 m depths for soils, and over $100\,m^2$ and to a 1 m depth for groundwater). These spatial scales also apply to deeper layers of soil. This particular usage, which identifies average soil and water volumes, differs from UK practice where attention has usually focused on the peak concentrations in soil and waters.

12.4.2 *Current practice*

Current Netherlands practice requires a three-stage evaluation process [4], i.e.

- soil investigation;
- exposure assessment for the contaminants found; and
- a clean-up investigation.

Soil investigations follow the same pattern as has been described for the UK (Chapter 3) and include desktop studies of existing historical information and data from later intrusive site investigations, sampling and analyses. Since a comprehensive range of intervention values exists, for both soils and waters, the identification of cases where future risk may arise is, of course, far simpler than it is in the UK.

The second – exposure assessment – evaluation is undertaken primarily to prioritize the urgency for remediation, and this inevitably calls for a consideration of the land use, since actual risks obviously are a consequence of which type of use is intended. Risks to human beings, to plants and animals, and to the wider environment, if contaminant immigration is possible, are each separately assessed.

The final, clean-up investigation, is carried out to determine whether a multifunctional clean-up standard is feasible, or whether an 'Isolation, Containment and Monitoring' (ICM) alternative has to be considered. Each option is usually examined in considerable detail to allow full comparisons. Netherlands practice, in multifunctional reclamations, is usually to excavate all suspect soil material and remove for off-site treatments, whereas in the ICM situations isolation by cut-off walls and top covers, and groundwater controls (to prevent uncontrolled migration of soluble contamination) is more likely to be recommended. Where an ICM solution is adopted it is usual to specify which future land uses are to be prohibited and the specific aftercare and maintenance programmes which have to be implemented. An example should make the point clearer – when groundwaters are contaminated by volatile chlorinated hydrocarbons, the spread of the water can be controlled by pumping or cut-off walls, but this would not preclude upward migration of evaporated contaminants. Thus either prohibiting any future domestic housing from such a site, or minimizing the vapour risks by hard covering the land (pavements and gas proof membranes in house foundations) to allow the presence of housing could be necessary. When an ICM solution is proposed, it often is helpful to design various options, based on different protective measures, to gain a fuller insight into the relationship between future land uses and the remediation costs which will be compelled, if residual risks are to be minimized adequately.

Netherlands' practices arise from the existence of a very comprehensive list of target and intervention values for soils and groundwaters. Such values of course have been derived from one default exposure scenario and consequently there will be many contaminated sites where site-specific conditions are less critical than the standard scenarios. Thus areas for debate with regulating authorities do exist.

Risk assessment is a far more developed judgemental tool in The Netherlands than has so far been the case in the UK, and is routinely undertaken to establish and cost necessary remediations and also to identify allowable re-use of land (Table 12.2).

Table 12.2   Use of risk assessment in The Netherlands

Aim	Method adopted	Details to be considered
Land quality	Comparison with published values	All possible exposure routes must be covered. Protection of species already on, or potentially present, must be considered
Urgency for remediation	Determine the risks to humans, to ecosystems, and if contamination migration occurs	Exposure assessments required* Site-specific ecosystem surveys required* Potential for migration has to be quantified
Establishing residual risks	Devise ICM variants. Formulate supplementary safeguards to reduce residual risks	Current and future land use. The aftercare and monitoring plans
Establishing allowable land re-uses	Compare soil quality requirements for different land uses	Various possible land uses. Risks possible for each land use

* If a decision is not possible on model predictions due to uncertainties.

THE CSOIL MODEL

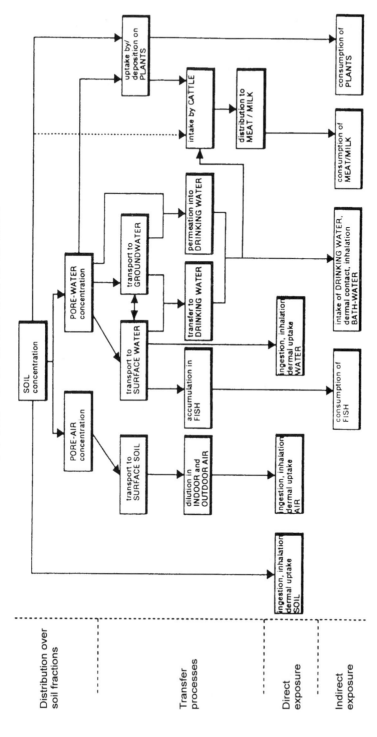

Figure 12.1 The Csoil model.

12.5 Designing a risk assessment

12.5.1 *Introduction*

As noted above human toxicological, ecotoxicological and contaminant migration assessments must separately be undertaken.

12.5.2 *Use of CSOIL model*

For risks to human health, use is generally made of an exposure model ('CSOIL'), which is comparable to the German UMS, the European HESP and the UK's CLEA models. A schematic of the Dutch CSOIL model indicates (Figure 12.1) the stages which are considered and that solid, water and gaseous contamination phases are each evaluated. The total site-specific exposure is taken to be the sum of exposures from these different contact routes.

Like all such models, CSOIL tends to overestimate the risks in certain situations and also is limited in the conditions it addresses. Mixtures of substances, such as to common association of PAHs and mineral oils below old production sites, cannot be assessed, and in consequence, a CSOIL variant assessment system was developed, for such substances as mineral oils, based on the products the oil mixture contains.

Among the simplifying assumptions built into the CSOIL model are:

- that the uptake of a contaminant by the human body equals the actual biological intake; as in fact, many contaminants are strongly absorbed onto soil particles and so non-bioavailable, this often results in significant over-estimations;
- that only two population groups, i.e. adults and children, need to be considered; since this overlooks some subpopulations that may be more susceptible (e.g. pregnant women) who could particularly be at risk from some contaminants, it can often be useful to make use of human subject parameters (body weight, exposure frequency, exposure duration, contaminant intake, etc.) which differ from those fixed by the model;
- that exposure routes are limited to ingesting soils, direct contact with soils, inhaling dusts, eating contaminant bearing crops, drinking water and bathing/showering. Milk, beef and fish intake exposure routes are not included, and could be important in some situations.

Because of these, and other limitations good practice should be to conduct confirmatory analyses in the contact media of concern. This is not difficult to do in many cases (e.g. establishing the actual contaminant concentration in the food crops grown in contaminated soils) and will give a more soundly based assessment of risk.

As with any other risk assessment process, the CSOIL model results will be far more credible if earlier site investigation has revealed an adequate body of truly representative contaminant concentration information. This implies that not only should results be numerous (50 analysed samples of the topsoils in gardens will obviously be more indicative than only two or three topsoil analytical results) but that they should be measurements of the important exposure route.

Soil analyses are entirely acceptable if soil ingestion is the route of concern, but quite inappropriate if gaseous or vapour exposures are important. In the latter case, attention would better be focused on analyses of soil vapours and consequently on inside and external atmospheres.

Model results are given in terms of total daily intakes at the site and when these exceed the maximum allowable doses a risk to human health is accepted. This, however, need not always be the case, since the safety factors included within the model may be large enough to preclude actual harm. While this could be taken as a criticism of the assessment approach (particularly when population health surveys have failed to identify any actual ill-effects) the essential point is that risk assessment, in Netherlands, practice, serves more as a support for policy decisions than as a precise indicator of the onset of health reductions.

12.5.3 *Ecotoxicological assessments*

Ecotoxicological risk assessments rest on the limits at which risks can arise, the extent and depth of contamination and the theory of island biogeography. This assumes that the complexity of an ecosystem is proportional to the size of the area involved.

So, in daily Dutch practice a risk assessment for the ecosystem is restricted to comparison of the concentrations in soil with the Hazardous Concentrations of the compounds involved. If the HC50 is exceeded the degree of the contamination will determine whether or not there is an 'actual' risk for the ecosystem. Ecotoxicological bioassays and field studies may be used afterwards to verify such an 'actual' risk, but the interpretation of the results of these studies is still subject to discussion. It is to be expected that relevance of soil contamination for the ecosystem will focus more on its influence in future development of natural reserves.

12.5.4 *Contamination migration*

The risks of uncontrolled spreading of contamination obviously are a serious matter, if for no other reason than the fact that remediating a larger area is more expensive than cleaning up the original smaller patch of contamination. In Netherlands' practice, and with the national need to protect groundwater resources, contaminant migration is most often considered as a water pollution problem. In the UK, with its lesser reliance on underground water reserves, the migration of gases and vapours has as much importance as the spreading of aqueous contamination.

For groundwater situations, classical hydrogeological methods are adopted. Factors such as the velocity of groundwater flow and the retardation of contaminants are important. This last point depends on the soil-water distribution co-efficient, which in turn arises from the soil type, its bulk density and porosity. Assessment models offer a range of soil property values, though these are average, and are better replaced with actual measurements taken on the land being assessed. Using such methods the volume of contaminant transported per year can be assessed, if enough hydrological information is available for accurate groundwater contours to be drawn. This approach is implemented in the Tauw-CSOIL computer model to allow meaningful assessments of groundwater contaminant migration to be made in most normal situations. From these and estimations of the dilution which a contaminant 'hotspot' will experience, it is usually possible to predict where a contaminated groundwater plume will reach in a specified time, and whether the resultant water quality deterioration will be enough to require off-site remediation work. As in all predictive model uses accuracies are much enhanced if actual water samples are obtained and analysed along the plume's migration pathway.

12.6 Remediation procedures

12.6.1 *Introduction*

If risk assessments have been conducted rigorously, situations which demand soil remediation will be identified.

Each year, the provinces draw up a programme which includes all cases where sites are to be investigated and/or remediated by private parties or the authorities. Private parties are required to report their investigation results to the competent authority, which then judges whether contamination is serious, whether remediation is urgent and then sets out an administrative decision. In this, the authority determines if:

- temporary safety measures are required;
- a change in land use is indicated;
- the submitted investigation results and/or remediation plan are acceptable; and
- the date by which remediation should be commenced.

The authorities really have two options when remediation seems to be urgent [2], i.e.

- to insist that it be commenced as soon as possible, with a maximum time limit of four years;

- to allow remediation to take place after four years from the issue of their decree, but within 25 years.

A similar procedure applies to investigations and remediations financed by the government (the so-called SPA cases), though approval of the remediation plan is not necessary in these situations, as the remediation plan is drawn up under the responsibilities of the authorities.

Unlike the UK the public is informed of contaminated land cases. Regional newspapers print accounts of the contaminated site investigations, the decisions of the statutory body, and the authorization of remediation work. Specific permits, however, are required for land remediation activities and include:

- a permit (under the Surface Water Pollution Act) to discharge contaminated groundwaters; a delay of six months will be caused by the necessary consideration of such an application;
- a permit for groundwater abstraction; provincial authorities consider these applications and decisions take three months; if abstraction rates are to be high notification and public involvement are likely, so delays could be longer;
- a permit (under the Environmental Management Act) for any necessary above-ground treatment facilities, which could be pollution sources;
- a permit for the discharge of groundwaters to the public sewers; this would be granted, after a month's delay, by the city council; and
- a permit for any planned tree felling; the city council would consider such requests and take a month to issue its decision.

This list of necessary permits, for an already approved land reclamation, is rather more onerous than current UK procedures.

12.6.2 Remediation plan and specifications

As would be expected from the greater levels of regulation and controls which typify Netherlands' practices, remediation plans and specifications have to be drawn up in more detail than is usual in the UK and are based on a national (RAW) specification. The relationship between the client and the contractor is also prepared on a uniform national basis (UAV 1989 – Uniform Administrative Conditions).

12.6.3 Regulation of excavated soils

All excavated contaminated soils have to be reported to the national Service Centre for Soil Treatment (SCG), which then distinguishes which soils are cleanable by available technologies and which are not. Soils containing in excess of 20% of finer particles (63 microns or smaller) would be uncleanable and a 'non-treatable declaration' would then be issued. Without this declaration, a cost surcharge of 26.2 guilders per tonne would be required when untreatable soils are taken into landfills.

The temporary storage of contaminated soils is managed by SCG which also sells cleaned soils as fill for building projects. Since the centre fixes the costs of soil treatment, provides advice on governmental clean-ups and is consulted on private-sector reclamations, SCG obviously has a dominant role in the Netherlands' practices on contaminated land clean-ups.

No other nation has adopted a centralized off-site soil cleaning system or the regulatory controls which SCG administers.

12.7 Soil remediation practices

12.7.1 *Introduction*

Netherlands' practices are continually evolving as innovative treatment solutions are devised. Unusually – compared to other nations – innovative reclamation solutions are viewed as opportunities rather than as risky endeavours.

Regulatory authorities support such approaches in the hope that cheaper and more effective outcomes will be achieved, and landowners generally are also supportive. Obviously if a trend setting solution is to be successful the parties involved (the authorities, the landowner(s), the consultants acting for both parties and the contractors who actually carry out the work) have to support and trust each other and be creative in resolving the inevitable problems which arise when technological innovation is attempted. The sadly all-too-common disputes and the employment of litigation – which increasingly has come to dominate the UK's reclamation endeavours – are not common and probably reflect the Dutch willingness to seek consensus and co-operation whenever this is justified.

Governmental funding for these innovative attempts is also available, as is the active participation of universities and research centres.

12.7.2 *Current trends*

Owing to the successful soil remediation operation the number of urgent sites will be greatly reduced in the next few years. Attention will shift from remediating the legacy of the past, towards soil quality management and prevention of soil pollution in the future.

The classic intensive soil remediation approach looks likely to become restricted to areas which are being developed and to cases where catastrophic leakage could create wider environmental degradation.

For sites which are still in active industrial use, the emphasis has moved to adopting longer time frame solutions which are cheaper and less disruptive.

Encouraging biological activities to increase degradation rates for organic contaminants and relying on natural decay processes are typical of those, as are the employment of active biological boundaries to preclude wider dispersion of aqueous pollution. In all such cases intensive monitoring is essential.

Bioremediation, which can be a slow in-situ process because of ineffective mixing and complex transfer phenomena in heterogeneous soils, will obviously become more widely employed as its low operating costs will often be more important than the slow achievement of soil cleanliness. An interesting example of a more innovative bio-remediation approach has recently been undertaken at a dry cleaning plant in south Holland. The tetrachloroethene (PCE) contamination is broken down in two consecutive steps, firstly PCE is dechlorinated to *cis*-dichloroethene with trichlorethene as an intermediate. In this first step, methanol is added to form a substrate for anaerobic dechlorination. Then the *cis*-dichlorethene is degraded aerobically by air sparging and phenol injection. This full scale trial is partially funded by the NOVEM programme and is a joint effort by the site owner, the consultant, the contractor and a participating university, all of whom have invested in the trial.

12.7.3 *Recent initiatives*

In October 1992, the Ministry of the Environment appointed a Soil Cleanup Working Group, known as the Welschen Committee. This was intended to translate strategic policy aims into practice and recommend operational solutions. By 1993, the committee had recommended:

- a sharing of costs between government and responsible parties for particular types of contaminated sites (such as old gasworks);
- a simplification of the administrative procedures when remediation has to take place; and
- an increase in the environmental benefit from remediation operations.

Many of the recommendations have been incorporated into the amended Soil Protection Act.

The problem of the blighting of strategic development projects as a consequence of soil contamination has also been addressed and it has become accepted that contaminated land impacts on land use planning, inner-city redevelopment and real estate transfers have to be given due attention. The current BEVER policy renewal process hopefully will remove many of the delays which are built into existing contaminated land procedures.

Active management of contaminated land, rather than total clean-up, is gaining in acceptance and this will – in the regional planning context – result in a stronger relationship between soil quality and suitable later uses of the land. Risk management will inevitably play an important role in this.

The reclamation market – which in The Netherlands is undeveloped by UK standards – is to be encouraged by increasing the opportunities for benefit both to government and the private sector. If successful this will offer the hope of additional private-sector injection of funds into contaminated land remediations.

12.8 Summary

Contaminated land practices in The Netherlands reflect the national determination to improve the Dutch environment. This has led to the establishment of a particularly comprehensive Soil Protection Act, accompanied by a substantial listing of soil and water quality values. For the prioritization of necessary clean-ups, risk assessments are routinely employed. The support and co-operation of Dutch industry has been a particular feature of the contaminated land reclamation successes which The Netherlands has achieved.

As noted earlier (Chapter 1) national concerns are the essential difference between clean-up policies in different countries and the contrast between The Netherlands' practices and the more market-controlled approaches which have grown up in the United Kingdom stem from this basic point.

The Netherlands – like the USA and the UK – is evolving its policy and seeking cheaper and less disruptive reclamation solutions. With other social needs demanding a share of scarce national resources, this is not surprising.

Technically, the distinction between approaches in The Netherlands and those in the United Kingdom include the willingness for government in Holland to support the trials of innovative treatment solutions and the co-operation of clients, regulators, consultants and contractors in making such treatments work. In contrast, UK practices seldom offer the opportunity for trials of innovative remediation methods. Thus Dutch practices do have access to a far wider range of technologies, which (while excessively expensive by the cost standards which have developed in the UK) are able to resolve a much wider scope of contamination problems.

13

Effective management of contaminated land reclamations

13.1 Introduction

A well-designed and managed reclamation is one where interested parties have no doubt that the reclaimed land is indeed fit for its intended purpose. In contrast, the qualities of ill-managed reclamations are questioned, regulators demand additional proof-testing work, and potential purchasers find the confirmation of the reclamation standards unconvincing. These doubts can cause the land resale values to fall and, in some worst cases, can leave the reclaimed land unusable.

These undesirable outcomes can be avoided by effective investigation and reclamation management, which only calls for a realization that the quality priorities differ in each of the investigation and reclamation stages.

The stages worthy of particular interest include:

- the pre-site work period;
- the site investigation stage;
- the reclamation work itself; and
- the post-reclamation proof-testing.

13.2 The pre-site work period

13.2.1 Introduction

This extends from the completion of the preliminary conceptual model (Chapter 3, section 3.4.4.) until the start of intrusive site investigations.

The priorities, at this point, are to appoint an investigation contractor who recognizes the problems which might arise, and who has taken reasonable precautions to prevent these actually occurring. The Health and Safety of investigation workers and the avoidance of environmental pollution are the concerns of greatest significance, though establishing co-operative contacts with regulators (Environment Agency, local environmental health officers, etc.) is also important.

Government legislation, and in particular the recent Construction (Design and Management) Regulations 1994 [1] intend to improve health and safety for the construction industry by placing new statutory duties on the client, the designer and the contractor. Site investigations where four or more people (including supervisors) are employed are subject to these rules, which demand that:

- the client appoints a competent planning supervisor, who has to ensure that a pre-tender health and safety plan and a health and safety file are prepared;
- the principal contractor (should more than one be engaged on different aspects of the work) has to co-ordinate health and safety throughout the construction period.

In cases where these regulations apply, it is necessary to issue a detailed description of the site conditions and of the hazards which would exist to all firms tendering to carry out a site investigation. The penalties for breaches of the regulations (i.e. up to £20 000 fines and six months' imprisonment in cases tried at Magistrates' Courts and unlimited fines and up to two years' imprisonment for these more serious breaches tried at Crown Court).

The basic requirements for all but very slightly contaminated sites are:

- to fully inform those bidding for the investigation work of the conditions and hazards which may exist;
- to be convinced that contractors are competent and have recognized the need for adequate health and safety precautions in their tendered prices;
- to confirm that necessary insurance cover is in existence;
- in cases where the CDM regulations apply that the principal contractor has properly developed the health and safety plan *before* any investigation work is commenced; and
- to liaise with those external bodies and regulators who will judge, at a later time, whether the site investigation has been adequate in scope and detail.

13.2.2 *Information to contractors*

Information which possible contractors require includes:

- a site zoning plan and sections (Chapter 3, section 3.4.5), attached to which is a listing of those different contamination assemblages which are anticipated in the various areas of the site;
- specific and emphasized advice on the possible hazards to the health and safety of contractor's work people; and
- advice on where ground disturbance could create unacceptable environmental impairment.

Contractors will also be supplied with tender documents, describing the works needed on more contaminated sites, and will usually price a detailed Bill of Quantities (Figure 13.1) in building up their final cost bid for the

Item	Description	Unit	Quantity	Rate	£
A 2.6.1	Establishing necessary perimeter air monitoring network (14 stations)	lump sum	–	–	
A. 2.6.2	Weekly monitoring of perimeter air quality and testing for asbestos fibres at each station	week	48		
A 2.6.3	Additional air monitoring in vehicle cabs and at excavation faces, as required	lump sum	–	–	
A 2.6.4	Weekly provision of air quality test reports	week	48		
A 2.6.5	Sampling and analysis of groundwaters as specified	–	50		
A 2.6.6	Sampling and testing of soils as specified	–	1000		

Figure 13.1 Typical extract from a land reclamation Bill of Quantities.

site investigation. In such cases – and where contractors have been warned of the existence of possible subsurface hazards – the specific health and safety and environmental precautions a contractor has priced into his bid can be very revealing, and are often quite enough to distinguish between competent and incompetent firms.

An illustrative example was on a derelict dock, where oil storage tanks had existed. Desk studies had suggested that these had leaked and that pools of impeded oils might have collected at rock head, where an impermeable marly sandstone band occurred. Below this were open permeable sandstones, widely used for public and industrial water supplies. If, in site investigations, oils oozed into excavations and pooled on top of the marly sandstone, further excavations would allow uncontrolled dispersal of pollutants and be a direct breach of water quality protection legislation (Water Resources Act 1991). An experienced contractor would appreciate the significance of this, and would add costs for the provision of pumps and waste liquid tankers to dry out excavations before penetrating the upper poorly permeable sandstone layer. Any contractor who failed to include such an extra cost either had not understood the warnings provided or was not interested in preventing environmental harm. In either case his or her employment would be unjustifiable.

Likewise, when tendering contractors had been told that health-affecting contaminants would be likely to exist at very high concentrations, concerns for staff safety should have led to a safety plan and the training of investigation workers being seen as essential. If the costs for these were not included in a contractor's tender, he or she might have been too inexperienced to recognize the significance of the supplied information, or possibly could have a casual disregard for health and safety regulations. Use of such a contractor could be distinctly imprudent, since the client would then (in cases where the CDM regulations apply) be exposed to the risk of criminal prosecution.

13.2.3 *Insurance cover*

Site investigation contractors necessarily carry insurance to cover accidental harm to their employees and to neighbouring public. To the direct employee cover and the insurance of third parties (including trespassers) has to be added insurance against accidental environmental impairment. Policies should be checked to confirm that they are still in force, and that no overly restricted exclusion clauses feature in the cover.

If an otherwise acceptable contractor is incompletely insured this need not be a problem, since specific cover for a particular contract can be obtained, and is only an extra cost against the work.

It often seems a thorough nuisance to have to check insurance policies and obtain extra cover, but in a litigious age it is worth the effort. Seeking recompense from the courts is far more common than was the case even five years ago. Trespassers who once recognized that their intrusion into land, in the process of investigation and reclamation, was both illegal and risky, now seem to feel that they have every right to rush to a solicitor if they suffer harm, and workmen who once took pride in being involved in dangerous work now seek legal remedies if their health suffers as a consequence.

Because of the defence costs and problems if litigation is commenced, it now makes sense to ensure that full insurance cover is in place before any site work is started.

13.2.4 *Contact with external bodies*

If health and safety, environmental protection and insurance concerns have been addressed, it only remains necessary to be sure that the investigation works will satisfy the various external bodies which will be consulted in the land development approval process. These bodies (Environment Agency, environmental health department, English Nature, etc.) will gain access to the site investigation reports and conclusions provided to the local planning authority, and may object to the granting of development approval if (in their view) investigations have been inadequate.

Given this, it can be useful to informally discuss the planned investigation details with these external agencies, and take on board more strongly voiced concerns. Quite often, these bodies do have specific data on past land uses and activities which the investigation planner might not have included.

While many practitioners prefer to avoid contact with regulatory authorities until site investigation has been completed, it is difficult to justify this view. These bodies *will* be those which will be consulted when development approval is sought, and it makes sense to investigate land well enough to remove their doubts and concerns.

13.2.5 *Conclusion*

Management, before intrusive site investigations are commenced, need not be an onerous task. All that is essential is to appoint a competent contractor who fully understands any predictable hazards, who has included necessary precautions, developed the pre-tender health and safety plan, and holds adequate insurances. If this is done, and if more important external agencies' views on site investigation adequacy are confirmed, then the reclamation work is probably off to a good start.

13.3 Health and Safety at Work Regulations and Legislation

13.3.1 *Introduction*

While Health and Safety safeguards in a site investigation will seldom be overly rigorous (because exposure durations to subsurface contaminants are usually only for very brief periods), much more comprehensive precautions are usually essential in the following reclamation stage. Thus it could be useful to highlight the essential features of the existing legislation and note more commonplace failings and deficiencies.

13.3.2 *Breaches of Health and Safety Regulations*

It is widely appreciated that the health and safety of workers is controlled by statute law [2,3], but the consequences of inadequate safeguards are not always fully understood.

Health and Safety inspectors can and do routinely issue Improvement Notices (S.21 of the 1974 Act) which demand specific corrective actions within specified times. In cases where failings are more serious, Prohibition Notices (S.22 of the 1974 Act) can require that particular activities are halted until revised health and safety precautions are fully implemented. Total cessation of all work on a site also can be enforced where more widespread failures of safety policies are identified after inspections, and – in worst cases – criminal prosecutions can be instigated.

Unlimited fines and/or periods of imprisonment can be imposed by courts for more serious breaches of the law.

A commonplace first reaction is that Health and Safety is a matter primarily for the appointed contractor. However, matters are not so simple as prosecutions can be directed against all the responsible parties, including the reclamation designers and consulting engineers, who have an obligation to insist on safety working practices. It makes little sense, of course, for a consultant to allow a contractor to work in a dangerous manner, since site activity could be halted, and adverse local publicity generated. Civil compensation claims often also follow successful Health and Safety prosecutions, and consulting engineers could be a personal financial risk if they have agreed to the use of hazardous working practices.

The introduction of the CDM Regulations only reinforces this view, health and safety is of importance to client, consultant and contractor and the force of law will be employed against those who ignore these requirements. Importantly, no time limit on prosecution features in these recent regulations and criminal prosecutions may be mounted whenever the Health and Safety Executive becomes aware that a breach of statute law had occurred. The latest strengthening of legal enforcement over health and safety issues has essentially levelled the playing field; it will no longer be the contractor's fault (no matter how poorly advised he was) when accidents occur, as clients and designers are equally at risk if they fail to adequately ensure that health and safety are protected.

13.3.3 *Consequences of information limitations*

One real difficulty is ensuring safe work on a contaminated land reclamation is that a full knowledge of possible hazards is never available. There could be undiscovered vats of flammable liquids in unexplored areas, and local pockets of (say) blue asbestos wastes could have been hidden within an otherwise quite clean area of a site. This is a far more difficult situation than that within a working factory, where only a restricted list of relatively easily identified hazardous substances and processes will occur.

Recently this was brought home to the author during reclamation of a 1960s domestic landfill for re-use as a retail park. The contractor had devised a safety policy (section 13.3.4) which focused on the predictable health risks from metal-enhanced ashy dusts and from the oxygen depletion and carbon dioxide occurrences in deep excavations. This adequately described the hazards indicated by site investigations, was fully explained to the reclamation workforce, and adequate personal protective equipment and decontamination facilities were provided [4,5, etc.]. Since the reclamation work consisted of no more than excavating ill-compacted ashy fills (the remnants left after biodegradation of the domestic wastes), screening out particular contaminated fractions, and relaying the screened ashes below a compacted clay capping, the consulting engineers visited the site infrequently, and left daily supervision to a junior resident engineer. It thus came as something of a shock when – towards the end of the reclamation – the resident engineer reported difficulty in having a 'pale blue granular powder' accepted by licensed waste disposal facilities. Investigation revealed that, within the very uniform ashy fills, someone in the past had carefully encapsulated pockets of cyanide wastes from an old town gas works. These, the resident engineer had had segregated and stockpiled, but had never chosen to have the blue granular material analysed or to seek advice on its safe retention and disposal. When analyses revealed that 60 mg/kg of free cyanide existed in samples from an uncovered and wind blowable stockpile (which workmen had treated with no concern since it did not feature on the site safety plan), a Health and Safety inspection was required by the local authority, and resulted in substantial fines being imposed *and* on the site being closed until the stockpile was fully protected from the atmosphere and disposal arrangements were devised. This example illustrates a particularly commonplace failing in most health and safety plans – i.e. that these *have* to be updated as additional information become available.

Other examples, such as the collapse of a workman because of oxygen deficiency in a sewer excavation, could be cited to stress how incomplete subsurface information will always be possible hazards to health. In the collapse case, landfill gas surveys had taken place, but had not been extended to one small area where chemical reactions (of slag wastes) had reduced local underground oxygen contents.

Contaminated sites exhibit a very wide range of possible hazards to workmen, i.e.

- of toxicity if metal rich dusts are inhaled on a daily basis;
- of cancer risks if high concentrations of asbestos fibres, benzenes and arsenic exist in soils;
- of disease if pathogen bearing sewage sludges are widespread;
- of asphyxiation, poisoning or explosion risks from a range of gases and vapours;
- of explosive combustion if heated materials are suddenly exposed to free air; and
- of respiratory ailments caused by continually dusty environments.

Given that these potential hazards are not infrequent, most practitioners insist on the provision of suitable personal protective equipment, fully equipped, decontamination facilities, and clean eating and smoking areas, and also impose controls over entry into any possibly ill-ventilated areas [4,5, etc.]. These safeguards, however, are not, in themselves, adequate to ensure health and safety.

13.3.4 *Regulatory requirements*

The Health and Safety at Work, etc. Act of 1974 specifically requires that employees:

- be provided with the information, training and supervision necessary for their continued well-being;
- be made aware of a specific safety policy which should be updated whenever new hazard information is gained (S.2(3) of the 1974 Act); and
- be consulted on the adequacy of the safety policy.

These legal requirements make it obvious that it is not the provision of safety facilities, but the management of the safety policy which is important. Cases such as the cyanide waste stockpile (section 13.3.3) demonstrate that the updating of a safety policy is practically imperative.

Thus the real necessity is for contaminated land practitioners to accept that health and safety should never be put at risk, establish a management structure to achieve this and nominate a specific individual to identify potential hazards and enforce and update the site safety policy. The complexities under many contaminated sites are such that access to specialist advice will often be required by the appointed safety manager.

The reasonableness of providing site safety instructions (Figure 13.2) and hazard data sheets (Figure 13.3) should not be a matter for debate. An educated workforce is less likely to make avoidable and hazardous work choices.

While Health and Safety can appear to be yet a further bureaucratic burden on the actual progress of reclamation work, there really is no need for time-consuming and unnecessary administration. In site investigation periods, a clear safety policy which is understood by the investigation workers and supplemented by specific hazard data sheets is probably all that is necessary. Later, in the main reclamation period, when exposure to subsurface hazards will be for much longer periods greater attention to detail is necessary, but even this can usually be included in the routine duties of site agents and resident engineers. The essential features of a reasonable reclamation site safety policy are clear descriptions of materials and conditions likely to be harmful, obtaining the acceptance of workmen that health and safety is a matter for everyone who works on a site, and insisting that anomalous materials and conditions be treated as suspect until specialist advice reveals that no risks need be voiced.

13.4 The site investigation period

13.4.1 *Introduction*

If adequate note has been paid to preserving the health and safety of workers and adjacent local inhabitants the main other quality criteria during site

1. The site was a sandstone rock quarry which operated until the 1920s. Later (1978 to 1983) the quarry void was filled with what should have been inert fills. These are too loosely compacted for housing foundations and have to be excavated, screened and relaid at maximum achievable compactions.

2. The safety risk on this site is due to the tipping of waste papers and sawdusts at depths of 4 to 5 m within the fills. These burnable materials were set on fire by vandals in 1981, and have continued to smoulder ever since. Underground temperatures are known to reach 85 °C and 95 °C at times, and the smouldering is also producing carbon monoxide gas on occasions.

3. Laboratory testing has shown that the sudden entry of atmospheric air into the heated papers and sawdust can cause explosive combustion to occur very rapidly.

4. The carbon monoxide gas is both toxic to people and very flammable. Because it is colourless and odourless its presence can only be proved by the use of carbon monoxide meters.

5. Reclamation can be conducted in an entirely safe manner if

    • no excavations take place until subsurface temperatures and gas contents have been measured and shown to be non-hazardous, in shallow (1 m deep) probe holes drilled at 5 m centres,

    • excavations then are limited, with only 300 mm layers being removed at any time,

    • repeat temperature and gas measurements are then conducted on the exposed excavated surface and show that no hazardous conditions exist immediately below this.

6. Should any subsurface temperatures in excess of 90 °C be encountered at 1 m depths, general excavation *must* be ceased and personnel *must* leave the excavation area. The specialized excavation team will then carefully expose the heated papers and sawdust and spread these to cool, before general site work can be recommenced. During this careful stripping, sand stockpiles will be available for use as fire blanketing if sudden combustion occurs. This careful stripping of unusually hot ground layers will be supervised by the Resident Engineer and must not be attempted without his direct instruction.

7. Should measurable carbon monoxide gas concentrations be found in any probe hole, site work must be terminated until the hazard has dispersed and this has been confirmed by repeated gas measurements conducted by personnel equipped with oxygen breathing equipment.

8. Because of the fire and explosion risks no smoking, fires or the use of non-intrinsically safe electrical equipment on site will be permitted. Any breach of this will result in immediate dismissal.

9. The mess facilities are the only locations where smoking and eating are permitted.

10. If you experience any headaches, giddiness or breathing difficulties you must immediately leave the working area and report your conditions to the Resident Engineer. These symptoms could indicate that you have been exposed to carbon monoxide gas and could be at risk if you remain in the working area.

24/3/95   Resident Engineer

Figure 13.2 Joycewood Site. Safety instructions for reclamation staff.

investigation is to obtain provably accurate and significant data on the site's conditions. This calls for a practical application of a basic quality assurance policy – without the annoying bureaucracy and delays which accompany many quality assurance schemes – which is really no more than a matter of common-sense.

SUBSTANCE PHENOL

UN No:	1671	CAS No: 108–95–2	Formula: C_6H_5OH	
OEs 8 h:	5 8 hppm	19 8 hmg/m^3	10 min 10 ppm	38 10 min mg/m

General risk: toxic.

Other effects: suspected or confirmed human carcinogen

Health surveillance: State: S: Vap.Den: 3.2;   Vap.Press:0.3mmHg;   Exp.limit 1.7–8.6%

BP	182 °C;	FIP	78 °C;	ignition temp;	605 °C

Reactivity:

Reacts with strong oxidizers; reacts with calcium hypochlorite; attacks Al, Zn and lead

Notes: On heating, emits toxic fumes; explosives mixtures formed > 78 °C; very rapid absorption through skin

RISK PHRASES

24/25	Toxic in contact with skin and if swallowed
34	Causes burns

SAFETY PHRASES:

2	Keep out of reach of children
28	After contact with skin, wash immediately with plenty of water
44	If you feel unwell, seek medical advice

HAZARD DATA

Haz. Chem:

2X	Use water fogs (fine water spray in absence of fog equipment)
	No danger of violent reaction or explosion
	Full protective clothing
	Contain

Fire extinguishant: CO_2, Halons, Powder, Alcohol-res., foam, water fog/spray

HEALTH PHRASES

1	May be absorbed via inhalation, ingestion and the skin
12	Corrosive to the eyes, skin and respiratory tract
39	Can affect the central nervous system, liver and kidney
30	Avoid all skin contact
50	Prevent dispersion of dust
60	Inhalation may cause lung oedema; serious cases may be fatal

FIRST-AID PHRASES

M1	Call doctor, or send/take to doctor/hospital; show medical staff substance data sheet or ensure information accompanies patient.
IH2	Inhalation: fresh air, rest and half upright position; especially in cases of irritating or corrosive substance
E1	Eyes: Rinse continuously with water for at least 10 min; transport to a doctor or hospital
S7	Skin: Place contaminated clothing in sealable thick plastic bag: wash skin with water and soap: rescuers must wear PPE

Figure 13.3 Hazard data sheet.

13.4.2 *Information to be collected*

An effective site investigation is not merely a precursor of site reclamation but an important element in the full body of information which will demonstrate the adequacy of the site reclamation and the suitability of the land for its intended re-use.

Detailing what should be produced from a competently managed site investigation is not difficult. Among the information wanted will be:

- precise locations for each trial pit and borehole to ensure that these can be relocated later;
- precise ground levels on each investigation point;
- accurate geological logs, which have been checked; personal comments can be most useful when these logs are later evaluated (Figure 13.4);
- precise depths and descriptions of materials taken for chemical analysis;
- accurate groundwater level depths, together with a note of the conditions in which these time variable measurements were made;
- a provable chain of custody set of arrangements to ensure that sample labelling, preservation and storage were adequate to ensure that analytical results are meaningful; and
- accurate and confirmed measurements of factors such as subsurface gassing conditions, soil density, etc.

Chapter 3, section 3.7.3, has earlier discussed many aspects of these issues and need not be repeated. Some additional emphasis, however, is useful to illustrate some more common deficiencies.

13.4.3 *Conclusion*

The positions of exploration holes are often fixed by approximate pacing from points on the site boundaries and the ground levels at the time of site investigation all too often are not established accurately. Both can create real difficulties in later reclamation work, though the lack of original ground levels is usually the more severe problem.

Depths to various horizons and contaminant occurrences are measured down from ground level. However, ground levels are usually altered by reclamation activities. Thus it can be difficult to relocate a suspect horizon, found in the earlier site investigation, if the ground levels with respect to Ordnance Datum have not been established. In some cases this can cause real difficulties. During the investigation of a former naval dockyard, several trial pits recorded the occurrence of visible blue asbestos fibres in a sandy layer within the near surface made ground capping of the site. During later site remediation the sandy soil was found to be chemically clean and so it was separated from other fills and retained for re-use as a subsoil on the surface of the reclaimed land. At this stage, a potential purchaser's advisers noted the earlier records of blue asbestos and required confirmation that the sandy material was asbestos free. As the investigation trial pits, where asbestos had been recorded, had all been excavated and because no original ground levels related to Ordnance Datum had been measured, it proved surprisingly

TRIAL PIT LOG			Project	Clarendon Phase 1
Trial Pit No 4			Job No	6447
Equipment JCB 3CX			Made By	T.C.
Ground Level 9.600 m Aod			Date	4 MAY 1992

Depth (m)	Reduced Level	Description of Strata	Legend	Remarks
0.9	8.700	Reinforced concrete slab surface with associated brick, concrete and sandstone foundation/ underbuildings		Heavy water ingress from 0.9 m
2.0	7.600	MADE GROUND Comprising very soft orange silty sandy clay, dense orange/ brown wet coarse clayey SAND with some gravel and loose black ash/coal material and gravel FILL		Sample 14A taken in FILL Initial water level 1.8 m
3.2	6.400	Dense grey wet clayey fine grained wet SAND with traces of peat and strong odour Sand becoming more loose and wet and peat increasing into bands with depth		Sample 14B taken in sand Groundwater sample taken
3.7	5.700	Medium dense grey/brown wet silty coarse sandy gravel with peat interspersed (strong odour)		
		Medium dense orange/ brown very sandy gravel. No peat evident although strong adour still identifiable		Trial pit terminated Descriptions and interfaces are approximate below water level

Weather: Overcast
Trench walls unstable from 1.0 m in fill
Trial Pit exposed for 1 hour
Water levelled at 1.6 m: walls very unstable

Figure 13.4 Trial pit log, example 1.

difficult to establish which actual horizon had held asbestos fibres and impossible to prove that these had been removed from site. The failure to establish accurate original ground levels on investigation points then compelled a thorough and quite expensive sampling and testing of the stockpiles of sandy subsoils to prove that these were free of asbestos contamination.

The logging of trial pits and boreholes is usually left to younger and less expensive site investigation staff, since the industry is highly competitive. This, if personnel are not properly trained, can result in a poorer information base than is necessary. Figure 13.4 is an example of the type of geological logging which is truly helpful. Original ground level, locations at which waters were encountered, depths to well described horizons, sample locations, ground stability and weather conditions are all included. Additionally useful personal observations are included. The 'odours' noted at the peaty bands near the base of trial pit would be enough to alert a supervisor that gas and vapour surveys ought to be conducted and would show where these were necessary. A particularly honest feature of this trial pit log is that the investigator stressed that descriptions and interface levels, below the groundwater level, were approximate and should not be interpreted with undue precision.

Figure 13.5 is in sharp contrast. Not only was the ground level and the occurrence of any groundwater ignored, but a sample for chemical analysis was taken from an unrecorded location in this pit. The description on the sample bag – 'ash 1.5 m to 3.1 m' – created doubts rather than providing useful information. Was the sample meant to be a bulk collection between the two recorded depths (and thus likely to indicate the average chemical condition of this thickness of 'ash') or was it in fact from a unique and probably atypical band lying somewhere between 1.5 m and 3.1 m (and so probably chemically unrepresentative of the bulk of the exposed 'ashes')? To make matters worse, inspection indicates that the strata description is false. The clue is the 'layer of clay bottles' (actually earthenware beer bottles of 1900s age). Re-investigation of the site was found to be necessary because of these doubts, and the 'ash' was identified to be decomposed domestic wastes, which were far from as uniform as had been suggested and emitted very obvious landfill gases and odours. On the basis of logs such as Figure 13.5, reclamation design could have been significantly misguided.

The quality assurance needed in site investigation work calls only for common-sense and an appreciation that site investigation data have to be as accurate and complete as is possible. If this is overlooked then risk assessments and reclamation designs will be very poorly based.

One final aspect of the routine quality assurance of site investigation work is that taking and preserving a large number of spare samples is always worthwhile. These will exist to allow later considerations of chemical analyses to actual materials and soil types and can themselves be put in for chemical examination if doubts exist over the completeness of the first set of chemical results.

TRIAL PIT LOG			Project	Cumbrian Way
Trial Pit No 7			Job No 794	
Equipment JCB 3C			Made By E.W.L.	
Ground Level Aod			Date 8 MAY 1992	
Depth (m)	Reduced Level	Description of Strata	Legend	Remarks
— 0.2		TOPSOIL		
		ASH of uniform appearance throughout pit. Layer of clay bottles 0.1 m thick at 1.5 m		Very easily excavated by machine
	4.0			
				Trial pit terminated
Rain/hail showers Walls fairly stable after 5 minutes				

Figure 13.5   Trial pit log, example 2.

13.5 Quality assurance of reclamations

13.5.1 *Introduction*

Potential purchasers of reclaimed sites have always been concerned not to pay too much for land, and this is today enhanced by fears of acquiring land which will bring with it unacceptable liabilities (Environment Act 1995). Thus there inevitably will be differences between the vendor and the purchaser over the land's value, its demonstrated fitness for particular re-use and whether remnant liabilities still persist despite the reclamation works.

Today's norm is for purchasers to use specialist consultants who will require access to:

- information of the pre-reclamation state of the land;
- the full site investigation reports and data;
- the risk assessment which was carried out;
- the reclamation specifications and their adequacy to reduce identified risks and liabilities;
- the reclamation records and testings which demonstrate the quality of what was achieved;
- the reclamation proof testing; and
- the confirmations that planning authorities and regulators have been fully satisfied by the works which were carried out.

Because of these predictable demands, current good practice is to produce an end-of-reclamation report (Figure 13.6), which is supplemented by all necessary chemical and other analytical results. This necessity, of course, forces the use of quality assurance systems during reclamations.

13.5.2 *Quality systems*

Avoiding time-consuming and expensive debates, re-investigations, and probable land value reductions indicates that the reclamation works should

1 Consultant's appointment terms
2 Desk study results
3 Site investigation results
4 Reclamation options and the reasons for the selected choice
5 Reclamation end quality specification
6 Reclamation history
7 Reclamation records and analyses which demonstrate the achievement of the necessary end quality standards
8 Proof-testing results, after reclamation has ended
9 Pre- and post-reclamation topographic survey plans
10 All relevant planning approvals and regulatory body requirements and approvals

Figure 13.6 Normal inclusions in an end-of-reclamation report.

be carried out to whatever defined and provable quality standards are appropriate for that site. Quality systems are, of course, far from new. The Ministry of Defence introduced quality assurance systems in the 1940s for the wartime munitions industries, and the British Standards Institute produced its standard BS5750: Quality Systems as early as 1978.

The publication, in 1981, of a guide for the use of BS5750, allowed interpretations that are clearly relevant to the construction industry. This guide [6] poses those critical questions that have to be addressed in any meaningful quality standard:

(a) *Quality system*
   - Has a system been specified for the necessary inspection and testing?
   - Does the system include the criteria for acceptance and rejection?

(b) *Inspection representative*
   - Has a named individual(s) been made clearly responsible for quality inspection?
   - Does he/she have the required authority, time, staffing, and facilities to carry out the specified duties?

(c) *Control of inspection and testing equipment*
   - Are the accuracies of all inspection and testing equipment known?
   - Have regular and routine calibration checks taken place?
   - Are these available to the inspection representative?

(d) *Records*
   - Are the collected test records adequate to prove that regular and adequate testing has taken place?
   - Do all records include the agreed acceptance standards?
   - Do the records list any corrective action that should have taken place?

(e) *Sampling procedures*
   - Have recognized sampling methods been used?
   - Are these the same as those defined in the reclamation specifications?
   - Do the chosen sampling methods give enough confidence that the required quality levels have been achieved?

(f) *Control of non-conforming materials*
   - Are non-conforming materials clearly specified?
   - If repair work has had to be carried out, is it in accordance with the original specifications?

(g) *Training*
   - Have all appropriate personnel been trained to the required standards?

Interpreting the British Standards guidance for use in contaminated land reclamations is not difficult.

Quality system.   While the essential information for defining an appropriate quality system has invariably been defined, i.e.

- The chosen reclamation strategy
- The different elements and activities that have to be included
- The programme of reclamation activities
- The standards that have to be achieved in each of the reclamation elements (e.g. compaction levels, allowable remnant chemical concentrations, etc.)

it is surprising how seldom this information is collected into a definitive quality system document, available for use by all the supervisory personnel.

The fact that this often is not done is probably the most significant reason for failure to achieve the required quality product. Site reclamations can last for two or more years, and so site supervisors can move on and be replaced. These newer staff can lack a full appreciation of what is important, particularly since they invariably have to react to short-term emergencies (e.g. dealing with the public reaction to closing a right of way rendered unsafe by the reclamation work, disputing a contractor's interpretation of the quantities detailed in the reclamation contract, obtaining a river water abstraction licence to wash gravel on-site, reacting to a Health and Safety Executive complaint, etc.). Thus it is far too commonplace for routine environmental monitoring to be delayed, or completed to only a partial level, simply because this work is routine and so appears to lack any real priority.

Inspection representative.   Appointing an individual to supervise the quality of particular activities is usually not difficult. The normal resident engineer arrangements generally ensure that the civil engineering inspections are carried out to a high standard. It is still quite common for consulting engineers to lack the in-house expertise to supervise the quality of the contamination and environmental aspects of reclamation work, but this can be covered by appointing an outside specialist. There is no special reason why the use of more than one inspection representative should be other than satisfactory, provided that a senior member of the consulting engineers' organization is able to co-ordinate the activities of these different inspectors, and support them appropriately, when circumstances require this.

An example of a failure in such support came to light when a former metal smelting works was to be reclaimed for a large retail complex, surrounded by public access landscaped areas. Planning approval to encapsulate some copper and zinc slags below the landscaped areas had been allowed. To avoid any contamination risks to the required vegetation cover, the reclamation design included a 1.5 m thick cover of soil, with high organic content and good water retaining properties, to encourage plant roots to remain in the clean materials.

A landscape architect was appointed to the resident engineer's team to supervise the landscaping. Part of this individual's duty was to test all imported soils. These tests soon revealed that no locally available soils matched the design specifications, and that all had very poor water-retaining characteristics. The landscape architect called for a halt in the landscaping work, until necessary redesign and plant species selection could be completed.

This, however, would have caused an overrun of the contract period, and the resident engineer (who saw environmental issues as very secondary to the contract problem) insisted on work continuing as planned.

The reclamation was completed to time and appeared initially satisfactory. However, the developer had had to agree to maintain the landscaped area for five years, before the local authority would accept ownership. Two very dry years occurred and abnormally high proportions (> 40%) of the planted species died off.

The local authority became concerned that plant roots might have migrated into the encapsulated slags, and insisted on contaminated studies of grass and leaf samples. These surveys showed no phytotoxic hazards. Further testing then revealed that the granular sandy soils, which had been used, dried out very easily and were inhospitable for the plant species selected. After a long and contentious dispute, the developer had to improve the soil's properties and replace the bulk of the already planted trees and shrubs.

In retrospect, this unnecessary expense resulted from an obvious failure to recognize the valid concerns of a specialist inspection representative.

Control of inspection and test equipment.   Contaminated site reclamations usually require two quite different systems of inspection and testing. One, for the physical properties of soils and construction materials is well defined in various British Standards and in the specifications issued by governmental departments (e.g. [7]), presents few if any problems, given the industry-wide expertise available in this area.

The other, for the chemical analysis of soil materials, liquids and gases, poses greater difficulties, partly because analytical expertise is not so widespread, and partly because use often has to be made of off-site analytical laboratories. Lord [8] considers this latter point and makes clear that since no officially approved catalogue of analytical methods for contaminated land reclamations yet exists, different laboratories can use methods of very variable sensitivity and accuracy. He also notes the very significant differences that can arise from the various methods adopted in laboratories to prepare samples for the analytical process. Thus, it is not possible to accept chemical analytical results as invariably and consistently accurate. Smith [9] confirms this view, and advocates that all analytical reports should include:

- a description of all analytical methods used;
- a listing of each method's accuracy and precision;
- a statement on the laboratory's quality control and quality assurance systems.

Since few, if any, analytical reports list of all the information that Smith rightly sees as essential, it is prudent to use analytical firms that can demonstrate a high level of analytical quality assurance, preferably by having been accredited by such organizations as NAMAS. It also is a sensible policy to ensure that analytical laboratories are supplied with the relevant data on a sample's origins, so that the most appropriate analytical methods can be employed. Giving a laboratory a sample of builder's rubble, which has a content of bitumen road surface debris, will lead to analytical results indicating high toluene extract levels, which then can be misinterpreted as due to dangerous tarry contamination. If the laboratory had been advised of the presence of the road surface debris, analytical costs might have been reduced, and concerns over a contamination hazard avoided.

Even if the laboratory service is fully advised and operates to a high analytical quality assurance, difficulties can very easily occur in testing for environmental hazards. A Midlands site had been a set of marl pits, which were later filled largely with inert wastes from the local pottery industries. Landfill gases were not thought to be a particular hazard, since biodegradable wastes were not encountered in the site exploration, but gas monitoring boreholes were installed around the perimeter of the site for routine gas monitoring over the two-year reclamation period. Monitoring results were taken monthly in the autumn and winter of 1988/1989 and in the winter of 1989/1990, by a junior member of staff using field portable gas monitoring meters. No results were available in the summer of 1989, since site stripping had destroyed the monitoring boreholes, which were not replaced until that autumn. When the site came to be sold on to a house builder in the summer of 1990 the buyer had his own advisers check the landfill gas conditions (Figure 13.7). The differences between the gas monitoring data collected during the reclamation work, and those obtained by the buyer's advisers, proved to result from a failure to calibrate the field portable gas meters properly and to ensure that gas monitoring took place in the hotter months, when biodegradation of the small proportion of wastes, that could produce landfill gases, was at its highest. The result of this very avoidable situation (which should have been prevented by giving the junior site supervisor a better level of training) was a very large reduction in the price obtained for the site.

Records.   Test records can be extremely numerous on any site of a reasonable size, and while these records are generally collected to a reasonable standard, dealing with them is often seen as a bureaucratic and boring task. The problem is really to store, check and analyse and present the records in a

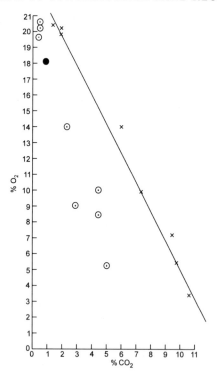

Figure 13.7 Oxygen/carbon dioxide relationships. Borehole 7, Marl pit site. (●) Gas meter reading, winter 1988/1989 and 1989/1990; (○) gas meter reading, summer 1990 (developer's equipment); (×) results from gas chromotographic analysis, summer 1990 (purchaser's adviser's data).

way that is not too time-consuming, and on sites where computer storage and graphical plotting systems are available, this should be simple. The increasing use of such systems as MOSS, for example, allows regular production of plans showing sampling locations, appropriate contouring and presentation.

Where the greatest failure often occurs is in the completeness of the records. It is still uncommon to find records that list the required quality standard and the actions that have been initiated to correct any unsatisfactory situations. A simple example occurred on a site in the north of England where oily contamination from a coal products works had polluted a near surface aquifer, which, in turn, was in hydraulic continuity with the adjacent river. As part of the planning approval, the requirement for works to prevent any off-site migration of oily contamination was included, and regular groundwater quality monitoring on the adjoining land had to be carried out at monthly intervals. One sample taken early in the reclamation project, on land immediately beside the river, proved to have an extremely high toluene extract value, which was not confirmed by any more specific analysis. Later sampling from the same location, however, indicated that no oily pollution had moved off-site. Late in the reclamation project, the planning authority suggested that the protection works, to prevent off-site migration of the oils, might not be fully satisfactory, and used the single off-site high toluene extract value to justify this concern. Since the monitoring records included no indication of any corrective or check action that had been taken when the high toluene figure was found, it proved extremely difficult to convince the statutory body that the protection works were in fact satisfactory, and that this single high pollution concentration was due to other causes, as in fact proved to be the case.

Sampling procedures.   This is perhaps the greatest source of confusion when contaminated land work is undertaken. Smith [9] has shown that minor position changes on a contaminated site can lead to very large variations in contamination levels, and it is also obvious that contamination levels vary with depth. Despite this, it is still all too common to read site investigation reports that list, for example, 'ashes and interbanded ashy clays – from 1.5 m to 3.0 m' and to have a single sample of this band analysed to produce cadmium, zinc and copper concentrations far higher than any safe levels. The obvious question that arises is whether the entire ashy band is contaminated to this level, or do the analytical results perhaps relate to a thin and atypical horizon, which attracted the sampler's attention.

Obviously, this sort of doubt is easily avoided (Chapter 3) by increasing the clarity of the site investigation reports and indicating clearly the level at which a sample was taken, and whether or not it was felt to typify the entire horizon. Internal quality control of this type is spreading in the site investigation industry, although a good deal more improvement has still to be achieved.

The example cited above led to quite serious problems, since the local planning authority had formed the view from the site investigation data provided to them, that the site was extremely contaminated and could only be safely developed if stringent and expensive analytical controls were exercised when the site was opened up. In fact, excavations indicated that the site investigation data had not properly typified the site, which proved to be only marginally contaminated. Convincing the planning authority of this proved difficult.

Control of non-conforming materials.   This is usually not a difficult matter, at least in the initial design phases of a reclamation. The civil engineering industry is familiar with the concept of identifying non-conforming materials and requiring contractors to deal with these in particular ways.

When repairs are needed later in the reclamation work after the debates and concerns that dominated the planning approval process have been forgotten, the situation need not be so foolproof. A former tar works in

South West England had heavy oil soaking of the demolition rubble and waste products surface layer, and its reclamation called for drainage and other works to reduce the oil content, after which the reduced (mainly metallic) contamination of the site would be counteracted by the installation of a clean cover. The local planning authority required considerable demonstration and testing to prove that the clean cover would be effective for at least 100 years and insisted that the entire site was capped before any redevelopment took place. This was to ensure that no airborne contamination could result. While this initial part of the work proved successful, redevelopment of the site was significantly delayed because of a downturn in the demand for housing, and parts of the site were sold on to different building contractors who had little understanding of the contamination risks that could arise if the clean cover were not maintained at its design quality. One of these contractors found it necessary to alter the previously installed drainage services, which had been included with the clean cover, trenched through the cover, and then backfilled his excavations with granular material obtained from a disused railway embankment.

Unfortunately this backfill material not only failed to meet the design physical properties of the original clean cover, but also proved to be contaminated with high concentrations of heavy metals. Obviously a failure in proper communication led to a very significant loss in quality assurance, which, in turn, had heavy financial penalties.

Training. Deficiencies in training are least apparent in those activities (e.g. land surveying or soil mechanics testing for civil engineering staff, and trial pitting for environmental scientists) that appeared as practical exercises in the site supervisor's formal education. This indicates that most employers still prefer to limit the professional development training they themselves provide, and expect newer staff to pick up necessary expertise by observation and experience.

In many cases (e.g. the landfill gas monitoring of the marl pit site and the sampling of the metal-contaminated ash site, noted above) this is not a cost-effective system, and greater note has to be taken of the 'abnormal' conditions that regularly occur on contaminated sites and the need to train staff to react to these.

13.5.3 *Appropriate quality systems*

The required quality standards will not always be the same and the extent of quality assurance work will be much less on some reclamations than on others. The obvious point here is that priorities have to be identified – some conditions are potentially more hazardous than others, and some quality monitoring activities can be carried out to less stringent standards. Simple examples of actual site reclamations should make this clear.

Housing redevelopment, Central Scotland. This small area of land (about 1 ha) is fairly typical of the marginally contaminated sites that occur within larger towns. Stiff clays originally surfaced the site and still extend to 7 m depths. No groundwater table is present in the upper 5 m of the site. Small-scale excavations in the centre of the site had taken place about 1940 and had resulted in a shallow depression up to 2.5 m deep, which was later infilled with post-war demolition rubble. At some later stage, a surfacing (up to 350 mm thick) of domestic ashes was tipped over most of the site.

Site investigation was by means of 24 trial pits, and chemical analyses of the samples revealed that, while the ashes contained the expected low to medium contamination of copper, nickel, zinc, lead and cadmium (Table 13.1), the underlying clays were uncontaminated, except where oil spills (from a car breaking operation) had taken place.

The presence of phytotoxic contaminants (mainly in low concentrations) was seen as unhelpful to the house sales, and the developer chose not to cap the site but to strip the top 400 mm, and redeposit the ashy material as a linear sound reducing mound, between the housing and an adjacent busy road. This sound mound was then capped with a clean cover, to prevent leaching of the metallic contamination (Chapter 5, section 5.5).

Quality standards for this reclamation proved to be extremely simple and consisted of no more than the following:

(1) *Site stripping process.* Ensuring that all stripped materials were removed to the sound reduction mound location. Recording the physical types of materials stripped off each area.

(2) *Site inspection post-stripping.* Visual confirmation that all ashy materials had been removed and that each housing plot was underlain by clean materials. Photographic records were taken to prove that ashes had been removed. Visual identification of any oil soaked

Table 13.1 Phytotoxic contaminants in the ashy band of a site in Central Scotland

Contaminant	Concentration found (mg/kg)	
	Range	*Mean*
Copper	53–310	140
Nickel	20–105	68
Zinc	110–4170	400
Lead	155–1020	490
Cadmium	2.0–5.7	2.4
Chromium	20–45	31
Arsenic	1–8	4
Phenol	< 1	
Toluene extract (oils)	1100–5000	1670
pH	6.21–7.81	
Sulphates	300–1500	645

ground that existed. If encountered, oil soaked materials were excavated, and the arisings removed to the sound reduction mound area
Oil soaked materials were tipped separately, from the ashy strippings,
to allow evaporation to occur. Testing of each house plot was limited to:

- one clay sample from each garden area analysed for phytotoxic contaminants; confirmation was noted that every sample was in fact chemically uncontaminated;
- spike tests were conducted under each house foot print, to ensure
that no landfill gases or oil vapours existed;
- occasional plate bearing tests were made on any clay materials that
visually seemed less dense than normal;
- plate bearing tests were completed on exposed demolition rubble fills
to confirm their suitability for foundation loads.

(3) *Sound reduction mound.* Oily materials were mixed with the granular
ashy fills, and then rolled prior to being capped with a clay surface capping, below which a drainage blanket was installed.

The staffing needs were limited to the normal site foreman, supplemented by
periodic inspection visits by the consulting engineer's staff.

The final records consisted only of site plans showing final and initial
ground levels, all test locations, photographic records and a summary of
the test results attained. These proved acceptable to the buyers' legal
representatives and to the National House Builders Council. No dispute
over the quality of the finished reclamation arose.

Housing redevelopment, South West England. Where a site contains mobile
contamination, in addition to relatively immobile metals, reclamation inevitably becomes more complex, and achieving the necessary quality standards is
a rather more significant task.

This particular site had been the location of oil and creosote storage tanks
for several decades, before which it was a tip for copper smelter slags. Leakages from the storage tanks had resulted in a high degree of oil saturation of
the fill deposits (demolition debris, ashes and the slags from a former copper
smelter), which averaged some 3 m in thickness. The existing site surface
sloped very steeply to a river, used for amenity and water supply purposes,
and whose water quality had to be ensured throughout the reclamation.

Site investigation by several sequences of trial pits, and deeper boreholes,
revealed the expected high metal levels, a very pervasive oil/tar/phenol contamination, and very high soil sulphate levels. The bearing capacities of the
site materials were such that piled foundations for the housing units proved
necessary.

The reclamation strategy adopted was:

(a) To break out the concrete slabs, which surfaced almost the entire site.

(b) To install land drainage at 10 m spacings, to draw off the free oily liquids. These drains were designed to act as permanent structures, to hold the site's groundwater level 1.5 m below the average level at which the site materials were oil contaminated. A central sump then collected the oily liquids, which were taken off-site for treatment and final disposal. No construction works took place until the drains had been in operation for an entire winter, to ensure that a maximum possible flushing out of the oil spillages was achieved (Table 13.2).

Table 13.2   Variation in oily contamination in one of the drainage trenches on a site in South West England

Week	pH	Phenols (mg/l)	Oils (mg/l)	
1	7.7	283	251	site breaking
4	7.2	96	100	out in
8	7.3	114	80	progress
16	7.4	197	135	
20	7.3	170	100	
24	7.5	50	60	
28	7.1	38	43	
32	7.1	15.5	25	
36	7.2	11.4	6.1	
40	7.1	11.0	6.0	
44	7.2	< 10	5.5	

(c) Foundation piles were then installed, after all exposed tarry contamination was removed for on-site encapsulation.
(d) A designed clean cover, and its overlying filter blanket, were laid to very tight particle size distribution and density specifications.
(e) The final ground levels were achieved by importing up to 4 m thicknesses of inert building rubble. This large-scale raising of the site levels was necessary to achieve acceptable road gradients and the required landscape topography.

The quality system needed for this project included:

(1) *Drainage works*
   • Recording the levels, gradients and positions of all drainage works; ensuring that these would not intersect the planned locations of pile foundations.
   • Monitoring the qualities and flow rates of liquids entering the central sump at weekly frequencies, and recording when particular zones of the site appeared to have drained.
   • Associated with the drainage monitoring, the river water quality at the upstream and downstream boundaries of the site were monitored weekly, to check if any oily pollution was occurring.

(2) *Off-site removal of oily liquids*
- The volumes of liquids removed for off-site treatment and disposal were recorded as a requirement of the waste disposal licence.

(3) *Tarry wastes*
- Locations of all exposed tarry wastes were recorded, as was the depth of excavation needed to remove these. Whenever tarry wastes occurred close to proposed pile locations, particular care was taken to remove the deposits. Where this proved impossible, note was made of the necessity to provide appropriate external protection for the particular concrete piles.
- All tarry wastes were stockpiled, air-dried, and then rolled, to highways specification, in a concrete tank base, which ultimately was covered to provide the on-site encapsulation cell.

(4) *Clean cover*
- All materials for use in the clean cover were derived from crushing concrete demolition waste. The particle size distribution curve of the crushed concrete was checked before each load was taken for use in the cover. Specific criteria for acceptability/rejection were devised and enforced.
- The cover was laid in 200 mm thick layers and the compaction of each layer was confirmed, by nuclear densiometer readings at 10 m centres.
- The filter blanket required above the clean cover was checked in an identical manner.

(5) *Imported inert fill*
- Each load of imported fill was visually checked to ensure that no biodegradable materials were included.
- A sample of each load was taken and analysed for its metal, toluene extract, and phenol contents. Criteria were enforced to ensure rejection of any material that failed to meet the necessary chemical quality.

These quality controls were such that a representative of the consulting engineer was permanently located on the site, during these initial reclamation stages, and this individual was supported by on-site and off-site analytical services. A full end-of-reclamation report was produced as detailed in Figure 13.6.

Industrial development, West Midlands. When a site has landfill gas problems in addition to mobile contamination, the quality system becomes even more demanding. This West Midlands site had been in part an iron foundry, manufacturing cast iron baths, and in part a refuse tipping area, which had been in intermittent operation for some 80 years. Coal mining occurred directly beneath the site, on which several mine shafts had existed. Whether these shafts had been properly filled and capped was not known. An infilled canal arm was also known to have existed.

Because of the coal mining and the possibility of land subsidence, the site exploration was particularly thorough. This exploration revealed that areas of high phenol concentrations occurred, together with oil soaking from demolished oil storage tanks, but the major contamination appeared to be from the lead and arsenic pigments, which had been used to glaze the cast iron baths. Apparently during the site demolition, the stores that had contained these hydroscopic salts had been bulldozed. Colliery wastes containing potentially combustible material were also widespread around the locations of the former mine shafts. Although the tipped wastes (up to 4 m thick) were all of a considerable age, landfill gases with methane concentrations of some 40% by vol. and carbon dioxide levels of up to 13% were frequently encountered in some gas monitoring boreholes.

The physical condition of the site made it necessary to improve its bearing capacity by vibro-compaction and the reclamation strategy was devised around this. The main reclamation phases were as follows:

(a) Zone the site into areas affected by particular potential hazards and deal with each in a specific controlled way.

(b) Install vent trenches at 10 m centres across the zone affected by landfill gases. Several months' monitoring and experimentation had revealed that passive venting and excavation of discrete areas of wastes would resolve the gas hazards.

(c) Collect all exposed patches of the lead/arsenic salts and remove these for off-site treatment. Excavations to about 2 m depths proved essential and this area was reinstated with a clean granular cover.

(d) Locate and remove all oil soaked ground. Stockpiling of this material allowed enough evaporation to give a marked reduction in the oily contamination, and only the more heavily contaminated material had to be removed to a licensed tip.

(e) Compacting the site surface to the levels needed to give the required bearing capacities and prevent enough air entering to allow any subterranean smouldering or burning in the more susceptible colliery spoil materials (Chapter 7).

Quality controls had to include:

(a) Weekly landfill gas monitoring (of both gas concentration and flow rates) to prove the effectiveness of the passive venting trenches, and to prove that the 10 m trench spacing was appropriate

(b) Chemical analysis to prove that the lead/arsenic levels had been reduced, to below the ICRCL guideline values, by the removal of the visible pigmentation wastes

(c) Chemical analysis on the oily soaked area to show that the remnant phenol, and oil concentrations were at the allowable final levels, once excavation of patches of oily ground had been completed
(d) Routine monitoring of the chemical qualities of the stockpiled oily arisings to determine if they could safely be introduced back into the site, or whether they would have to be removed to a licensed tip
(e) Regular monitoring of the site compaction, to prove that no subterranean smouldering hazard would be likely
(f) Routine confirmation that the necessary ground bearing capacities had been achieved
(g) Routine updating of site plans, sample location plans and analytical records

This proved to be very successful. The work load, however, called for almost the full time attention of a senior member of the consulting engineer's organization, supported by the scientific personnel from the analytical firm.

The examples noted above all are derived from the engineering-biased reclamations that have so far dominated UK reclamation practices. With the availability of chemical and microbial treatment methods offered by specialist contractors who are now able to enter into commercial contracts that include end-quality achievement levels, it is obvious that a rather different pattern of land reclamation strategies will soon become the norm. These are likely to include appropriate mixes of various on-site treatments, on-site encapsulations, and a reduced emphasis on clean covers.

In quality assurance terms this inevitably will lead to more complex situations. Different organizations are likely to be involved on a single site, at much the same time, and the reclamation techniques each is employing will have very different requirements and achievable qualities.

13.6 Summary

Contaminated land is invariably reclaimed for re-use and/or resale, and this is done in a legislative climate in which any adverse effects the land might present to human health or the local environment could attract criminal charges.

Because of this, it should be apparent that the entire investigation and reclamation effort will be scrutinized by other parties, and that failures (or voiced suspicions of inadequacies) will – at best – cause land values to diminish.

It is possible to avoid these undesirable outcomes by adequate reclamation management. The emphases which are essential do differ in the various stages of site reclamation, but none calls for excessive bureaucracy or – in fact – for other than good professional practices.

Appendix I
Soil guidelines (UK) and Dutch Intervention Values

UK Advisory Guidelines for Soils

1 Since 1990, the UK's Department of the Environment has sought to confirm or revise its existing threshold trigger level guidelines (ICRCL 59/83 *Guidance on the assessment and redevelopment of contaminated land*). This was compelled by a widespread criticism of the limited coverage of contaminants included in ICRCL 59/83, and the lack of "action" values which would necessitate land quality remediation.

2 It might have been expected that a comprehensive listing of critical contaminant concentrations, similar to that devised by the Netherlands' authorities (Netherlands Ministry of Housing, Physical Planning and Environmental Conservation - Report HSE 94.021 of 9th May 1994) would have resulted.

3 However, as discussed in Chapter 2 (Section 2.4) listings such as the latest Dutch standards are relevant only in respect of the targets (ecosystems and human health are the Netherlands' targets of importance) chosen by those devising the listings, and of the methodologies by which critical contamination concentrations have been identified.

4 The UK authorities recognise that a wide range of risks can feasibly arise from contaminated land (Chapter 2, Table 2.6) and so have accepted that different contaminant listings and concentrations should be devised for each risk separately. Obviously this imposes greater difficulties on those charged with producing the advisory soil guidelines, particularly since the Netherlands' acceptance of the results of laboratory studies on experimental animals, exposed to high doses of contamination, is not favoured by the UK. This difference in viewpoint is acknowledgement of the inevitable uncertainly in extrapolating from the results of high dose exposures on rats to the consequences likely when human beings come into contact with much smaller contaminant doses in the complexities of the modern environment.

5 Given the scale of the task undertaken by the UK authorities, it is perhaps unsurprising that no replacements for ICRCL 59/83 have yet been published.

6 The first–addressing human health risks when contaminated soils and dusts are ingested and inhaled–is anticipated shortly. This is likely to list only a small number of contaminants (probably benzene and the zootoxic contaminants cited in ICRCL 59/83) and be relevant only to contamination in near surface soils (300 mm to 500 mm depths). Thus its applicability will be limited, particularly on sites where remediation has not employed any type of clean cover.

7 Other guidance, to give acceptable safeguards to water qualities, plant populations and homes which could be at risk from gas migrations or structural failure caused by contaminant attack, are likely to take longer to appear.

8 The reasonable expectation is, thus, that the ICRCL 59/83 guidance will continue in use as the main indicator of contaminant issues of interest for the foreseeable future, until gradually superseded by more defensible and relevant soil quality guidances.

9 The evolving UK approach to guidances on soil contamination arguably is intellectually superior to the approach accepted by the Netherlands' authorities. However, it does call for a less mechanistic usage of guidance values, and so will force a more educated approach on those who employ the guidances.

ICRCL 59|83 (pp 305 to 307, First Edition)

Dutch Intervention Values (9 May 1994)

1. The Netherlands abandoned the usage of its 'A-B-C' soil and groundwater values on 9 May 1994 (Ministry of Housing, Physical Planning and Environmental Conservation report HSE 94.021)
2. Intervention Values are now available to identify 'serious contamination' of soils, sludges and groundwaters and to indicate when remediation is necessary.
3. These new Intervention Values

    - have been devised from both human and ecotoxicological risk studies and are deemed to be more defensible than the former 'C' values;
    - should be applied against the *average chemical concentrations* of typical garden areas (7 m by 7 m and to a depth of 0.5 m) and of smaller (100 m^3 groundwater volumes;
    - allow for variations in soil type (particularly different concentration of organic carbon and clay in soils); soil 'correction factors' are specified.

4. Additionally 'Target Values' are specified to indicate desirable uncontaminated chemical contents of both soils and waters. These represent the ultimate national soil quality objectives but *not* clean-up criteria for the foreseeable future.
5. The former 'B' value (at which further consideration of soil and/or groundwater contamination must take place to establish whether risks actually are likely) is now replaced by the average of the Intervention and Target Values, or (where no Target Value is listed) by half the Intervention Value.
6. Intervention values are not cited for those broad groups of substances (e.g. herbicides) whose individual substances have toxicological properties which vary appreciably.
7. It should also be noted that – irrespective of the cited Intervention Values – The Netherlands' authorities acknowledge that hazardous conditions may occur where contaminant mobility is especially high even if Intervention Values are not exceeded.
8. For heavy metals, the Target and Intervention Values are dependent on clay/silt and organic matter contents and the standard soil values (Table A.1) must be modified by the formula

$$Ib = Is \frac{(A + B\% \text{ clay/silt } + C\% \text{ organic matter})}{A + 25B + 10C}$$

    where Ib = *Intervention Value for a particular soil*

    Is = *Intervention Value for a standard soil*

    A, B and C = *compound dependent constants (see Table A.2)*

9. For organic contamination, Target and Intervention Values are affected by a soil's organic matter content in accordance with the formula

$$Ib = Is \frac{(\% \text{organic matter})}{10}$$

Table A.1 Target and Intervention Values for micro-contaminations for a standard soil (10% organic matter and 25% 'clay'). Soil/sediment in mg/kg, ground water in µg/l; unless stated differently

Substance	Soil/sediment (mg/kg *dry matter*)		Ground water (µg/l)	
	Target value	*Intervention value*	*Target value*	*Intervention value*
I Metals				
Arsenic	29	55	10	60
Barium	200	625	50	625
Cadmium	0.8	12	0.4	6
Chromium	100	380	1	30
Cobalt	20	240	20	100
Copper	36	190	15	75
Mercury	0.3	10	0.05	0.3
Lead	85	530	15	75
Molybdenum	10	200	5	300
Nickel	35	210	15	75
Zinc	140	720	65	800
II Inorganic substances				
Cyanides (free)	1	20	5	1500
Cyanides compl. (pH < 5)1	5	650	10	1500
Cyanides compl. (pH ≥ 5)	5	50	10	1500
Thiocyanates (sum)		20		1500
III Aromatic substances				
Benzene	0.05(d)	1	0.2	30
Ethyl benzene	0.05(d)	50	0.2	150
Phenol	0.05(d)	40	0.2	2000
Cresols (sum)		5	(d)	200
Toluene	0.05(d)	130	0.2	1000
Xylene	0.05(d)	25	0.2	70
Catechol		20	(d)	1250
Resorcinol		10		600
Hydrochinon		10		800
IV Polycyclic Aromatic Hydrocarbons (PAH)				
PAH (sum 10)2,11	1	40	–	–
Naphthalene			0.1	70
Anthracene			0.02	5
Phenanthrene			0.02	5
Fluoranthene			0.005	1
Benzo(a)anthracene			0.002	0.5
Chrysene			0.002	0.05
Benzo(a)pyrene			0.001	0.05
Benzo(ghi)perylene			0.002	0.05
Benzo(k)fluoranthene			0.001	0.05
Indeno(1,2,3-cd)pyrene			0.0004	0.05
V Chlorinated hydrocarbons				
1,2-Dichloroethane		4	0.01(d)	400
Dichloromethane	(d)	20	0.01(d)	1000
Tetrachloromethane	0.001	1	0.01(d)	10
Tetrachloroethene	0.01	4	0.01(d)	40
Trichloromethane	0.001	10	0.01(d)	400
Trichloroethene	0.001	60	0.01(d)	500

Vinyl chloride		0.1		0.7
Chlorobenzenes (sum)3,11		30		–
Monochlorobenzene	(d)	–	0.01(d)	180
Dichlorobenzene	0.01	–	0.01(d)	50
Trichlorobenzenes (sum)	0.01	–	0.01(d)	10
Tetrachlorobenzenes (sum)	0.01	–	0.01(d)	2.5
Pentachlorobenzene	0.0025	–	0.01(d)	1
Hexachlorobenzene	0.0025	–	0.01(d)	0.5
Chlorophenols (sum) 3,11		10		–
Monochlorophenols (sum)	0.0025	–	0.25	100
Dichlorophenols (sum)	0.003	–	0.08	30
Trichlorophenols (sum)	0.001	–	0.025	10
Tetrachlorophenols (sum)	0.001	–	0.01	10
Pentachlorophenol	0.002	5	0.02	3
Chloronaphthalene		10		6
PCBs (sum) 5	0.02	1	0.01(d)	0.01
VI Pesticides				
DDT/DDE/DDD6	0.0025	4	(d)	0.01
Drins 7		4		0.1
Aldrin	0.0025		(d)	
Dieldrin	0.0025		0.02 ng/l	
Endrin	0.001		(d)	
HCH (sum)8		2		1
α-HCH	0.0025		(d)	
β-HCH	0.001		(d)	
γ-HCH	0.05 µg/kg		0.2 ng/l	
Carbaryl		5	0.01 (d)	0.1
Carbofuran		2	0.01 (d)	0.1
Maneb		35	(d)	0.1
Atrazine	0.05 µg/kg	6	0.0075	150
VII Other contaminants				
Cyclohexanone	0.1	270	0.5	15 000
Phthalates (sum)9	0.1	60	0.5	5
Mineral oil	50	5000	50	600
Pyridine	0.1	1	0.5	3
Styrene	0.1	100	0.5	300
Tetrahydrofurane	0.1	0.4	0.5	1
Tetrahydrothiophene	0.1	90	0.5	30
(d) = detection limit				

Notes for Table A.1.

1. Acidity: pH (0.01 M CaCl$_2$). The 90-percentile of the measured values is taken for the determination of pH larger or equal to 5 and pH smaller than 5.
2. PAH (sum of 10) equals the sum of anthracene, benzo(a)anthracene, benzo(k)fluoranthene, benzo(a)pyrene, chrysene, phenanthrene, indeno(1,2,3-cd)pyrene, naphthalene, benzo(ghi)perylene.
3. Chlorobenzenes (sum) equals the sum of all chlorobenzenes (mono-, di-, tri-, tetra-, penta-, and hexachlorobenzene).
4. Chlorophenols (sum) equals the sum of all chlorophenols (mono-, di-, tri-, tetra-, and pentachlorophenol).
5. The Intervention Value for polychlorinated biphenyls (PCB) equals the sum of PCB 28, 52, 101, 118, 138, 180. The Target Value equals the PCB sum not including PCB 118.
6. DDT/DDD/DDE equals the sum of DDT, DDD and DDE.
7. Drins equals the sum of aldrin, dieldrin and endrin.
8. HCH-compounds equals the sum of α-HCH, β-HCH, γ-HCH and δ-HCH.

9. Phthalates equals the sum of all phthalates.
10. Mineral oil equals the sum of linear and branched alkanes. In the case of contamination with mixtures (e.g. gasoline or heating oil), both aromatic and polycyclic aromatic hydrocarbons have to be analysed. This sum parameter is included for practical reasons. Further toxicological and chemical differentiation is being studied.
11. The sum parameter for polycyclic aromatic hydrocarbons, chlorophenols and chlorobenzenes in soil/sediment is valid for the total concentration of all compounds within one group. In the case of only one contaminant being present, the value given is the Intervention Value for that compound. For two or more contaminants the sum of the concentrations is used. For soil/sediment the effects are directly additive (i.e. 1 mg of substance A has the same effect as 1 mg of substance B), therfore sum parameters can be used by adding concentrations for groups of compounds (for further information concerning additivity see Technische Commissie Bodembescherming (1989)[1]. Effects for groundwater are indirect, and added as a fraction of the individual Intervention Values.

Table A.2 Compound related constants for metals

	A	B	C
Arsenic	15	0.4	0.4
Barium	30	5	0
Cadmium	0.4	0.007	0.021
Chromium	50	2	0
Cobalt	2	0.28	0
Copper	15	0.6	0.6
Mercury	0.2	0.0034	0.0017
Lead	50	1	1
Molybdenum	1	0	0
Nickel	10	1	0
Zinc	50	3	1.5

[1] Technische Commissie Bodembescherming (1989) 'Advies beoordeling van bodemverontreiniging met polycyclische aromaten'. TCB A89/03. (Advice evaluation soil contamination with PCB.)

Appendix II
Semi-quantified risk assessment methodology (after reference 2.13)

1. Most semi-quantified risk assessments use a numerical system which puts all information and situations into a common (albeit an arbitrary) scale by assigning scores to those factors relevant to the source–pathway–target scenarios of concern.
2. Since a considerable number of risks are possible when contaminated land is re-used for sensitive end purposes (Table 2.6), the semi-quantified approach forces assessors to consider all possible risks and then rank these in order of significance.
3. A number of risk ranking and site reclamation prioritization (semi-quantified) schemes exist, including the Hazard Ranking Scheme developed by the US Environmental Protection Agency, the Canadian National Classification System and the rather simpler New South Wales scheme.
4. The method adopted, in reference 2.13, covers all seven possible environmental risk situations and employs a three-part assessment process:

   Part A – establishing the potential for future risk. Historical and archival source material is employed as discussed in Chapter 3, section 3.4 – without any attempt to ascribe numerical estimates of risks – as a screening process to establish which of the potential risk scenarios are so improbable that further consideration of these can be omitted. An assessor's decision to terminate consideration of a particular risk must, however, be fully justified and limitations are deliberately included to restrict individual bias.

   Part B – This focuses on those risks not excluded by the initial screening, and employs site investigation results to allow the relative quantification of various risk situations. Deficiencies of site investigation information (whether relating to contaminant occurrences, mobilities and concentrations or to the pathways which could allow contaminant migration) are identified and penalized. 'Liabilities' are scored as negative values which reflect the relative magnitudes of the various environmental risks in the site-specific conditions which occur. This allows a more precise identification of which reclamation method should be selected, since highest negative scores indicate the most important of the risks which could arise in future years, and so those which have to be removed by site remediation.

   Part C – This focuses on how effective site reclamation has been, and since achievements (which must be demonstrated by appropriate proof-testing and monitoring) are scored as positive values. The subtraction of these from the Part B risk assessments demonstrates whether or not reclamation has actually removed the concerns which became apparent when site investigation data was assessed.

5. Simple check-lists are provided for the assessment and use is made of the lists and hazardous concentrations of those contaminants significant for a particular risk category.
6. The assessment method is intended to offer the rigour of a fully quantified risk assessment while still retaining the flexibility necessary to overcome scientific data deficiencies. Thus it goes beyond the systems adopted in the USA and Canada and provides for judgements on risk reduction and for the relative appropriateness of remediation alternative methods.
7. In the current position of scientific uncertainties and the necessity to tailor risk assessments to affordable budgets, the method described in reference 2.13 has the particular advantage of consistency and relative cheapness.

Appendix III
Other useful guidelines and standards

Environmental reference values

DoE Circular 7/89: Environmental quality objectives

List I substances
Mercury
    0.3 mg/kg (wet weight) in a representative sample of fish flesh
    1 µg/l (annual mean) total mercury in inland surface waters
    0.5 µg/l (annual mean) dissolved mercury in estuary waters
    0.3 µg/l (annual mean) dissolved mercury in marine waters
Cadmium
    5 µg/l (annual mean) total cadmium in inland surface waters
    5 µg/l (annual mean) dissolved cadmium in estuary waters
    2.5 µg/l (annual mean) dissolved cadmium in marine waters
Hexachlorocyclohexane (HCH)
    0.1 µg/l total HCH (annual mean) in inland surface waters
    0.02 µg/l total HCH (annual mean) in estuary and marine waters
Carbon tetrachloride
    12 µg/l (annual mean) in all waters
DDT
    0.01 µg/l (annual mean) for the isomer *para-para*-DDT in all waters
    0.025 µg/l (annual mean) for total DDT in all waters
Pentachlorophenol (PCP)
    2 µg/l (annual mean in all waters)
The drins' (aldrin, dieldrin, endrin and isodrin)
    0.03 µg/l (annual mean) total drins for all waters, with a maximum of 0.005 µg/l for endrin
    From 1 January 1994
    0.01 µg/l (annual mean) aldrin for all waters
    0.01 µg/l (annual mean) dieldrin for all waters
    0.005 µg/l (annual mean) endrin for all waters
    0.005 µg/l (annual mean) isodrin for all waters
Hexachlorobenzene (HCB)
    0.03 µg/l (annual mean) in all waters
Hexachlorobutadiene (HCBD)
    0.1 µg/l (annual mean) in all waters
Chloroform
    12 µg/l (annual mean) in all waters

List II substances (a)

Quality objective (e)		Lead (f) TR 208*	Chromium TR207	Zinc TR 209	Copper (g) TR210	Nickel TR211	Arsenic TR212
Fresh water							
Direct	A1(b)	50PT	50PT	3000PT	20PT	50PT	50PT
abstraction to	A2(b)	75MT	75MT	5000PT	50PT	50PT	50PT
potable supply							
	Total hardness (as mg/l CaCO$_3$)						
Protection of	0–50	4AD	5AD	8AT(30P)	1AD(5P)	50AD	50AD
sensitive	50–100	10AD	10AD	50AT(200P)	6AD(22P)	100AD	50AD
aquatic life	100–150	10AD	20AD	75AT(300P)	10AD(40P)	150AD	50AD
(e.g. salmonid	150–200	20AD	20AD	75AT(300P)	10AD(40P)	150AD	50AD
fish (c)	200–250	20AD	50AD	75AT(300P)	10AD(40P)	200AD	50AD
	250+	20AD	50AD	125AT(500P)	28AD(112P)	200AD	50AD
Protection of	0–50	50AD	150AD	75AT(300P)	1AD(5P)	50AD	50AD
other aquatic	50–100	125AD	175AD	175AT(700P)	6AD(22P)	100AD	50AD
life (e.g.	100–150	125AD	200AD	250AT(1000P)	10AD(40P)	150AD	50AD
cyprinid fish)	150–200	250AD	200AD	250AT(1000P)	10AD(40P)	150AD	50AD
	200–250	250AD	250AD	250AT(1000P)	10AD(40P)	200AD	50AD
	250+	250AD	250AD	500AT(2000P)	28AD(112P)	200AD	50AD
Salt water							
Protection of							
salt water life	25AD	15AD	40AD	5AD	30AD	25AD	

*WRC Report reference number.
All values are given as µg/l. A, Annual average; P, 95% of samples (d); M, maximum allowable concentration; D, dissolved; T, total.

List II substances (a)

Quality objective (e)		Boron (h) TR256*	Iron (h)(i) TR258	pH TR259	Vanadium TR253	
Fresh water						
Direct abstraction	A1(b)	1000PT	300PD	6.5–8.5P		
to potable supply	A2(b)	1000PT	2000PD	5.5–9.0P		
					Total hardness (as mg/l CaCO$_3$)	
Protection of sensitive aquatic life (e.g. salmonid fish) (c)		2000AT	1000AD	6.0–9.0P	0–200 200+	20AT 60AT
Protection of other aquatic life (e.g. cyprinid fish)		2000AT	1000AD	6.0–9.0P	0–200 200+	20AT 60AT
Salt water						
Protection of salt water life		7000AT	1000AD	6.0–8.5P(k)	100AT	

* WRC Report reference number.
All values given as µg/l, except pH where 95% of samples must lie within the range shown.
A, annual average; P, 95% of samples (d); D, dissolved; T, total.

List II substances (a)

Quality objective (e)		Triorganotin Compounds TR255*	
		Tributyltin	Triphenyltin
Fresh water			
Direct abstraction to potable supply	A1(b)	0.02MT	0.09MT
	A2(b)	0.02MT	0.09MT
Protection of sensitive aquatic life (e.g. salmonid fish) (c)		0.02MT	0.02MT
Protection of other aquatic life (e.g. cyprinid fish)		0.02MT	0.02MT
Salt water			
Protection of salt water life		0.002MT(1)	0.008MT(1)

* WRC Report reference number.
All values given as µg/l, P, 95% of samples M, maximum allowable concentration; T, total.

List II substances (a)

Quality objective (e)		Mothproofing agents TR261*				
		PCSDs	Cyfluthrin	Sulcofuron	Fulcofuron	Permethrin
Fresh water						
Direct abstraction	A1(b)		0.001PT			0.01PT
to potable supply	A2(b)		0.001PT			0.01PT
Protection of sensitive aquatic life (e.g. salmonid fish) (c)		0.05PT	0.001PT	25PT	1.0PT	0.01PT
Protection of other aquatic life (e.g. cyprinid fish)		0.05PT	0.001PT	25PT	1.0PT	0.01PT
Salt water						
Protection of salt water life		0.05PT	0.001PT	25PT(j)	1.0PT(j)	0.01PT(j)

*WRC Report reference number.
All values given as µg/l. P, 95% of samples (d); T, total.

Notes
  (a) These quality standards are set for the purpose of controlling discharges of dangerous substances under Directive 76/464/EEC. A number of other EC directives set down standards for some of the substances listed here in respect of particular uses of water. For the purposes of implementing these directives, certain provisions may apply (for example in relation to definitions, sampling frequency or possible derogations) which are not set out here, and authorities should consult those directives separately (see also paragraph 45 of this circular).
  (b) These categories correspond to those defined in Directive 75/440/EEC. Waters in category A1 should be suitable for abstraction for drinking after simple physical treatment and disinfection; waters in category A2 for drinking after normal physical treatment, chemical treatment and disinfection.

(c) In some cases more stringent values may be appropriate locally to protect particularly sensitive flora or fauna (see appropriate WRC Report).

(d) Notwithstanding the advice in paragraph 45 of this circular, where the values shown for certain substances are applied as 90 percentiles under other EC directives, authorities may apply them here also on the basis that 90% of samples should comply. Otherwise 95% of samples should fall within the quality standard shown.

(e) Other standards may be applicable for other particular water uses, notably irrigation of crops and livestock watering. In some cases these are identified in the appropriate WRC report and in published advice from ADAS. In other cases authorities should consult ADAS as necessary.

(f) Where breeding populations of rainbow trout are present the quality standard for lead should be 50% of that recommended for sensitive aquatic life. The standards given also assume that the lead present is almost entirely inorganic; if a significant proportion of organic lead is present, more stringent standards may be necessary.

(g) Higher concentrations of copper may be acceptable where the presence of organic matter may lead to complexation.

(h) Certain crops are particularly sensitive to these substances and may require especially stringent standards for irrigation (see appropriate WRC report).

(i) The toxicity of iron increases at pHs below 7, and authorities may need to set more stringent quality standards, especially where pH is below 6.5.

(j) These standards may need to be reviewed when more adequate data become available.

(k) A more restricted range of 7.0–8.5 should be applied for the protection of shellfish.

(l) Further analytical development is likely to be needed before these standards could be verifiable in receiving waters. They can, however, be used in calculating acceptable concentrations in effluents.

Standards for various water uses

EC Directive 75/440 EEC: Characteristics of surface water intended for the abstraction of drinking water

	Parameters		A1 G	A1 I	A2 G	A2 I	A3 G	A3 I
1	pH		6.5–8.5		5.5–9		5.5–9	
2	Coloration (after simple filtration)	mg/l Pt scale	10	20 (0)	50	.100 (O)	50	200 (O)
3	Total suspended solids	mg/l SS	25					
4	Temperature	°C	22	25 (O)	22	25 (O)	22	25 (O)
5	Conductivity	µs/cm⁻¹ at 20 °C	1000		1000		1000	
6	Odour	(dilution factor at 25°C)	3		10		20	
7*	Nitrates	mg/l NO_3	25	50 (O)		50 (O)		50 (O)
8¹	Fluorides	mg/l F	0.7–1	1.5	0.7–1.7		0.7–1.7	
9	Total extractable organic chlorine	mg/l Cl						
10*	Dissolved iron	mg/l Fe	0.1	0.3	1	2	1	
11*	Manganese	mg/l Mn	0.05		0.1		1	
12*	Copper	mg/l Cu	0.002	0.05 (O)	0.05		1	
13	Zinc	mg/l Zn	0.5	3	1	5	1	5
14	Boron	mg/l B	1		1		1	
15	Beryllium	mg/l Be						
16	Cobalt	mg/l Co						
17	Nickel	mg/l Ni						
18	Vanadium	mg/l V						
19	Arsenic	mg/l As	0.01	0.05	0.05		0.05	0.1
20	Cadmium	mg/l Cd	0.001	0.005	0.001	0.005	0.001	0.005
21	Total chromium	mg/l Cr		0.05		0.05		0.05
22	Lead	mg/l Pb		0.05		0.05		0.05
23	Selenium	mg/l Se		0.01		0.01		0.01
24	Mercury	mg/l Hg	0.0005	0.001	0.0005	0.001	0.0005	0.001
25	Barium	mg/l Ba		0.1		1		1
26	Cyanide	mg/l Cn		0.05		0.05		0.05

(Table cont.)

#	Parameters		A1 G	A1 I	A2 G	A2 I	A3 G	A3 I
27	Sulphates	mg/l SO$_4$	150	250	150	250 (O)	150	250 (O)
28	Chlorides	mg/l Cl	200		200		200	
29	Surfactants (reacting with methyl blue)	mg/l (laurylsulphate)	0.2		0.2		0.5	
30*2	Phosphates	mg/l P$_2$O$_3$	0.4		0.7		0.7	
31	Phenols (phenol index) paranitraniline 4 aminoantipyrine	mg/l C$_6$H$_5$OH						
32	Dissolved or emulsified hydrocarbons (after extraction by petroleum ether)	mg/l		0.05		0.2	0.5	1
33	Polycyclic aromatic hydrocarbons	mg/l		0.0002		0.0002		0.001
34	Total pesticides (parathion, BHC, dieldrin)	mg/l		0.001		0.0025		0.005
35*	Chemical oxygen demand (COD)	mg/l O$_2$					30	
36*	Dissolved oxygen saturation rate	% O$_2$	> 70		> 50		> 30	
37*	Biochemical oxygen demand (BOD) (at 20°C without nitrification)	mg/l O$_2$	< 3		< 5		< 7	
38	Nitrogen by Kjeldahl method (except NO$_3$)	mg/l N	1		2		3	
39	Ammonia	mg/l NH$_4$	0.05		1	1.5	2	4 (O)
40	Substances extractable with chloroform	mg/l SEC	0.1		0.2		0.5	
41	Total organic carbon	mg/l C						
42	Residual organic carbon after flocculation and membrane filtration (5 μm) TOC	mg/l C						
43	Total coliforms 37°C	/100 ml	50		5000		50 000	
44	Faecal coliforms	/100 ml	20		2000		20 000	
45	Faecal streptococci	/100 ml	20		1000		10 000	
46	Salmonella			Not present in 5000 ml		Not present in 1000 ml		

I, mandatory; G, guide; O, exceptional climatic or geographical conditions; *, see Article 8(d).

1 The values given are upper limits set in relation to the mean annual temperature (high and low).

2 This parameter has been included to satisfy the ecological requirements of certain types of environment.

Article 8(d). In the case of surface water in a shallow lake or virtually stagnant surface water, for parameters marked with an asterisk in the table in Annex II, this derogation being applicable only to lake with a depth not exceeding 20 m, with an exchange of water slower than 1 year, and without a discharge of waste water into the water body.

EC Directive 76/160 EEC : Quality requirements for bathing water

Parameters	G	I	Minimum sampling frequency	Method of analysis and inspection
Microbiological:				
1 Total coliforms/ 100 ml	500	10000	Fortnightly (1)	Fermentation in multiple tubes. Subculturing of the positive tubes on a
2 Faecal coliforms / 100 ml	100	2000	Fortnightly (1)	confirmation medium. Count according to MPN (most probable number) or membrane filtration and culture on an appropriate medium such as Tergitol lactose agar, endo agar, 0.4% Teepol broth, subculturing and identification of the suspect colonies
				In the case of 1 and 2, the incubation temperature is variable according to whether total or faecal coliforms are being investigated
3 Faecal streptococci /100 ml	100	–	(2)	Litsky method. Count according to MPN (most probable number) or filtration on membrane. Culture on an appropriate medium
4 *Salmonella* /1 1	–	0	(2)	Concentration by membrane filtration. Inoculation on a standard medium. Enrichment, subculturing on isolating agar, identification
5 Entero viruses (PFU/10 l)	–	0	(2)	Concentrating by filtration. flocculation or centrifuging and confirmations

(Table cont.)

Parameters	G	I	Minimum sampling frequency	Method of analysis and inspection
Physico-chemical				
6 pH	–	6–9(0)	(2)	Electrometry with calibration at pH 7 and 9
7 Colour	–	No abnormal change in colour (0)	Fortnightly (1)	Visual inspection or photometry with standards on the Pt.Co scale
	–	–	(2)	
8 Mineral oils (mg/l)	–	No film visible on the surface of the water and no odour	Fortnightly (1)	Visual and olfactory inspection extraction using an adequate volume and weighing the dry residue
	≤ 0.3	–	(2)	
9 Surface-active substances reacting with methylene blue (mg/litre, laurylsulfate)	–	No lasting foam	Fortnightly (1)	Visual inspection or absortion spectro-photometry with methylene blue
	≤ 0.3	–	(2)	
10 Phenols (phenol indices) (mg/l C_4H_5OH)	–	No specific odour	Fortnightly (1)	Verification of the absence of specific odour due to phenol or absorption
	≤ 0.005	≤ 0.05	(2)	spectrophotometry 4-aminoantipyrines (4 AAP) method
11 Transparency (m)	2	1 (0)	Fortnightly (1)	Secchi's disc
12 Dissolved oxygen (% saturation O_2)	80–120	–	(2)	Winkler's method or electrometric method (oxygen meter)
13 Tarry residues and floating materials such as wood, plastic articles, bottles, containers of glass, plastic, rubber or any other substance; waste or splinters	Absence		Fortnightly (1)	Visual inspection
14 Ammonia (mg/l NH_4)			(3)	Absorption spectrophotometry, Nessler method, or indophenol blue method

15 Nitrogen Kjeldahl (mg/l N)	(3)	Kjeldahl method
Other substances regarded as indications of pollution		
16 Pesticides (parathion, HCH, dieldrin) mg/l	(2)	Extraction with appropriate solvents chromatographic determination
17 Heavy metals such as: Arsenic (mg/l As) Cadmium (mg/l Cd) Chrome VI (mg/l Cr^{vi}) Lead (mg/l Pb) Mercury (mg/l Hg)	(2)	Atomic absorption possibly preceded by extraction
18 Cyanides (mg/l Cn)\	(2)	Absorption spectrophotometry using a specific reagent
19 Nitrates (mg/l NO_3) Phosphates (mg/l PO_4)	(2)	Absorption spectrophotometry using a specific reagent

G, guide; I, mandatory

(0) Provision exists for exceeding the limits in the event of exceptional geographical or meterorological conditions.

(1) When a sampling taken in previous years produced results which are appreciably better than those in this Annex and when no new factor likely to lower the quality of the water has appeared, the competent authorities may reduce the sampling frequency by a factor of 2.

(2) Concentration to be checked by the competent authorities when an inspection in the bathing area shows that the substance may be present or that the quality of the water has deteriorated.

(3) These parameters must be checked by the competent authorities when there is a tendency towards the eutrophication of the water.

EC 78/659 EEC : *Quality of fresh water to support fish life*

Annex 1 List of parameters

Parameter	Salmonid waters		Cyprinid waters		Method of analysis or inspection	Minimum sampling and measuring frequency	Observations
	G	I	G	I			
1. Temperature (°C)	1. Temperature measured downstream of a point of thermal discharge (at the edge of the mixing zone) must not exceed the unaffected temperature by more than				Thermometry	Weekly, both upstream and downstream of the point of thermal discharge	Over-sudden variations in temperature shall be avoided
		1.5°C		3°C			
	Derogations limited in geographical scope may be decided by Member States in particular conditions if the competent authority can prove that there are no harmful consequences for the balanced development of the fish population						
	2. Thermal discharges must not cause the temperature downstream of the point of thermal discharge (at the edge of the mixing zone) to exceed the following:						
	21.5 (0)		28 (0)				
	10 (0)		10 (0)				
	The 10°C temperature limit applies only to breeding periods of species which need cold water for reproduction and only to waters which may contain such species						
	Temperature limits may, however, be exceeded for 2% of the time						

						Methods of analysis	Minimum sampling and measuring frequency
2. Dissolved oxygen (mg/l O₂)	50% ≥ 9 100% ≥ 7	50% ≥> 9 When the oxygen concentration falls below 6 mg/l, Member States shall implement the provisions of Article 7 (3). The competent authority must prove that this situation will have no harmful consequences for the balanced development of the fish population	50% ≥ 8 100% ≥ 5	50% ≥ 7 When the oxygen concentration falls below 4 mg/l, Member States shall implement the provisions of Article 7 (3). The competent authority must prove that this situation will have no harmful consequences for the balanced development of the fish population		Winkler's method or specific electrodes (electro-chemical method)	Monthly, minimum one sample representative of low oxygen conditions of the day of sampling However, where major daily variations are suspected, a minimum of two samples in one day shall be taken
3. pH	6−9 (0)[1]			6−9 (0)[1]		Electrometry calibration by means of two solutions with known pH values, preferably on either side of, and close to the pH being measured	Monthly

(*Annex I Cont.*)

Parameter	Salmonid waters		Cyprinid waters		Method of analysis or inspection	Minimum sampling and measuring frequency	Observations
	G	I	G	I			
4. Suspended solids (mg/l)	≤ 25(0)		≤ 25(0)		Filtration through a 0.45 μm filtering membrane, or centrifugation (five minutes minimum, average acceleration of 2800 to 3200 g) drying at 105°C and weighing		The values shown are average concentrations and do not apply to suspended solids with harmful chemical properties Floods are liable to cause particularly high concentrations
5. BOD₅(mg/l O₂)	≤ 3		≤ 6		Determination of O₂ by the Winkler method before and after 5 days incubation in complete darkness at 20 ± 1°C (nitrification should not be inhibited)		

6. Total phosphorus (mg/l P)		Molecular absorption spectrophotometry	In the case of lakes of average depth between 18 and 300 m, the following formula could be applied: $$L \leq 10 \frac{Z}{T_w}\left(1 + \sqrt{T_w}\right)$$ where L is the loading expressed as mg P per m² of lake surface in 1 year, Z is the mean depth of lake (m) and, T_w is the theoretical renewal time of lake water (years)
			In other cases limit values of 0.2 mg/l for salmonid and of 0.4 mg/l for cyprinid waters, expressed as PO_4, may be regarded as indicative in order to reduce eutrophication
7. Nitrites (mg/l NO_2)	≤ 0.01	≤ 0.03	Molecular absorption spectrophotometry

(Annex I Cont.)

Parameter	Salmonid waters		Cyprinid waters		Method of analysis or inspection	Minimum sampling and measuring frequency	Observations
	G	I	G	I			
8. Phenolic compounds (mg/l C_6H_5OH)[2]		2		2	By taste		An examination by taste shall be made only where the presence of phenolic compounds is presumed
9. Petroleum hydrocarbons[3]		3		3	Visual By taste	Monthly	A visual examination shall be made regularly once a month, with an examination by taste only where the presence of hydrocarbons is presumed
10. Non-ionized ammonia (mg/l NH_3)	≤ 0.005	≤ 0.025	≤ 0.005	≤ 0.025	Molecular absorption spectrophotometry using indophenol blue or Nessler's method associated with pH and temperature determination	Monthly	Values for non-ionized ammonia may be exceeded in the form of minor peaks in the daytime
	In order to diminish the risk of toxicity due to the non-ionized ammonia, of oxygen consumption due to nitrification and of eutrophication, the concentrations of total ammonium should not exceed the following:						
11. Total ammonium (mg/l NH_4)	≤ 0.04	≤ 1[4]	≤ 0.2	≤ 1[4]			

Parameter	G	I	Method of analysis	Minimum sampling and measuring frequency	Observations
12. Total residual chlorine (mg/l HOCl)	≤ 0.005	≤ 0.005	DPD-method (diethyl-*p*-phenylenediamene)	Monthly	The I-values correspond to pH = 6; higher concentrations of total chlorine can be accepted if the pH is higher
13. Total zinc (mg/l Zn)	≤ 0.3	≤ 1.0	Atomic absorption spectrometry	Monthly	The I-values correspond to a water hardness of 100 mg/l CaCO$_3$; for hardness levels between 10 and 500 mg/l corresponding limit values can be found in Annex II
14. Dissolved copper (mg/l Cu)	≤ 0.04	≤ 0.04	Atomic absorption spectrometry		The G-values correspond to a water hardness of 100 mg/l CaCO$_3$; for hardness levels between 10 and 300 mg/l corresponding limit values can be found in Annex II

[1] Artificial pH variations with respect to the unaffected values shall not exceed ±0.5 of a pH unit within the limits falling between 6.0 and 9.0 provided that these variations do not increase the harmfulness of other substances present in the water.

[2] Phenolic compounds must not be present in such concentrations that they adversely affect fish flavour.

[3] Petroleum products must not be present in water in such quantities that they: form a visible film on the surface of the water or form coatings on the beds of water-courses and lakes; impart a detectable 'hydrocarbon' taste to fish; produce harmful effects in fish.

[4] In particular geographical or climatic conditions, and particularly in cases of low water temperature and of reduced nitrification or where competent authority can prove that there are no harmful consequences for the balanced development of the fish population, Member States may fix values higher than 1 mg/l.

General observation: It should be noted that the parametric values listed in this Annex assume that the other parameters, whether mentioned in this Annex or not, are favourable. This implies, in particular, that the concentrations of other harmful substances are very low. Where two or more harmful substances are present in mixture, joint effects (additive, synergic or antagonistic effects) may be significant. G, guide; I, mandatory; (0), derogations are possible in accordance with Article 11.

Annex II: Particulars regarding total zinc and dissolved copper

Zinc concentrations[1] (mg/l Zn) for different water hardness values between 10 and 500 mg/l CaCO$_3$

	Water hardness (mg/l CaCO$_3$)			
	10	50	100	500
Salmonid waters (mg/l Zn)	0.03	0.2	0.3	0.5
Cyprinid water (mg/l Zn)	0.3	0.7	1.0	2.0

[1] See Annex I, No. 13, Observations column.

Dissolved copper[1] concentrations (mg/l Cu) for different water hardness values between 10 and 300 mg/l CaCO$_3$

	Water hardness (mg/l CaCO$_3$)			
	10	50	100	300
mg/l Cu	0.005 (2)	0.022	0.04	0.112

[1] See Annex I, No. 14, observations column.
[2] The presence of fish in waters containing higher concentrations of copper may indicate a predominance of dissolved organo-cupric complexes.

SI 1989/2286 : The Surface Water (Dangerous Substances) (Classification) Regulations

Schedule 1: Regulation 3(1)

Classification of inland waters (DSI)

Substance	Concentration (μg/l) (annual mean)
Aldrin, dieldrin, endrin and isodrin	(i) 0.03 for the four substances in total (ii) 0.005 for endrin
Cadmium and its compounds	5 (total cadmium, both soluble and insoluble forms)
Carbon tetrachloride	12
Chloroform	12
DDT (all isomers)	0.025
para-para-DDT	0.01
Hexachlorobenzene	0.03
Hexachlorobutadiene	0.1
Hexachlorocyclohexane (all isomers)	0.1
Mercury and its compounds	1 (total mercury, both soluble and insoluble forms)
Pentachlorophenol and its compounds	2

Schedule 2: Regulation 3(2)

Classification of coastal waters and relevant territorial waters (DS2)

Substance	Concentration (μg/l) (annual mean)
Aldrin, dieldrin, endrin and isodrin	(i) 0.03 for the four substances in total (ii) 0.005 for endrin
Cadmium and its compounds	2.5 (dissolved cadmium)
Carbon tetrachloride	12
Chloroform	12
DDT (all isomers)	0.025
para-para-DDT	0.01
Hexachlorobenzene	0.03
Hexachlorobutadiene	0.1
Hexachlorocyclohexane (all isomers)	0.02
Mercury and its compounds	0.3 (dissolved mercury)
Pentachlorophenol and its compounds	2

SI 1989/1148 : The Surface Water (Classification) Regulations

Schedule: Regulation 3

Criteria for the classification of waters (the limits set out below are maxima)

No. in Annex II to 75/440/EEC	Parameters		DW1	DW2	DW3
2	Coloration (after simple filtration)	mg/l Pt scale	20	100	200
4	Temperature	°C	25	25	25
7	Nitrates	mg/l NO_3	50	50	50
8(1)	Fluorides	mg/l F	1.5		
10	Dissolved iron	mg/l Fe	0.3	2	
12	Copper	mg/l Cu	0.05		
13	Zinc	mg/l Zn	3	5	5
19	Arsenic	mg/l As	0.05	0.05	0.1
20	Cadmium	mg/l Cd	0.005	0.005	0.005
21	Total chromium	mg/l Cr	0.05	0.05	0.05
22	Lead	mg/l Pb	0.05	0.05	0.05
23	Selenium	mg/l Se	0.01	0.01	0.01
24	Mercury	mg/l Hg	0.001	0.001	0.001
25	Barium	mg/l Ba	0.1	1	1
26	Cyanide	mg/l Cn	0.05	0.05	0.05
27	Sulphates	mg/l SO_4	250	250	250
31	Phenols (phenol index) paranitraniline 4-aminoantipyrine	mg/l C_6H_5OH	0.001	0.005	0.1
32	Dissolved or emulsified hydrocarbons (after extraction by petroleum ether)	mg/l	0.05	0.2	1
33	Polycyclic aromatic hydrocarbons	mg/l	0.0002	0.0002	0.001
34	Total pesticides (parathion, BHC, dieldrin)	mg/l	0.001	0.0025	0.005
39	Ammonia	mg/l NH_4		1.5	4

Note: (1) The value given is an upper limit set in relation to the mean annual temperature (high and low).

Explanatory note (this note is not part of the Regulations). These Regulations pre-scribe the system of classifying the quality of inland waters (as defined in Section 103(1)(c) of the Water Act 1989) according to their suitability for abstraction by water undertakers for supply (after treatment) as drinking water.

The classifications DW1, DW2 and DW3 reflect the mandatory values assigned by Annex II to Council Directive 75/440/EEC (OJ No. L 194, 25.7.75, p. 26) (concerning the quality required of surface water intended for the abstraction of drinking water) to the parameters listed in the Schedule to the Regulations.

The classifications are relevant for the purposes of setting water quality objectives for rivers, lakes and other inland waters under Section 105 of the Act and for ascer-taining the treatment to which the water is to be subjected before it is supplied for public use, in accordance with Part VI of the Water Supply (Water Quality) Regula-tions 1989 (S.I. 1989/1147).

Standards proposed for classifying wastes

C190/33 EC : Proposal for a Council Directive on the Landfill of Waste Control Criteria for Assessing Leachate Potential.

Treatment of the samples. The original structure of the sample used should be main-tained as far as possible; large parts should be crushed. The proposed analytical method is DIN 38414-S4 (October 1984 issue) with the following additions and/or simplifications:

A wide-necked glass bottle (10 cm diameter) should be used
Shake, rotating bottle by 180° once/min for 24 h
Centrifuge, 250 μl filter syringes with 0.45 μ filters should be used for sampling

Assignment values. This table fixes the ranges by which wastes will be characterized for the purpose of landfilling according to the composition of their eluates:

Wastes whose eluate concentration is in the range fixed for hazardous wastes will be considered as such with respect to landfilling; for eluate concentrations higher than the maximum values fixed, hazardous wastes will have to be treated prior to landfill, unless compatible for joint disposal with municipal waste, or, if treatment is not possible, destinated to a mono-landfill

Wastes whose eluate concentration is not above the maximum values fixed for inert wastes will be considered as such

Wastes whose eluate concentration falls in the range between inert wastes and the minimum value for hazardous wastes will be considered non-hazardous

		Hazardous waste range	Inert waste
1.01	pH value	4–13	4–13
1.02	TOC	40–200 mg/l	< 200 mg/l
1.03	Arsenic	0.2–1.0 mg/l	< 0.1 mg/l
1.04	Lead	0.4–2.0 mg/l	The total of these metals: < 5 mh/l[1]

1.05	Cadmium	0.1–0.5 mg/l	
1.06	Chromium	0.1–0.5 mg/l	The total of these
1.07	Copper	2–10 mg/l	metals: < 5 mg/l
1.08	Nickel	0.4–2.0 mg/l	
1.09	Mercury	0.02–0.1 mg/l	
1.10	Zinc	2–10 mg/l	
1.11	Phenols	20–100 mg/l	< 10 mg/l
1.12	Fluoride	10–50 mg/l	< 5 mg/l
1.13	Ammonium	0.2–1.0 mg/l	< 50 mg/l
1.14	Chloride	1.2–6.0 mg/l	< 0.5 g/l
1.15	Cyanide[2]	0.2–1.0 g/l	< 0.1 g/l
1.16	Sulphate[3]	0.2–1.0 g/l	< 1.0 g/l
1.17	Nitrite	6–30 mg/l	< 3 mg/l
1.18	AOX[4]	0.6–3.0 mg/l	< 0.3 mg/l
1.19	Solvents[5]	0.02–0.10 mgCl/l	< 10 μg Cl/l
1.20	Pesticides[5]	1–5 μg Cl/l	< 0.5 μg Cl/l
1.21	Lipophilic substances	0.4–2.0 mg/l	< 1 mg/l

[1] And no single value above the minimum fixed for hazardous water.
[2] Readily released.
[3] If possible < 500 mg/l.
[4] Adsorbed organically bound halogens.
[5] Chlorinated.

Notes
(1) For characterization purposes the components to be analysed in the eluates shall be chosen in function of the qualitative composition of the waste.
(2) In addition to these eluate criteria, a determination of asbestos on a representative sample of the crude inert waste shall be performed, according to the annexes of the Council Directive 87/217/EEC on the prevention and reduction of environmental pollution by asbestos.

Analytical methods. The following ISO or DIN methods are proposed as reference methods. Any equivalent method after a certification procedure based on the use of a certified reference material will be accepted. In case of discrepancy of the results the proposed methods will be used as reference.

1.01	pH	ISO-DP 10 523 or DIN 38404-C5-84
1.02	TOC in eluate	DIN 38409-H3-85
1.03	Arsenic	ISO 6595-1982 or DIN 38405-E6-81
1.04	Lead	ISO 8288-1985 or DIN 38406-E6-81
1.05	Cadmium	ISO 8288-1985 or DIN 38406-E19-80
1.06	Chromium(VI)	ISO-DIS 9174-88 or DIN 38405-D24-87
1.07	Copper	ISO 8288-1985 or DIN 38406-E21-80
1.08	Nickel	ISO 8288-1985 or DIN 38406-E21-80
1.09	Mercury	ISO 5666-1/3-88 or DIN 38406-E12-80
1.10	Zinc	ISO 8288-1985 or DIN 3840-E8-85
1.11	Phenols	ISO 6439-1990 or DIN 38409-H16-84
1.12	Fluoride	ISO-DP 359-1 or DIN 38406-D4-85
1.13	Ammonium	ISO 7150-1983 or DIN 38406-E5-83
1.14	Chloride	ISO-DIS 9297 or DIN 38405-D1-85
1.15	Cyanide	DIN 38405-D14-88
1.16	Sulphate	ISO-DIS 9280-1 or DIN 38405-D5-85
1.17	Nitrite	ISO 6777-1983 or DIN 38405-D10-81
1.18	AOX	ISO-DIS 9562 or DIN 38409-H14-85
1.19	Chlorinated solvents[1]	ISO-DP 10 301 or GC head-space
1.20	Chlorinated pesticides[2]	GC (capillary column)
1.21	Extractible lip. substances[3]	cf. param. 27, Directive 80/778/EEC

[1] Needs 2 ml of eluate.
[2] After extraction of 1 l of eluate.
[3] Needs 250 ml of eluate; chloroform extract, results in 'dry residue' (mg/l).

SI 1989/1156 : Trade Effluents (Prescribed Processes and Substances) Regulations 1989

Section 74 of the 1989 Act (control of exercise of trade effluent functions in certain cases) shall apply to trade effluent in which any of the substances listed in Schedule 1 to these Regulations is present in a concentration greater than the background concentration.

Schedule 1: Regulation 3

Prescribed substances

Cadmium and its compounds
γ-Hexachlorocyclohexane
DDT
Pentachlorophenol
Hexachlorobenzene
Hexachlorobutadiene
Aldrin
Dieldrin
Endrin
Carbon tetrachloride
Polychlorinated biphenyls
Dichlorvos
1,2-Dichloroethane
Trichlorobenzene
Atrazine
Simazine
Tributyltin compounds
Triphenyltin compounds
Trifluralin
Fenitrothion
Azinphos-methyl
Malathion
Endosulfan

Salient guidance on landfill gas concentrations

Waste Management Paper No. 27 : references to gas concentrations

Clause	
1.17(a) and 7.9	Monitoring of tip to continue until flammable gas < 1%, CO_2 < 1.5% for 24 months
3.3	Flammability range: methane 5–15%; hydrogen 4–74%
3.4	Asphyxiation: no one to enter confined space where oxygen < 18% by vol. Carbon dioxide: short-term exposure limit 1.5% by vol (over 10 min); occupational exposure standard (over 8 h) 0.5% by vol.
3.5	Toxicity: hydrogen sulphide: short-term exposure limit 15 ppm (over 10 min.); occupational exposure standard 10 ppm (over 8 h)

6.19	Natural concentrations of carbon dioxide up to 2 m deep may occur up to 7% by vol.
7.21	Trigger values in buildings: evacuation requirement when flammable gas $> 1\%$ by vol. and/or $> 20\%$ of LEL; carbon dioxide $> 1.5\%$ by vol.
C.16.2	Ventilation of rooms required when: methane $> 0.25\%$ by vol. and/or $> 5\%$ of LEL; carbon dioxide $> 0.5\%$ by vol.

National Rivers Authority 1994 (Interim Leaching Test Method Guidance)

1. Return samples to laboratory in a sealed container.
2. Do not dry or sieve samples. Inert materials, such as stones, should be removed and their percentage recorded.
3. Size reduction is only allowable on materials coarser than 5 mm.
4. On no account must fine grinding take place.
5. Cone and quarter sample. Remove three samples from the same quarter for

   - the leaching test
   - 'total' contaminant soils analysis
   - moisture content test.

6. Fill a container with de-ionized/distilled water and leave to stand overnight. pH should be approximately 5.6 by morning.
7. No other pH adjustment is required.
8. Place 100 g of the sample at its natural moisture content in one litre volume of the test water. Record sample weight.
9. Agitate the flask for 24 hours. Avoid excessive agitation which may abrade particles.
10. Settle flask for 15 minutes. Filter liquid through a 0.45 μm filter. Do not rinse the filter.
11. Store filtered leachate at 4 °C in the dark until analysis takes place. Discard solid materials and filters.
12. Analyse filtered leachate and express results in *either* mg per litre of leachate (useful for comparison against water quality standards) *or* as mg of contaminant per kilogram of soil (more useful as an indicator of probable environmental impact of the contaminated site).

References

Chapter 1

1. Evelyn, J. (1661) *Fumifugium: or inconvenience of the air and smoke of London.*
2. Trevelyan, G.M. (1952) *Illustrated English Social History*, Vol. 4, Longmans, London.
3. Carson, R. (1963) *Silent Spring*, Hamish Hamilton.
4. Rorsch, A. (1985) Foreword, in E.W. Assink and W.J. Van Der Brink (eds), *Proceedings of the International TNO Conference on Contaminated Soils*, Martinus Nijhoff, Dordrecht.
5. Netherlands Law on Soil Protection 1994, Ministry of Housing, The Hague, Netherlands.
6. Beckett, M.J. and Simms D.L. (1984) The development of contamination land, in *Conference on Hazardous Waste Disposal and the Re-use of Contaminated Land*, SCI, London.
7. Environmental Act (1995) HMSO, London.
8. Ross, S.M. (1994) Retention, transformation and mobility of toxic metals in soils, in S.M. Ross (ed.), *Toxic Metals in Soil-Plant Systems*, John Wiley & Sons, Chichester.
9. Thorton, I. (1985) Metal contamination of soils in UK urban gardens: implications to health, in *1st TNO International Conference on Contaminated Soils*, Martinus Nijhoff, Dordrecht, 203–10.
10. Van Wijnen, J.H. (1985) Health risk assessment, population survey and contaminated soil, in *1st TNO International Conference (op.cit.)*, 181–90.
11. Soczo, E. and Meeder, T. (1992) Clean-up of contaminated sites in Europe and the USA, in *Eureco '92'* (European Urban Regeneration Conference), Birmingham, UK.
12. de Bruijn, P. (1992) Biotreatment in soil remediation, in *Contaminated Land – Policy, Regulation and Technology*, IBC Technical Services, London.
13. Edelman, T. and de Bruin, M. (1985) Background values of 32 elements in Dutch topsoils, in *1st TNO International Conference (op.cit.)*, 89–99.
14. Leonard, M. and Privett, K. (1991) Environmental assessments of reclaimed land in the USA, in *Land Reclamation: An End to Dereliction?*, Elsevier Applied Science, London, 235–40.
15. Harris, M.R. (1987) Recognising the problem, in T. Cairney (ed.), *Reclaiming Contaminated Land*, Blackie & Sons, Glasgow.
16. Swanson, A.E. (1992) Legal considerations and liabilities, in *Contaminated Land – Policy Regulation and Technology (op.cit.)*.
17. Sanning, D.E. (1992) The NATO/CCMS Pilot Study Program, in *Contaminated Land – Policy, Regulation and Technology (op.cit.)*.

Chapter 2

1. Royal Commission on Environmental Pollution (1984) *10th Report: Tackling Pollution – Experience and Prospects*, Cmnd 9194, HMSO, London.
2. Barry, D. (1991) Hazards in land recycling, in G. Fleming (ed.), *Recycling Derelict Land*, Thomas Telford, London.
3. Beckett, M.J. (1993) Land contamination, in T. Cairney (ed.), *Contaminated Land: Problems and Solutions* (1st edn), Blackie Academic and Professional, London, Glasgow, New York.
4. House of Commons Select Committee on the Environment (1990) *First Report on Contaminated Land,* HMSO, London, 170–1.
5. Department of the Environment (1990) *Contaminated Land, the Government's Response*, Cm 1161, HMSO, London.

6. Environmental Protection Act (1990) HMSO, London.
7. *Environment News Release* (24 November 1993).
8. Ross, S.M. (1994) Retention, transformation and mobility of toxic metals in soils, in S.M. Ross (ed.), *Toxic Metals in Soil – Plant Systems*, John Wiley & Sons Ltd, Chichester.
9. Griffin, R.P. (1988) *Principles of Hazardous Materials Management*, Lewis Publishers, USA.
10. Angus Environmental Ltd. (1991) *Review and Recommendations for Canadian Interim Environmental Quality Criteria for Contaminated Sites*, Report to the CCME Subcommittee on Environmental Quality Criteria.
11. Interdepartmental Committee on the Redevelopment of Contaminated Land (ICRCL) (1987) Guidance Notes 59/83, *Guidance on the Assessment and Redevelopment of Contaminated Land*, Department of the Environment, London.
12. Ministère de l'Environment du Québec (1988) *Contaminated Sites Rehabilitation Policy*, Direction des Substances Dangereuses, Québec, Canada.
13. Cairney, T. (1995) *The Re-use of Contaminated Land: A Handbook of Risk Assessment*, John Wiley & Sons Ltd, Chichester.
14. LaGoy, P.K., Nisbet, I.C.T. and Schulz, C.O. (1989) Endangerment assessment for the Smuggler Mountain Site – Pitkin County, Colorado, in D.J. Paustenback (ed.), *Risk Assessment of Environmental Hazards*, John Wiley & Sons Ltd, Chichester.
15. Traves, L. (1992) Applying risk assessment concepts to evaluate alternative uses of contaminated industrial properties, *Proc. of Decommissioning, Decontamination and Demolition Conference*, IBC Technical Services, London.
16. Aspinwall & Company (1994) *A Framework for Assessing the Impact of Contaminated Land on Groundwater and Surface Water*, Contaminated Land Research Report, Department of the Environment, London.
17. Welsh Development Agency (1993) *Manual on the Remediation of Contaminated Land*.
18. Construction Industry Research and Information Association (1995) *Remedial Treatment for Contaminated Land – Volume III Site Investigations*, CIRIA, Storeys Gate, Westminster, London.
19. Thompson, H. (1993) Environmental liability – the risks for financial organisations and the implications for borrowers, *5th Annual Risk Management and Insurance Conference on Pollution, Environmental Impairment and Waste*, IBC Financial Focus Ltd, London.
20. British Standards Institute (1988) *Draft for Development DD175 – Code of Practice for the Identification of Potentially Contaminated Land*, BSI, London.
21. National Rivers Authority (1994) *Leaching Test Method for the Assessment of Contaminated Land: Interim N.R.A. Guidance*, National Rivers Authority, Bristol, UK.
22. de Bruijn, P.J. and Verheul, J.H.A.M. (1992) Biotreatment in soil remediation, in *Contaminated Land – Policy, Regulation and Technology*, IBC Technical Services Ltd, London.
23. Netherlands Ministry of Housing (1994) *Dutch Intervention Values for Soil Remediation*, Report HQ 94–021 Environmental Quality Objectives in the Netherlands, Ministry of Housing, The Hague, Netherlands.

Chapter 3

1. British Standards Institute (1988) Draft for Development DD 175: 1988, *Code of Practice for the Identification of Potentially Contaminated Land and its Investigation*, BSI, London.
2. Environment Act (1995) HMSO, London.
3. Interdepartmental Committee for the Reclamation of Contaminated Land (ICRCL) (1987) *Guidance on the Assessment and Redevelopment of Contaminated Land*, Guidance Note 59/83, 2nd edn, Department of the Environment, London.
4. Environmental Protection Act (1990) HMSO, London.
5. Water Resources Act (1991) HMSO, London.
6. Hobson, D.M. (1991) Planning for reclamation, in M.C.R. Davies (ed.), *Land Reclamation : an End to Dereliction*, Elsevier Applied Science, London, 75–81.
7. Ground Board of the Institution of Civil Engineers (1991) *Inadequate Site Investigation*, Thomas Telford, London.
8. ICRCL (1990) *Notes on the Restoration and Aftercare of Metalliferous Mining Sites for Pastures and Grazing*, Guidance Note 70/90, Department of the Environment, London.

9. ICRCL (1978) *Notes on the Redevelopment of Landfill Sites*, Guidance Note 17/78, Department of Environment, London.
10. ICRCL (1979) *Notes on the Redevelopment of Gasworks Sites*, Guidance Note 18/79. Department of Environment, London.
11. ICRCL (1979) *Notes on the Redevelopment of Sewage Works and Farms*, Guidance Note 23/79, Department of Environment, London.
12. Kirk Othmer (1978 to 1984) *Encyclopaedia of Chemical Technology* (3rd edn), John Wiley & Sons, Chichester, UK.
13. Department of the Environment (1996) *Industry Profiles*, DoE Publications.
14. Market, B. and Thornton, J. (1990) Multi-element analysis of English peat bog soil, *Water, Air and Soil Pollution*, **49**, Kluwer Academic Publishers, pp. 113–23.
15. Environmental Resources Ltd (1987) *Problems Arising from the Redevelopment of Gas Works and Similar Sites* (2nd edn), Department of the Environment, London.
16. Bell, R.M., Gildon, A. and Parry, G.D.R. (1983) Sampling strategy and data interpretation for site investigation of contaminated land, in *Reclamation of Former Iron and Steel Works Sites*, GP Double-day, Durham.
17. Ferguson, C.C. (1992) The statistical basis for spatial sampling of contaminated land, *Ground Engineering*, June 1992.
18. Smith, M.A. and Ellis, A.C. (1986) An investigation into methods used to assess gas work sites for reclamation, *Reclamation and Revegetation Research*, **4**, 183–209.
19. Welsh Development Agency (1993) *Manual on the Remediation of Contaminated Land*, Ecotec Research and Consulting Ltd, Birmingham.
20. Thornton, I. (1980) Background levels of heavy metals in soils and plants, in *Reclamation of Contaminated Land*, Society of Chemical Industry, London, C5/1–C5/12.
21. Borrow, M.C. and Burridge, J.C. (1980) Inorganic pollution and agriculture, in *Ministry of Agriculture, Fisheries and Food Reference Book 326*, HMSO, London, 159–83.
22. National Rivers Authority (1992) *Methodology for Monitoring and Sampling of Groundwater 2*, R & D Note 126, National Rivers Authority, Bristol.
23. British Standards Institute (1990) BS 1377, *Methods of Test for Soils for Civil Engineering Purposes*, BSI, London.
24. British Standards Institute (1983) BS 6068, *Water Quality – Part 2. Physical, Chemical and Biochemical Methods*, BSI, London
25. British Standards Institute (1977–1984) BS 1016, *Methods for the Analysis and Testing of Coal and Coke*, BSI, London.
26. Health & Safety Executive (no date) *Methods for the Determination of Hazardous Substances*, MDHS Series, HSE.
27. Agriculture Development and Advisory Service (1986) *The Analysis of Agricultural Materials: A Manual of Analytical Methods Used by ADAS*, HMSO, London.
28. HMSO (no date) *Methods for the Examination of Waters and Associated Materials*, HMSO, London.
29. United States Environmental Protection Agency (1986) *Test Methods for Evaluating Solid Waste: Physical/Chemical Methods*, SW – 846, 3rd edn, Office of Solid Waste and Emergency Response, Washington, USA.
30. American Public Health Association, American Water Works Association and Water Pollution Control Federation (1989) *Standard Methods for the Examination of Water and Waste Water*, 17th edn, APHA.
31. Cairney, T., Clucas, R.C. and Hobson, D.M. (1990) Evaluation of subterranean fire risks on reclaimed sites, in *Reclamation, Treatment and Utilisation of Coal Mining Wastes*, Balkerina, Rotterdam, 237–43.
32. National Rivers Authority (1993) *Pollution Potential of Contaminated Sites*, R & D Note 181, NRA Bristol, UK.
33. Building Research Establishment (1991) *Sulphate and Acid Resistance of Concrete in the Ground*, Digest No. 363, BRE, Garston, UK.
34. Lord, D.W. (1987) Appropriate site investigations, in T. Cairney (ed.), *Reclaiming Contaminated Land*, Blackie & Son, Glasgow.
35. *Construction Plant and Equipment Annual*, Blackheath Publishing Ltd., Morgan Grampian.
36. Russel, A. and Gee, R. (1990) Use of the dynamic probe on polluted and marginal sites, in M Forde (ed.), *Proc. Int. Conf. on Construction on Polluted and Marginal Land*, Glasgow.

37. British Standards Institute (1981) *BS 5930, Code of Practice for Site Investigations*, BSI, London.
38. Smith, M.A. (1991) Data analysis and interpretation, in G. Fleming (ed.), *Recycling Derelict Land*, Thomas Telford, London.
39. Neilsen, D.M. (1991) *Practical Handbook of Groundwater Monitoring*, Lewis, New York.

Chapter 4

1. Construction Industry Research and Information Association (1995) *Remedial Treatment for Contaminated Land – Volume IV – Classification and Selection of Remedial Methods*, Special Publication 104, CIRIA, Westminster, London.
2. ICRCL (1983) *Guidance Note 59/83*, 2nd edn, HMSO, London.
3. Anon. (1989) Cleaning up sites with on-site process plants, *Environmental Science and Technology*, **23**, 912–16.
4. Environmental Business (1995) Landfill tax set at £7.00/tonne, *Environment Business*, 6 Dec., Information for Industry, London.
5. Anon. (1991) Contaminated land treatment technology, *Croner's Environmental Management*, **3**, 325–38.
6. House of Commons Select Committee on the Environment (1990) *First Report on Contaminated Land*, HMSO, London.
7. Burnett-Hall, R. (1990) Legal aspects, in *Contaminated Land, Policy, Regulation and Technology*, IBC Technical Services, London.
8. National Rivers Authority (1992) *Policy and Practice for the Protection of Groundwater*, National Rivers Authority, Bristol, UK.
9. Petts, J. (1990) Contaminated land – is the UK cleaning up or covering up?, in *Contaminated Land, Policy Regulation and Technology (op. cit.)*.
10. *Texaco News* (1991) Hi-tech clean up at Aberdeen terminal, Issue 142.
11. LaGrega, M.D., Ball, R.O. and Anzia, M.J. (1995) Optimal remedies selection: the United States operative procedure, in *Remediation of Contaminated Soils*, European Conference on the Environment, 29/11/95–1/12/95, Palermo, Sicily, CIPA, 323–58.
12. Bull, M. and Toksuad, T. (1995) The problem of leaking tanks and pipelines, in *European Conference on the Environment (op. cit.)*, 359–78.
13. Hinsenveld, M. (1990) Alternative physico-chemical and thermal cleansing technologies for contaminated land, in *Contaminated Land, Policy, Regulation and Technology (op. cit.)*.
14. Haiges, L. *et al.* (1989) Evaluation of underground fuel spill clean-up technologies, in *Haztech International Conference*, San Francisco.
15. Anon. (1985) Getting to grips with waste solidification, *ENDS Report*, **120**, 11–13.
16. Hubbert, S.J. (1990) Practical examples of the U.S.E.P.A. S.I.T.E. (Superfund Innovative Technology Evaluation) program, in *Contaminated Land, Policy, Regulation and Technology (op. cit.)*.
17. Boelsing, F. (1988) *D.C.R. Technology*, Ministry of Economics, Technology and Traffic, Hanover, Germany.

Chapter 5

1. Thorburn, S. and Buchanan, N.W. (1987) Building on chemical waste, in *Building on Marginal and Derelict Land*, Thomas Telford, London, 281–96.
2. Collins, S.P. *et al.* (1987) Rehabilitation of the Old Palace gasworks site, in *Building on Marginal and Derelict Land*, Thomas Telford, London, 449–96.
3. Mills, G. and Clark, J.C. (1987) The redevelopment of the Wandsworth gasworks site, in *Building on Marginal and Derelict Land*, Thomas Telford, London, 497–520.
4. House of Commons Select Committee on the Environment (1990) *First Report on Contaminated Land*, HMSO, London.

5. Jones, A.K. (1980) Monitoring of reclaimed contaminated sites. Unpublished report to Department of the Environment.
6. Lord, A. (1991) Options available for problem solving, in G. Fleming (ed.), *Recycling Derelict Land*, Thomas Telford, London, 145–90.
7. Cairney, T. (1987) Soil cover reclamations, in T. Cairney (ed.), *Reclaiming Contaminated Land*, Blackie, Glasgow, 144–69
8. ICRCL (1987) Guidance Note 59/83, *Guidance on the Assessment and Redevelopment of Contaminated Land*, Department of the Environment.
9. Sharrock, T. (1986) Methods of evaluating soil cover materials and quantifying design proposals, *Civil Engineering Technology*, 9(8), 2–11.
10. Rawls, W.J. and Brakensiek, D.L. (1982) Estimating soil water retention from soil properties, *Journal of the American Society of Civil Engineers*, 108(IR2), 166–71.
11. Russo, D. (1988) Determining soil hydraulic properties by parameter estimation: on the selection of a model for hydraulic properties, *Water Resources Research*, 24(3), 453–9.
12. Bloemen, G.W. (1980) Calculation of steady state capillary rise from a groundwater table and through multi-layered soils, *Zeitschrift für Pilanzenern*, 143, 701–19.
13. Anders, I.J. (1989) Evaluation of the soil cover reclamation method for chemically contaminated land, PhD thesis (unpublished), Liverpool Polytechnic.
14. Bhuiyan, S.I. *et al.* (1971) Dynamic simulation of vertical infiltration into unsaturated soils, *Water Resources Research*, 7, 1597–1605.
15. Al Saeedi, A. (1992) Irrigation design in arid environments, PhD thesis (unpublished), Liverpool Polytechnic.
16. Driscoll, R. (1983) The influence of vegetation on the swelling and shrinking of clay soils in Britain, *Geotechnique*, 33, 93–105.
17. Waters, P. (1980) Comparison of the ceramic plate and the pressure membrane to determine the 15 bar water content of soils, *Journal of Soil Science*, 31, 443–6.
18. Heilman, P. (1981) Root penetration of Douglas Fir seedlings into compacted soil, *Forestry Science*, 27, 660–6.
19. Cairney, T. (1983) Accelerated techniques for predicting the movement of contaminants in soils, Unpublished research report, EC Environmental Programme, ENV/675/UK(H).
20. Department of the Environment (1986) *Landfilling Wastes*, Waste Management Paper No. 26, HMSO, London.
21. Knox, K. (1991) Water management at landfills: water balance and practical aspects. Lecture notes (unpublished) for NAWDC Course, Coventry.
22. National Rivers Authority (1994) *Leaching Test Method for the Assessment of Contaminated Land – Interim NRA Guidance*, National Rivers Authority, Bristol, UK.
23. Bouma, J., Belmans, C.F.M. and Dekker, L.W. (1982) Water infiltration in a silt loam subsoil with vertical worm channels, *Soil Science Society of America*, 46, 917–21.
24. Hillel, D. (1980) *Application of Soil Physics*, Academic Press, New York.

Chapter 6

1. Chipp, P. (1990) Geotechnical processes for the prevention and control of pollution, in *Symposium on Management and Control of Waste Fill Sites*, Leamington Spa.
2. European Communities (1991) Proposal for a Council Directive on the landfill of waste, 91/C190/01, *Official Journal of the European Communities*, C190/33, EC.
3. Her Majesty's Inspectorate of Pollution (1989) *Waste Management Paper 27, The Control of Landfill Gas*, HMSO, London.
4. Haxo, H. E. *et al.* (1985) Liner materials for hazardous and toxic wastes and municipal solid waste leachate, *Pollution Technology Review No. 124*, Noyes Publications.
5. Mitchell, J. K. (1991) Conduction Phenomena, 31st Rankine Lecture, *Geotechnique*, September.
6. Anon. (1985) First Stent wall installed at Kingston upon Thames, *Ground Engineering*, October, 27–31.
7. Xanthakos, P. (1979) *Slurry Walls*, McGraw-Hill, New York.
8. Philipp Holtzman Aktiengesellschaft (1991) *Innovative Glastechnologie für Deponiedichtwande*.

9. D'Appolonia, D. J. (1980) Soil-bentonite slurry trench cut-offs, *Journal of the Geotechnical Engineering Division*, ASCE, **106**(4), 399–417.
10. Krause, R. (1989) *New Developments and Trends in Ground Water Protection with Flexible Membrane Liners*, LT Lining Technology GmbH.
11. Hass, H. J. and Hitze, R. (1986) All-round encapsulation of hazardous wastes by means of injection gels and cut-off materials resistant to aggressive agents, in *ESME3 Seminar on Hazardous Waste*, Bergamo, Italy.
12. Oil Companies Materials Association (1973) *Specification No. DFCP 4, Drilling Fluid Materials Bentonite*.
13. Brice, G. J. and Woodward, J. C. (1984) Arab potash solar evaporation system: design and development of a novel membrane cut-off wall, *Proceedings of the Institution of Civil Engineers*, Part 1, **76**, 185–250.
14. Jefferis, S. A. (1985) Discussion on the Arab potash solar evaporation system, *Proceedings of the Institution of Civil Engineers*, Part 1, 641–6464.
15. Jefferis S. A. (1992) Contaminant-grout interaction, in *ASCE Specialty Conference, Grouting, Soil Improvement and Geosynthetics*, New Orleans.
16. Howsam, P. (ed.) (1990) *Microbiology in Civil Engineering*, E & F. N. Spon, London.
17. Jefferis, S. A. and Mangabhai, R. J. (1989) The divided flow permeameter, in *Materials Research Society, Symposium on Pore Structure and Permeability of Cementitious Materials*, Vol. 137.

Chapter 7

1. Beever, P. F. (1989) Subterranean fires in the U.K. – the problem, *BRE Information Paper, IP3/89*.
2. Redpath, P. G. (1989) Containment, spread and effect of an industrial site fire, in *BRE Research Colloquium*.
3. Drake, D. (1987) *Subterranean Heating at Oakthrope Village*, Institute of Mining Engineering, S. Staffs. Branch.
4. Rainbow, A. K. M. (ed.) (1990) Reclamation, treatment and utilization of coal mining wastes, in *Proceedings of the Third International Symposium on the Reclamation, Treatment and Utilization of Coal Mining Wastes*, Glasgow.
5. Fardell, P. J. and Lukas, C. (1987) Understanding Fire, *Chemistry in Britain*, March.
6. Bowes, P. C. (1984) *Self Heating: Evaluating and Controlling the Hazards*, HMSO, London.
7. Street, P. J., Smalley, J. and Cunningham, A. T. S. (1975) Hydrogen as an indicator of the spontaneous combustion of coal, *Journal of the Institute of Fuel*, September.
8. British Standards Institution (1975) BS476, *Fire Tests on Building Materials and Structures*, London.
9. British Standards Institution (1973) BS1016, *Method for the Analysis and Testing of Coal and Coke, Part 3. Proximate Analysis of Coal*, BSI London.
10. Ball, D. F. (1964) Loss-on-ignition as an estimate of organic matter and inorganic carbon in non-calcareous soils, *Journal of Soil Science*, **15**, 84–92.
11. British Standards Institution (1973) BS1016, *Method for the Analysis and Testing of Coal and Coke, Part 5. Gross Calorific Value of Coal and Coke*, London.
12. Davies, C. (1970) *Calculations in Furnace Technology*, Pergamon Press, Oxford.
13. Chigier, N. A. (1981) *Energy, Combustion and Environment*, McGraw-Hill, New York.
14. Rose, J. W. and Cooper, J. R. (eds) (1977) *Technical Data on Fuel*, 7th edition, The British National Committee World Energy Conference, Scottish Academic Press, Edinburgh.
15. Smith, M. A. (1991) in S. Fleming (ed.), *Recycling Derelict Land*, published for the Institute of Civil Engineers by Thomas Telford, London.
16. Beever, P. F. (1982) Spontaneous combustion, *BRE Information Paper IP 6/82*.
17. Beever, P. F. (1982) Spontaneous combustion – isothermal test methods, *BRE Information Paper IP 23/82*.
18. Baker, B. (1989) Subterranean fires, *BRE Research Colloquium*.
19. Sebastian, J. J. S. and Mayers, M. A. (1937) Coke reactivity. Determination by a modified ignition method, *Journal of Industrial and Engineering Chemistry*, **29**(10).
20. British Standards Institution (1975) BS1377, *Methods of Testing for Soils for Civil Engineering Purposes*, London.

21. Nagata, N. (1988) Air permeability of undisturbed soils, *Bulletin of the Faculty of Agriculture*, Mie University, Japan, pp. 35–55.
22. CP3 (1972) *Wind Loads*, HMSO, London, Chap. V, Part 2.

Chapter 8

1. Staff, M.G., Sizer, K.E. and Newson, S.R. (1991) The potential for surface emissions of methane from abandoned mine workings, *Proc. Symp. on Methane – 'Facing the Problems'*, Paper 1.1, Nottingham, March 1991.
2. Orr, W.E., Wood, A.M., Beaver, J.J., Ireland, R.J. and Beagley, D.P. (1991) Abbeystead outfall works: background to repairs and modifications and lessons learned, *J. Inst. Water Env. Manag.*, 5 Feb., 7–22.
3. Williams, G.M. and Aitkenhead, N. (1989) The gas explosion at Loscoe, Derbyshire, *Proc. Symp. on Methane – 'Facing the Problems'* (*op. cit.*), Paper 3–6.
4. Health and Safety Executive (1991) *Protection of Workers and the General Public during Development of Contaminated Land*, HMSO, London.
5. County Surveyors Society (1982) *Gas Generation from Landfill Sites*, Special Activity Group No. 7, London, UK.
6. Crowhurst, D. (1987) *Measurement of Gas Emissions from Contaminated Land*, Building Research Establishment, Fire Research Station, Boreham Wood, Herts, UK.
7. Institute of Wastes Management (1990) *Monitoring of Landfill Gas*, IWM, Nottingham, UK.
8. Crowhurst, D. and Manchester, S.J. (1993) *The Measurement of Methane and Associated Gases from the Ground*, CIRIA Special Publication, Construction Industry Research and Information Association, London, UK.
9. Rowan, S. and Raybould, J.G. (1993) *Procedures for Investigation of Sites for Methane and Associated Gases in the Ground*, CIRIA Special Publication, Construction Industry Research and Information Association, London, UK.
10. Jefferis, S.A. (1993) In-ground barriers, Chapter 6 in T. Cairney (ed.), *Contaminated Land: Problems and Solution*, Blackie Academic and Professional, London, Glasgow, New York.
11. Cairney, T. (1995) Chapter 9 Risks to site users from gases and vapours, in *The Re-use of Contaminated Land*, John Wiley & Sons, Chichester, New York, Queensland, Canada.
12. Ward, R.S., Williams, G.M. and Hill, C.C. (1993) Changes in landfill gas composition during migration, in *Proceedings of 'Discharge Your Obligations'*, Institute of Waste Management, Kenilworth, 381–92.
13. Barry, D. (1991) Hazards in land recycling, Chapter 3 in G. Fleming (ed.), *Recycling Derelict Land*, Thomas Telford, London.
14. Bauer, L.D., Gardner, W.H. and Gardner, W.R. (1972) *Soil Physics*, 4th edn, John Wiley & Sons, New York.
15. Hooker, P.J. and Bannon, M.P. (1993) *Methane: Its Occurrence and Hazards in Construction*, CIRIA Special Publication, Construction Industry Research and Information Association, Westminster, London, UK.
16. Card, G.B. (1993) *Protecting Development from Methane*, CIRIA Report CP/8, Construction Industry Research and Information Association, Westminster, London, UK.
17. Rees, J.F. and Grainger, J.M. (1982) Rubbish dump or fermenter, *Process Biochemistry*, Nov/Dec., 1982, 41–4.
18. Barry, D. (1987) Hazards from methane (and carbon dioxide), Chapter 11 in T. Cairney (ed.), *Reclaiming Contaminated Land*, Blackie & Son, Glasgow and London.
19. Evans, O.D. and Thompson, G.M. (1986) Field and interpretative techniques for delineating subsurface petroleum hydrocarbon spills using soil gas analysis, *Proc. NWWA/API Petroleum Hydrocarbons and Organic Chemicals in Groundwater – Prevention Detection and Restoration Conference*, National Water Well Association, Dublin.
20. ICRCL (1990) *Notes on the Redevelopment of Landfill Sites*, Circular 17/87, 8th edn, Department of the Environment, London, UK.
21. Emberton, J.R. and Parker, A. (1987) The problems associated with building on landfill sites, *Waste Management and Research*. **5,** 473–82.
22. Department of the Environment (1985) *Building Regulations, Part C – Site Preparation and Resistance to Moisture*, HMSO, London, UK.

23. Department of the Environment (1987) *Landfill Sites: Development Control*, Circular 17/87, HMSO, London, UK.
24. Department of the Environment (1986) *Landfilling Wastes. Waste Management Paper. No. 26*, HMSO, London, UK.
25. Her Majesty's Inspectorate of Pollution (1989) *Waste Management Paper No. 2, The Control of Landfill Gas*, revised March 1990, HMSO, London, UK.
26. Lord, J.A. (1991) Recycling landfill and chemical waste sites, in 'Containment of pollution and redevelopment of closed landfill sites', Paper 6.1, *Proc. Leamington Spa Conference.*
27. Smith, M.V. (1993) Landfill gases, Chapter 8 in T. Cairney (ed.), *Contaminated Land: Problems and Solutions*, Blackie Academic and Professional, London and Glasgow.

Chapter 9

1. Gasson, P.E. and Cutler, D.F. (1990) Tree root plate morphology, *Arboricultural Journal*, **14**, 193–264.
2. Strahler, A.N. (1960) *Physical Geography*, Wiley, New York.
3. Roberts, R.D. and Roberts, J.M. (1986) The selection and management of soils in landscape schemes, *Ecology and Design in Landscape*, 99–126.
4. Samuel, P. (1991) Revegetation of reclaimed land, in *Proceedings of Land Reclamation – An End to Dereliction*, Cardiff, 366–76.
5. Brophy Organic Products (1991) Having your cake and selling it, *Landscape Industry International*, **9** (5), 24–5.
6. Scullion, J. (1991) Re-establishing earthworm populations on former open-cast coalmining land, in *Proceedings of Land Reclamation – An End to Dereliction*, Cardiff, 377–86.
7. Gemmell, R.P. (1985) Wildlife habitats created by mining and tipping, *Land and Mineral Surveying*, August, 422– 31.

Chapter 10

1. Department of the Environment (1992) *Development Plans and Regional Planning Guidance*, PPG Note 12, Department of the Environment.
2. Department of the Environment/Welsh Office (1992) *Planning Policy Guidance on Planning and Pollution Controls*, Consultation paper, June 1992.
3. Department of the Environment/Welsh Office (1985) *The Use of Conditions in Planning Permissions*, Circular 1/85, HMSO, London.
4. *The Waste Management Licensing Regulations 1994* (1994) HMSO, London (SI 1056).

Chapter 11

1. *Cleaning Up the Nation's Waste Sites: Markets and Technology Trends*, EPA 542-R-92-012 (Washington, DC: EPA, Office of Solid Waste and Emergency Response, Technology Innovation Office, 1993).
2. *Innovative Treatment Technologies: Annual Status Report*, 7th edn, EPA-542-R-95-008, No. 7 (Washington, DC: EPA, Office of Solid Waste and Emergency Response, Technology Innovation Office, 1995).
3. J.R. Conner, *Chemical Fixation and Solidification of Hazardous Wastes* (New York: Van Nostrand Reinhold, 1990).
4. US EPA Best Demonstrated Available Technology (BDAT) Background Document for K061, Washington, DC, 1988.
5. C.A. Wentz, *Hazardous Waste Management* (New York: McGraw Hill Publishing Company, 1989).
6. Conner, Jesse R., Chemical stabilization of contaminated soils, chapter in D.J. Wilson and Ann N. Clarke, (eds), *Hazardous Waste Site Soil Remediation*: Marcel Dekker, Inc., New York, 1994.

7. US EPA, *Federal Register 55 (61): 11798–11877* (29 March 1990).
8. US EPA, *Federal Register 59: 47982*
9. Loehr, R.C., Changing Times for Environmental Management, *Environmental Engineer*, **32**(1), January 1996.
10. LaGrega, M.D., R.O. Ball and M.J. Anzia, Optimal remedies selection: The United States operative procedure, *Proceedings of the European Conference of the Environment – Remediation of Contaminated Soils*, Palermo, Italy, 29 November–1 December 1995, 323–57.
11. Risk assessment methods for deriving cleanup levels, *The Hazardous Waste Consultant*, **9**(3), May/June 1991, 1.1–1.6.
12. P. Chrostouski, Risk assessment and accepted regulatory cleaning levels, *Remediation* **4**(4), Autumn 1994, 383–98.
13. Padovani, S.J., Mastrolanardo, M.S., and M.J. Freibert, Conducting an accelerated cleanup at a Superfund site in a mixed residential and industrial area, *Proceedings of the Superfund XV Conference, Washington, DC* 29 November through 1 December, 1994, 1079–86.
14. RCRA corrective actions and CERCLA remediations at federal facilities: a comparison, *The Hazardous Waste Consultant*, **12**(6), November/December, 1994, 2.1–2.8.

Chapter 12

1. Ministry of Housing, Physical Planning and Environment (1994) *Circular on intervention values for soil cleanup* (*Circulaire interventiewaarden bodemsanering*), DBO/07494013, Staatscourant, 24 May, 95.
2. Commission on Soil Cleanup of Operational Industrial Sites (1991) *Final Report*, SDU, Vitgeverij Koninginnesgracht, The Hague.
3. Holtkamp, A.B. and Gravestegn L.J.J. (1993) Large scale voluntary soil cleanup operation for contaminated sites in the Netherlands now on its way, in F. Arendt *et al.* (eds), *Contaminated Soil '93*, Kluwer Academic Publishers, Dordrecht, 27–34.
4. Visser, W.J.F. (1993) *Contaminated Land Policies in Some Industrialised Countries*, Report TCB RO2, Technical Soil Protection Committee, The Hague.

Chapter 13

1. *The Construction (Design and Management) Regulations 1994* (1994) SI 1994, No. 3140, HMSO, London.
2. HMSO (1974) *Health & Safety at Work, etc. Act*, HMSO, London.
3. Health & Safety Executive (1991) *Protection of Workers and the General Public during Development of Contaminated Land*, HMSO, London.
4. Health & Safety Executive (no date) *Respiratory Protective Equipment: A Practical Guide for Users*, HSE publication HS (G) 53, Health & Safety Executive Library Services, Broad Lane, Sheffield or HMSO, London.
5. Health & Safety Executive (no date) *Entry into Confined Spaces*, Guidance Note GS5, HMSO, London.
6. British Standards Institute (1981) *B.S. 5750, Part 6*, HMSO, London.
7. Department of Transport (1986) Specifications for Highway Works – Part 2, HMSO, London.
8. Lord, D.W. (1987) Appropriate site investigations, in T. Cairney (ed.), *Reclaiming Contaminated Land*, Blackie & Son, Glasgow and London.
9. Smith, M.A. (1991) Data analysis and interpretation, in G. Fleming (ed.), *Recycling Derelict Land*, Thomas Telford. London.

Index

Additive interactions of
    contaminants 15
Analytical guidances for testing
    samples 59
Analytical packages offered by
    laboratories 25
Analytical quality control 318
Antagonistic interactions of
    contaminants 15

Background contamination levels 54
Building Regulations 203, 257

Calorific value (significance of) 163
Clean covers
    appropriate applications 126
    design decisions 100
    design lives 101
    functions of different layers 123
    materials employed 102
    material properties 103
    failure mechanisms 108
    quantification 107
    types 100, 112, 113, 119
Combustion
    British Standard test 162
    calorific value testing 163
    effect of air permeability 171
    Fire Research Station method 165
    loss-on-ignition test 163
    potential test 165
    processes 158
    tests available 162
Contaminant availability (variations
    in) 15
Contaminant mobility (variations in) 16
Contaminant speciation 14
Contamination
    as a concept 11
    due to earlier land usage 40
    hazards 11
    pathways 12
    risks 12, 19, 248

significance 3, 248
    standards and criteria 13, 16
Cut-off barriers 128
    gas impermeability 131
    potentials affecting integrity 132
    requirements 130
    slurry trench process 135
    slurry trench specifications 150
    types 129, 133, 135, 138

Duty of Care (waste disposal) 258

Environmental legislation
    (Netherlands) 7, 25, 288
    consequences 7, 26, 300
    decision process 291–296
    regulatory clean-up standards
        Appendix I 289
    Soil Clean-up Interim Act 288
    Soil Protection Act 289
Environmental legislation (UK)
    case law 237
    common law 237, 246
    EU influences 238
    Environmental Protection Act
        (1990) 239, 243
    Environment Act (1995) 13, 20, 239,
        244, 247–256, 259
    guidance notes and Codes of
        Practice 238
    subordinate legislation 237
    Water Resources Act (1991) 244
Environmental legislation (USA)
    applicable, relevant and appropriate
        requirements 85, 263
    Best Demonstrated and Available
        Technologies 265
    consequences of regulations 8, 268,
        275, 286
    C.E.R.C.L.A. 8, 261, 263, 275
    C.E.R.C.L.A. remediation
        process 275–82
    land bans 262 264

Environmental legislation (contd)
Potentially Responsible Parties 263
R.C.R.A. 261, 283
S.A.R.A. 262
target compound list 263
toxicity characteristic rule 266
underlying policy 85
Universal Treatment Standards 270
Environment Act (1995) 13 20, 239, 244, 247–256, 259
Environment Agency (England and Wales) 21, 250, 260
Environmental Protection Agency (USA) 8, 18, 261, 265

Gases
barometric influence on results 191, 192
barriers to entries 200
concentrations 178, 186
consistency of collected data 197, 206
controls on redevelopment 202
emission predictions 204
factors affecting collected data 188, 191, 210
flow rates 186, 212
investigations 177–214
pressures 178
principal gases encountered 177
properties of principal gases 180–181, 186
risks from 177
source identification 184, 201
solutions to gas hazards 199
temperature effects on results 190
venting 200
Gas and vapour monitoring
adequacy of surveys 196
borehole installations 179
multi-point boreholes 183, 185
probehole installations 182
spike test surveys 183, 204
worst case conditions 188, 195
Gassing categories 197, 199
Groundwater investigations 55
Groundwater quality protection 82

Hazard-pathway-target scenario 11, 250
Health and safety
Construction (Design and Management) Regulations 301, 306
enforcement 305

hazard data sheets 308, 310
Improvement Notices 305
legislation (UK) 305
management of safety policies 308
Prohibition Notices 305
responsibilities of various parties 306
safety policies required 308
site safety instructions 308
training required for site staff 308
types of risks on contaminated sites 307
Hydraulic conductivity (clean cover design) 104, 106

I.C.R.C.L. assessment approach 20, 22
I.C.R.C.L. guidance, Appendix I 41, 52, 60

Land contamination
clean-up levels 6, 7, 25 Appendix I; complexity of soil systems, 3, 26
definition difficulty 5, 11
differences in natural reactions 2, 4, 85
extent 5, 12
legislation (Netherlands) 2 7, 288
legislation (UK) 3, 239, 243–244, 247–256
legislation (USA) 8, 261–263, 275–282
legislation (other nations) 4
remediation options 87–98
registers of land 12
significance 3, 248
systematic approach to assessment 23
Landscape maintenance 232
Leachability tests for soils 60, 357

Multifunctionality (for land reclamations) 4, 25

Natural regeneration of vegetation cover 234
Netherlands soils quality standards, Appendix I 330
Nitrogen purging (gas flow measurement) 212

Pathways for contaminant migration 12, 49, 85
Plant requirements 216
Pollution (in contrast to contamination) 11

Quality assurance
  effect on land values 315
  necessary records 311, 313, 315, 319, 325, 327
  of reclamations 315–322
  systems likely to be successful 316–322, 322–328

Reclamation
  bioremediation 80, 93
  challenges to UK approaches 81
  chemical stabilisation 80, 97
  clean covers 90, 99–127
  dilution of contaminants 89
  engineering based methods 87
  excavation and disposal 87
  innovative techniques 81, 84, 93
  oil soaked ground 93
  oils on groundwater 95
  on-site encapsulation 88, 122, 124
  screening techniques 88, 90
  site specific restrictions 80
  soil washing 80, 90, 94, 97
  thermal 80, 94, 96
  vacuum extraction 94
Reclamation objectives (UK) 78, 256
Remediation Notice (Environment Act 1995) 252, 255
Risk assessment 11, 17
  methods 17, 23, Appendix II
Risks from soil contaminants 19, 30, 31
Risks, site specific variations 27

Sampling
  random 50
  targeted 53
Sensitivities of land re-use 30
Site investigation
  aims 47
  background contamination conditions 54
  British Standards draft guidance 33, 37, 41, 51
  costs 86, 282
  data required 311
  design 48
  exploration methods 62
  gas monitoring 58

  groundwater investigation 55
  investigation grid patterns 51
  level of detail required 33
  multistage 34, 49, 55
  preliminary investigations 36
  preliminary model 44
  proof testing 24
  random sampling 50
  rational approach 28
  reports 72, 73
  safety precautions 70, 307–308
  sample protection and chain of custody 66, 69
  sample sizes and types 60, 321
  sampling strategy 49
  site history 37
  site zoning 45, 73, 302
  specification 66
  strategy 35
  targeted sampling 53
Slurry
  necessary properties 140–142
  preparation 139
Slurries (for cut-off walls) 136–138
Soils
  density effects on root penetration 110
  fertilities 226
  moisture retention 210
  nutrient levels 221
Soil covers, desirable properties 126
Soil quality ameliorants 224
Subterranean combustion
  critical air flow rate 171
  ignition temperature 168–171
  incidences 158
  runaway combustion 169
  tests 162–171
Synergistic interaction of contaminants 15

Targets at risk from contaminants 19, 30, 248
Town and Country Planning Acts 20, 239, 243

Waste Management Papers 203
Waste management powers 257, 260